T0213457

Encyclopedia Of Mathematics And Its Applications

Edited By G.-C. Rota

Volume 39

Combinatorial Matrix Theory

ENCYCLOPEDIA OF MATHEMATICS AND ITS APPLICATIONS

ENCYCLOPEDIA OF MATHEMATICS AND ITS APPLICATIONS

Combinatorial Matrix Theory

RICHARD A. BRUALDI
University of Wisconsin

HERBERT J. RYSER

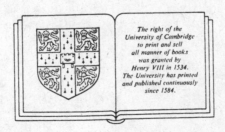

*The right of the
University of Cambridge
to print and sell
all manner of books
was granted by
Henry VIII in 1534.
The University has printed
and published continuously
since 1584.*

CAMBRIDGE UNIVERSITY PRESS
Cambridge
New York   Port Chester   Melbourne   Sydney

CAMBRIDGE
UNIVERSITY PRESS

32 Avenue of the Americas, New York NY 10013-2473, USA

Cambridge University Press is part of the University of Cambridge.

It furthers the University's mission by disseminating knowledge in the pursuit of
education, learning and research at the highest international levels of excellence.

www.cambridge.org
Information on this title: www.cambridge.org/9781107662605

First published 1991
First paperback edition 2013

A catalogue record for this publication is available from the British Library

Library of Congress Cataloguing in Publication data
Brualdi, Richard A.
Combinatorial matrix theory / Richard A. Brualdi, Herbert J. Ryser
p. cm. – (Encyclopedia of mathematics and its applications ; 39.)
Includes bibliographical references and index.
ISBN 0-521-32265-0
1. Matrices. 2. Combinatorial analysis.   I. Ryser, Herbert John.
II. Title. III. Series.
QA188.B78   1991
512.9′434 – dc20   90-20210
                                                          CIP

ISBN 978-1-107-66260-5 Paperback

CONTENTS

PREFACE

It was on March 20, 1984, that I wrote to Herb Ryser and proposed that we write together a book on the subject of combinatorial matrix theory. He wrote back nine days later that "I am greatly intrigued by the idea of writing a joint book with you on combinatorial matrix theory. ... Ideally, such a book would contain lots of information but not be cluttered with detail. Above all it should reveal the great power and beauty of matrix theory in combinatorial settings. ... I do believe that we could come up with a really exciting and elegant book that could have a great deal of impact. Let me say once again that at this time I am greatly intrigued by the whole idea." We met that summer at the small Combinatorial Matrix Theory Workshop held in Opinicon (Ontario, Canada) and had some discussions about what might go into the book, its style, a timetable for completing it, and so forth. In the next year we discussed our ideas somewhat more and exchanged some preliminary material for the book. We also made plans for me to come out to Caltech in January, 1986, for six months in order that we could really work on the book. Those were exciting days filled with enthusiasm and great anticipation.

Herb Ryser died on July 12, 1985. His death was a big loss for me.[1] Strange as it may sound, I was angry. Angry because Herb was greatly looking forward to his imminent retirement from Caltech and to our working together on the book. In spite of his death and as previously arranged, I went to Caltech in January of 1986 and did some work on the book, writing preliminary versions of what are now Chapters 1, 2, 3, 4, 5 and 6. As I have been writing these last couple of years, it has become clear that the book we had envisioned, a book of about 300 pages covering the basic results and methods of combinatorial matrix theory, was not realistic. The subject, at least as I view it, is too vast and too rich to be compressed into 300 or so pages. So what appears in this volume represents only a

---

[1] My article "In Memoriam Herbert J. Ryser 1923–1985" appeared in the *Journal of Combinatorial Theory, Ser. A*, Vol. 47 (1988), pp. 1–5.

portion of combinatorial matrix theory. The choice of chapters and what
has been included and what has been omitted has been made by me. Herb
contributed to Chapters 1, 2 and 5. I say all this not to detract from his
contribution but to absolve him of all responsibility for any shortcomings.
Had he lived I am sure the finished product would have been better.

As I have written elsewhere,[2] my own view is that "*combinatorial matrix theory* is concerned with the use of matrix theory and linear algebra in
proving combinatorial theorems and in describing and classifying combinatorial constructions, and it is also concerned with the use of combinatorial
ideas and reasoning in the finer analysis of matrices and with intrinsic combinatorial properties of matrix arrays." This is a very broad view and it
encompasses a lot of combinatorics and a lot of matrix theory. As I have
also written elsewhere[3] matrix theory and combinatorics enjoy a symbiotic
relationship, that is, a relationship in which each has a beneficial impact
on the other. This symbiotic relationship is the underlying theme of this
book. As I have also noted[4] the distinction between matrix theory and
combinatorics is sometimes blurred since a matrix can often be viewed as
a combinatorial object, namely a graph.

My view of combinatorial matrix theory is then that it includes a lot of
graph theory and there are separate chapters on matrix connections with
(undirected) graphs, bipartite graphs and directed graphs, and in addition,
a chapter on special graphs, most notably strongly regular graphs. In order
to efficiently obtain various existence theorems and decomposition theorems for matrices of 0's and 1's, and more generally nonnegative integral
matrices, I have included some of the basic theorems of network flow theory. In my view latin squares form part of combinatorial matrix theory and
there is no doubt that the permanent of matrices, especially matrices of
0's and 1's and nonnegative matrices in general, is of great combinatorial
interest. I have included separate chapters on each of these topics. The final and longest chapter of this volume is concerned with generic matrices
(matrices of indeterminates) and identities involving both the determinant
and the permanent that can be proved combinatorially.

Many of the chapters can be and have been the subjects of whole books.
Thus I have had to be very selective in deciding what to put in and what to
leave out. I have tried to select those results which I view as most basic. To
some extent my decisions have been based on my own personal interests.
I have included a number of exercises following each section, not viewing
the exercises as a way to further develop the subject but with the more

---

[2] "The many facets of combinatorial matrix theory," *Matrix Theory and Applications*,
C. R. Johnson ed., Proceedings of Symposia in Applied Mathematics, Vol. 40, pp. 1–35,
Amer. Math. Soc., Providence (1990).
[3] "The symbiotic relationship of combinatorics and matrix theory," *Linear Algebra
and Its Applications*, to be published.
[4] Ibid.

modest goal of providing some problems for readers and students to test their understanding of the material presented and to force them to think about some of its implications.

As I mentioned above and as the reader has no doubt noticed in my brief chapter description, the topics included in this volume represent only a part of combinatorial matrix theory. I plan to write a second volume entitled "Combinatorial Matrix Classes" which will contain many of the topics omitted in this volume. A tentative list of the topics in this second volume incudes: nonnegative matrices; the polytope of doubly stochastic matrices and related polytopes; the polytope of degree sequences of graphs; magic squares; classes of (0,1)-matrices with prescribed row and column sums; Costas arrays, pseudorandom arrays (perfect maps) and other arrays arising in information theory; combinatorial designs and solutions of corresponding matrix equations; Hadamard matrices and related combinatorially defined matrices; combinatorial matrix analysis including, for instance, the role of chordal graphs in Gaussian elimination and matrix completion problems; matrix scalings of combinatorial interest; and miscellaneous topics such as combinatorial problems arising in matrices over finite field, connections with partially ordered sets and so on.

It has been a pleasure working these last several years with David Tranah of Cambridge University Press. In particular, I thank him and the series editor Gian-Carlo Rota for their understanding of my desire to make a two-volume book out of what was originally conceived as one volume. I prepared the manuscript using the document preparation system LaTeX, which was then edited by Cambridge University Press. I wish to thank Wayne Barrett and Vera Pless for pointing out many misprints. My two former Ph.D. students, Tom Foregger and Bryan Shader, provided me with several pages of comments and corrections. During the nearly five years in which I have worked on this book I have had, and have been grateful for, financial support from several sources: the National Science Foundation under grants No. DMS-8521521 and No. DMS-8901445, the Office of Naval Research under grant No. N00014-85-K-1613, the National Security Agency under grant No. MDA904-89 H-2060, the University of Wisconsin Graduate School under grant No. 160306 and the California Institute of Technology. For the last 25 years I have been associated with the Department of Mathematics of the University of Wisconsin in Madison. It's been a great place to work and I am looking forward to the next 25.

I wish to dedicate this book to the memory of my parents:

Mollie Verni Brualdi
Ulysses J. Brualdi

Richard A. Brualdi
Madison, Wisconsin

1

Incidence Matrices

1.1 Fundamental Concepts

Let

$$A = [a_{ij}], \quad (i = 1, 2, \ldots, m; j = 1, 2, \ldots, n)$$

be a matrix of m rows and n columns. We say that A is of *size m by n*, and we also refer to A as an m by n matrix. In the case that $m = n$ then the matrix is square of *order n*. It is always assumed that the entries of the matrix are elements of some underlying field F. Evidently A is composed of m row vectors $\alpha_1, \alpha_2, \ldots, \alpha_m$ over F and n column vectors $\beta_1, \beta_2, \ldots, \beta_n$ over F, and we write

$$A = \begin{bmatrix} \alpha_1 \\ \alpha_2 \\ \vdots \\ \alpha_m \end{bmatrix} = \begin{bmatrix} \beta_1 & \beta_2 & \cdots & \beta_n \end{bmatrix}.$$

It is convenient to refer to either a row or a column of the matrix as a *line* of the matrix. We use the notation A^T for the transpose of the matrix A. We always designate a zero matrix by O, a matrix with every entry equal to 1 by J, and the identity matrix of order n by I. In order to emphasize the size of these matrices we sometimes include subscripts. Thus $J_{m,n}$ denotes the all 1's matrix of size m by n, and this is abbreviated to J_n if $m = n$. The notations $O_{m,n}$, O_n and I_n have similar meanings. In displaying a matrix we often use $*$ to designate a submatrix of no particular concern. The $n!$ permutation matrices of order n are obtained from I_n by arbitrary permutations of the lines of I_n. A permutation matrix P of order n satisfies the matrix equations

$$PP^T = P^T P = I_n.$$

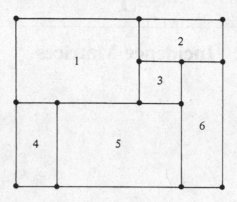

Figure 1.1

In most of our discussions the underlying field F is the field of real numbers or the subfield of the rational numbers. Indeed we will be greatly concerned with matrices whose entries consist exclusively of the integers 0 and 1. Such matrices are referred to as (0,1)-*matrices*. and they play a fundamental role in combinatorial mathematics.

We illustrate this point by an example that reformulates an elementary problem in geometry in terms of (0,1)-matrices. Let a rectangle R in the plane be of integral height m and of integral length n. Let all of R be partitioned into t smaller rectangles. Each of these rectangles is also required to have integral height and integral length. We number these smaller rectangles in an arbitrary manner $1, 2, \ldots, t$. An example with $m = 4$, $n = 5$ and $t = 6$ is illustrated in Figure 1.1.

We associate with the partitioned rectangle of Figure 1.1 the following two (0,1)-matrices of sizes 4 by 6 and 6 by 5, respectively:

$$X = \begin{bmatrix} 1 & 1 & 0 & 0 & 0 & 0 \\ 1 & 0 & 1 & 0 & 0 & 1 \\ 0 & 0 & 0 & 1 & 1 & 1 \\ 0 & 0 & 0 & 1 & 1 & 1 \end{bmatrix},$$

$$Y = \begin{bmatrix} 1 & 1 & 1 & 0 & 0 \\ 0 & 0 & 0 & 1 & 1 \\ 0 & 0 & 0 & 1 & 0 \\ 1 & 0 & 0 & 0 & 0 \\ 0 & 1 & 1 & 1 & 0 \\ 0 & 0 & 0 & 0 & 1 \end{bmatrix}.$$

The number of 1's in column i of X is equal to the height of rectangle i, and all of the 1's in column i occur consecutively. The topmost and bottommost 1's in column i of X locate the position of rectangle i with respect to the top and bottom horizontal lines of the full rectangle. The matrix Y behaves in much the same way but with respect to rows. Thus the number of 1's in row j of Y is equal to the length of rectangle j, and all the 1's in row j of Y occur consecutively. The first and last 1's in row j of Y locate the position of rectangle j with respect to the left and right vertical lines of the full rectangle. It follows from the definition of matrix multiplication that the product of the matrices X and Y satisfies

$$XY = J. \tag{1.1}$$

The matrix J in (1.1) is the matrix of 1's of size 4 by 5.

This state of affairs is valid in the general case. Thus the partitioning of a rectangle in the manner described is precisely equivalent to a matrix equation of the general form (1.1) with the following constraints. The matrices X and Y are (0,1)-matrices of sizes m by t and t by n, respectively, and J is the matrix of 1's of size m by n. The 1's in the columns of X and the 1's in the rows of Y are required to occur consecutively. If the original problem is further restricted so that all rectangles involved are squares, then we must require in addition that $m = n$ and that the sum of column i of X is equal to the sum of row i of Y, $(i = 1, 2, \ldots, t)$.

We next describe in more general terms the basic link between (0,1)-matrices and a wide variety of combinatorial problems. Let

$$X = \{x_1, x_2, \ldots, x_n\}$$

be a nonempty set of n elements. We call X an n-set. Now let $X_1, X_2, \ldots, X_m$ be m not necessarily distinct subsets of the n-set X. We refer to this collection of subsets of an n-set as a *configuration* of subsets. Vast areas of modern combinatorics are concerned with the structure of such configurations. We set $a_{ij} = 1$ if $x_j \in X_i$, and we set $a_{ij} = 0$ if $x_j \notin X_i$. The resulting (0,1)-matrix

$$A = [a_{ij}], \quad (i = 1, 2, \ldots, m; j = 1, 2, \ldots, n)$$

of size m by n is the *incidence matrix* for the configuration of subsets $X_1, X_2, \ldots, X_m$ of the n-set X. The 1's in row α_i of A display the elements in the subset X_i, and the 1's in column β_j display the occurrences of the element x_j among the subsets. Thus the lines of A give us a complete description of the subsets and the occurrences of the elements within these subsets. This representation of our configuration in terms of the (0,1)-matrix A is of the utmost importance because it allows us to apply the powerful techniques of matrix theory to the particular problem under investigation.

Let A be a (0,1)-matrix of size m by n. The *complement* C of the incidence matrix A is obtained from A by interchanging the roles of the 0's and 1's and satisfies the matrix equation

$$A + C = J.$$

We note that the matrices O and J of size m by n are complementary and correspond to the configurations with the empty set repeated m times and the full n-set repeated m times, respectively. A second incidence matrix associated with the $(0,1)$-matrix A of size m by n is the transposed matrix A^T of size n by m. The configuration of subsets associated with a transposed incidence matrix is called the *dual* of the configuration.

Suppose that we have subsets $X_1, X_2, \ldots, X_m$ of an n-set X and subsets $Y_1, Y_2, \ldots, Y_m$ of an n-set Y. Then these two configurations of subsets are regarded as the same or *isomorphic* provided that we may relabel the subsets $X_1, X_2, \ldots, X_m$ and the elements of the n-set X so that the resulting configuration coincides with the configuration $Y_1, Y_2, \ldots, Y_m$ of the n-set Y. This means that our original configurations are the same apart from the notation in which they are written.

The above isomorphism concept for configurations of subsets has a direct interpretation in terms of the incidence matrices that represent the configurations. Thus suppose that A and B are two $(0,1)$-matrices of size m by n that represent incidence matrices for subsets $X_1, X_2, \ldots, X_m$ of an n-set X and for subsets $Y_1, Y_2, \ldots, Y_m$ of an n-set Y, respectively. Then these configurations of subsets are isomorphic if and only if A is transformable to B by line permutations. In other words the configurations of subsets are isomorphic if and only if there exist permutation matrices P and Q of orders m and n, respectively, such that

$$PAQ = B.$$

In many combinatorial investigations we are often primarily concerned with those properties of a $(0,1)$-matrix that remain invariant under arbitrary permutations of the lines of the matrix. The reason for this is now apparent because such properties of the matrix become invariants of isomorphic configurations.

If two configurations of subsets are isomorphic, then their associated incidence matrices of size m by n are required to satisfy a number of necessary conditions. Thus the row sums including multiplicities are the same for both matrices and similarly for the column sums. The ranks of the two matrices must also coincide. The incidence matrices may be tested for invariants like these. But thereafter it may still be an open question as to whether or not the given configurations are isomorphic. Suppose, for example, that A is a $(0,1)$-matrix of order n such that all of the line sums of A are equal to the positive integer k. We may ask if the configuration associated with this incidence matrix is isomorphic to its dual. This will be the case if and only if there exist permutation matrices P and Q of order n such that

$$PAQ = A^T.$$

We now look at some examples of the isomorphism problem which are readily solvable. Suppose that the configuration is represented by a permutation matrix of order n. Then clearly the configuration may be represented equally well by the identity matrix I_n, and thus any two such configurations are isomorphic.

Suppose next that the configuration is represented by a $(0,1)$-matrix A of order n such that all of the line sums are equal to 2. This restriction on A allows us to replace A under line permutations by a direct sum of the form

$$A_1 \oplus A_2 \oplus \cdots \oplus A_e.$$

Each of the components of this direct sum has line sums equal to 2, and is "fully indecomposable" in the sense that it cannot be further decomposed by line permutations into a direct sum. Each fully indecomposable component is itself normalized by line permutations so that the 1's appear on the main diagonal and in the positions directly above the main diagonal with an additional 1 in the lower left corner. For example, a normalized fully indecomposable component of order 5 is given by the matrix

$$\begin{bmatrix} 1 & 1 & 0 & 0 & 0 \\ 0 & 1 & 1 & 0 & 0 \\ 0 & 0 & 1 & 1 & 0 \\ 0 & 0 & 0 & 1 & 1 \\ 1 & 0 & 0 & 0 & 1 \end{bmatrix}.$$

We have constructed a canonical form for A in the sense that if

$$B_1 \oplus B_2 \oplus \cdots \oplus B_f$$

is a second decomposition for A, then we have $e = f$ and the A_i are equal to the B_j in some ordering. The essential reasoning behind this is as follows. We first label all of the 1's in A from 1 to $2n$ in an arbitrary manner. Under line permutations two labeled 1's in a line always remain within the same line. Consider the component A_1 and its two labeled 1's in the $(1,1)$ and $(1,2)$ positions of A_1. These labeled 1's occur in some row of a component of the second decomposition, say in component B_i. But then the labeled 1 in the $(2,2)$ position of A_1 also occurs in B_i and similarly for the labeled 1 in the $(2,3)$ position of A_1. In this way we see that all of the labeled 1's in A_1 occur in B_i. But then A_1 and B_i are equal because both matrices are fully indecomposable and normalized. We may then identify the labeled 1's in the component A_2 with the labeled 1's in another component B_j of the second decomposition and so on.

The above canonical form for A implies that we have devised a straightforward procedure for deciding the isomorphism problem for configurations whose incidence matrices have all of their line sums equal to 2. The situation for $(0,1)$-matrices that have all of their line sums equal to 3 is vastly more complicated. Such matrices may already possess a highly intricate internal structure.

In the study of configurations of subsets two broad categories of problems emerge. The one deals with the structure of very general configurations, and the other deals with the structure of much more restricted configurations. In the present chapter we will see illustrations of both types of problems. We will begin with the proof of a minimax theorem that holds for an arbitrary (0,1)-matrix. Later we will also discuss certain (0,1)-matrices of such a severely restricted form that their very existence is open to question.

1.2 A Minimax Theorem

We now prove the fundamental minimax theorem of König[1936]. This theorem has a long history and many ramifications which are described in detail in the book by Mirsky[1971]. The theorem deals exclusively with properties of a (0,1)-matrix that remain invariant under arbitrary permutations of the lines of the matrix.

Theorem 1.2.1.   *Let A be a $(0,1)$-matrix of size m by n. The minimal number of lines in A that cover all of the 1's in A is equal to the maximal number of 1's in A with no two of the 1's on a line.*

Proof. We use induction on the number of lines in A. The theorem is valid in case that $m = 1$ or $n = 1$. Hence we take $m > 1$ and $n > 1$. We let ρ' equal the minimal number of lines in A that cover all of the 1's in A, and we let ρ equal the maximal number of 1's in A with no two of the 1's on a line. We may conclude at once from the definitions of ρ and ρ' that $\rho \leq \rho'$. Thus it suffices to prove that $\rho \geq \rho'$. A minimal covering of the 1's of A is called *proper* provided that it does not consist of all m rows of A or of all n columns of A. The proof of the theorem splits into two cases.

In the first case we assume that A does not have a proper covering. It follows that we must have $\rho' = \min\{m, n\}$. We permute the lines of A so that the matrix has a 1 in the $(1,1)$ position. We delete row 1 and column 1 of the permuted matrix and denote the resulting matrix of size $m - 1$ by $n - 1$ by A'. The matrix A' cannot have a covering composed of fewer than $\rho' - 1 = \min\{m - 1, n - 1\}$ lines because such a covering of A' plus the two deleted lines would yield a proper covering for A. We now apply the induction hypothesis to A' and this allows us to conclude that A' has $\rho' - 1$ 1's with no two of the 1's on a line. But then A has ρ' 1's with no two of the 1's on a line and it follows that $\rho \geq \rho'$.

In the alternative case we assume that A has a proper covering composed of e rows and f columns where $\rho' = e + f$. We permute lines of A so that these e rows and f columns occupy the initial positions. Then our permuted matrix assumes the following form

$$\begin{bmatrix} * & A_1 \\ A_2 & O \end{bmatrix}.$$

In this decomposition O is the zero matrix of size $m - e$ by $n - f$. The matrix A_1 has e rows and cannot be covered by fewer than e lines and the matrix A_2 has f columns and cannot be covered by fewer than f lines. This is the case because otherwise we contradict the fact that $\rho' = e + f$ is the minimal number of lines in A that cover all of the 1's on A. We may apply the induction hypothesis to both A_1 and A_2 and this allows us to conclude that $\rho \geq \rho'$. $\square$

The maximal number of 1's in the $(0,1)$-matrix A with no two of the 1's on a line is called the *term rank* of A. We denote this basic invariant of A by

$$\rho = \rho(A).$$

We next investigate some important applications of the König theorem. Let $X_1, X_2, \ldots, X_m$ be m not necessarily distinct subsets of an n-set X. Let

$$D = (a_1, a_2, \ldots, a_m)$$

be an ordered sequence of m distinct elements of X and suppose that

$$a_i \in X_i, (i = 1, 2, \ldots, m).$$

Then the element a_i *represents* the set X_i, and we say that our configuration of subsets has a *system of distinct representatives* (abbreviated SDR). We call D an SDR for the ordered sequence of subsets $(X_1, X_2, \ldots, X_m)$. The definition of SDR requires $a_i \neq a_j$ whenever $i \neq j$, but X_i and X_j need not be distinct as subsets of X.

The following theorem of P. Hall[1935] gives necessary and sufficient conditions for the existence of an SDR. We derive the Hall theorem from the König theorem. We remark that one may also reverse the procedure and derive the König theorem from the Hall theorem (Ryser[1963]).

Theorem 1.2.2.   *The subsets $X_1, X_2, \ldots, X_m$ of an n-set X have an SDR if and only if the set union $X_{i_1} \cup X_{i_2} \cup \cdots \cup X_{i_k}$ contains at least k elements for $k = 1, 2, \ldots, m$ and for all k-subsets $\{i_1, i_2, \ldots, i_k\}$ of the integers $1, 2, \ldots, m$.*

Proof. The necessity of the condition is clear because if a set union $X_{i_1} \cup X_{i_2} \cup \cdots \cup X_{i_k}$ contains fewer than k elements then it is not possible to select an SDR for these subsets.

We now prove the reverse implication. Let A be the (0,1)-matrix of size m by n which is the incidence matrix for our configuration of subsets. Suppose that A does not have the maximal possible term rank m. Then by the König theorem we may cover the 1's in A with e rows and f columns, where $e + f < m$. We permute the lines of A so that these e rows and f columns occupy the initial positions. Then our permuted A assumes the form

$$\begin{bmatrix} * & A_1 \\ A_2 & O \end{bmatrix}.$$

In this decomposition O is the zero matrix of size $m - e$ by $n - f$. The matrix A_2 of size $m - e$ by f has $m - e > f$. But then the last $m - e$ rows of the displayed matrix correspond to subsets of X whose union contains fewer than $m-e$ elements, and this is contrary to our hypothesis. Hence the matrix A is of term rank m, and this in turn implies that our configuration of subsets has an SDR.   □

Let $A = [a_{ij}]$ be a matrix of size m by n with elements in a field F and suppose that $m \leq n$. Then the *permanent* of A is defined by

$$\text{per}(A) = \sum a_{1i_1} a_{2i_2} \cdots a_{mi_m},$$

where the summation extends over all the m-permutations $(i_1, i_2, \ldots, i_m)$ of the integers $1, 2, \ldots, n$. Thus $\text{per}(A)$ is the sum of all possible products of m elements of A with the property that the elements in each of the products lie on different lines of A. This scalar valued function of the matrix A occurs throughout the combinatorial literature in connection with various enumeration and extremal problems. We remark that $\text{per}(A)$ remains invariant under arbitrary permutations of the lines of A. Furthermore, in the case of square matrices $\text{per}(A)$ is the same as the determinant function apart from a factor ± 1 preceding each of the products in the summation. In the case of square matrices certain determinantal laws have direct analogues for permanents. In particular, the Laplace expansion for determinants has a simple counterpart for permanents. But the basic multiplicative law of determinants

$$\det(AB) = \det(A) \det(B)$$

is flagrantly false for permanents. Similarly, the permanent function is in general greatly altered by the addition of a multiple of one row of a matrix to another. These facts tend to severely restrict the computational procedures available for the evaluation of permanents.

We return to the (0,1)-matrix A of size m by n, and we now assume that $m \leq n$. Then it follows directly from the definition of $\text{per}(A)$ that per$(A) > 0$ if and only if A is of term rank m. The following theorem is also a direct consequence of the terminology involved.

Theorem 1.2.3. *Let A be the incidence matrix for m subsets $X_1, X_2, \ldots,$ X_m of an n-set X and suppose that $m \leq n$. Then the number of distinct SDR's for this configuration of subsets is* per(A). □

The permanent function is studied more thoroughly in Chapter 7.

We have characterized a configuration of subsets by means of a $(0,1)$-matrix. The choice of the integers 0 and 1 is particularly judicious in many situations, and this is already exemplified by Theorem 1.2.3. But the configuration could also be characterized by a $(1, -1)$-matrix or for that matter by a more general matrix whose individual entries possess or fail to possess a certain property. For example, the following assertion is entirely equivalent to our formulation of the König theorem. *Let A be a matrix of size m by n with elements from a field F. The minimal number of lines in A that cover all of the nonzero elements of A is equal to the maximal number of nonzero elements in A with no two of the nonzero elements on a line.* In what follows we apply the König theorem to nonnegative real matrices.

A matrix of order n is called *doubly stochastic* provided that its entries are nonnegative real numbers and all of its line sums are equal to 1. The $n!$ permutation matrices of order n as well as the matrix of order n with every entry equal to $1/n$ are simple instances of doubly stochastic matrices. The following theorem on doubly stochastic matrices is due to Birkhoff[1946].

Theorem 1.2.4. *A nonnegative real matrix A of order n is doubly stochastic if and only if there exist permutation matrices $P_1, P_2, \ldots, P_t$ and positive real numbers $c_1, c_2, \ldots, c_t$ such that*

$$A = c_1 P_1 + c_2 P_2 + \cdots + c_t P_t \tag{1.2}$$

and

$$c_1 + c_2 + \cdots + c_t = 1. \tag{1.3}$$

Proof. If the nonnegative matrix A satisfies (1.2) and (1.3) then

$$AJ = JA = J$$

and A is doubly stochastic.

We now prove the reverse implication. We assert that the doubly stochastic matrix A has n positive entries with no two of the positive entries on a line. For if this were not the case, then by the König theorem we could cover all of the positive entries in A with e rows and f columns, where $e + f < n$. But then since all of the line sums of A are equal to 1, it follows that $n \leq e + f < n$, and this is a contradiction. Now let P_1 be the permutation matrix of order n with 1's in the same positions as those occupied by the n positive entries of A. Let c_1 be the smallest of these n positive entries. Then $A - c_1 P_1$ is a scalar multiple of a doubly stochastic matrix, and at least one more 0 appears in $A - c_1 P_1$ than in A. Hence we may iterate the argument on $A - c_1 P_1$ and eventually obtain the desired decomposition (1.2). We now multiply (1.2) by J and this immediately implies (1.3).   □

Corollary 1.2.5. *Let A be a $(0, 1)$-matrix of order n such that all of the line sums of A are equal to the positive integer k. Then there exist permutation matrices $P_1, P_2, \ldots, P_k$ such that*

$$A = P_1 + P_2 + \cdots + P_k.$$

Proof. The $(0,1)$-matrix A is a scalar multiple of a doubly stochastic matrix. This means that the same arguments used in the proof of Theorem 1.2.4 may be applied directly to the matrix A. But we now have each $c_i = 1$ and the entire process comes to an automatic termination in k steps.   □

Corollary 1.2.5 has the following amusing interpretation. A dance is attended by n boys and n girls. Each boy has been previously introduced to exactly k girls and each girl has been previously introduced to exactly k boys. No further introductions are allowed. Is it possible to pair the boys and the girls so that the boy and girl of each pair have been previously introduced? We number the boys $1, 2, \ldots, n$ in an arbitrary manner and similarly for the girls. Then we let $A = [a_{ij}]$ denote the $(0,1)$-matrix of order n defined by $a_{ij} = 1$ provided boy j has been previously introduced to girl i and by $a_{ij} = 0$ in the alternative situation. Then A satisfies all of the requirements of Corollary 1.2.5, and each of the k permutation matrices P_i gives us a desired pairing of boys and girls. The totality of all of the permitted pairings of boys and girls is equal to per(A). But it should be noted that per(A) depends not only on n and k, but also on detailed information involving the structure of the previous introductions.

Exercises

1. Derive Theorem 1.2.1 from Theorem 1.2.2.
2. Suppose in Theorem 1.2.2 the set union $X_{i_1} \cup X_{i_2} \cup \cdots \cup X_{i_k}$ always contains at least $k + 1$ elements. Let x be any element of X_1. Show that the sets $X_1, X_2, \ldots, X_m$ have an SDR with the property that x represents X_1.
3. Let A be a $(0,1)$-matrix of order n satisfying the equation $A + A^T = J - I$. Prove that the term rank of A is at least $n - 1$.
4. Let A be an m by n $(0,1)$-matrix. Suppose that there exist a positive integer p such that each row of A contains at least p 1's and each column of A contains at most p 1's. Prove that per$(A) > 0$.
5. Let $X_1, X_2, \ldots, X_m$ and $Y_1, Y_2, \ldots, Y_m$ be two partitions of the n-set X into m subsets. Prove that there exists a permutation $j_1, j_2, \ldots, j_m$ of $\{1, 2, \ldots, m\}$ such that

$$X_i \cap Y_{j_i} \neq \emptyset, \quad (i = 1, 2, \ldots, m)$$

if and only if each union of k of the sets $X_1, X_2, \ldots, X_m$ contains at most k of the sets $Y_1, Y_2, \ldots, Y_m, (k = 1, 2, \ldots, m)$.

6. Let $x = (x_1, x_2, \ldots, x_n)$ and $y = (y_1, y_2, \ldots, y_n)$ be two *monotone* real vectors:

$$x_1 \geq x_2 \geq \cdots \geq x_n; \qquad y_1 \geq y_2 \geq \cdots \geq y_n.$$

Assume that there exists a doubly stochastic matrix S of order n such that $x = yS$. Prove that

$$x_1 + \cdots + x_k \leq y_1 + \cdots + y_k, \quad (k = 1, 2, \ldots, n)$$

with equality for $k = n$. (The vector x is said to be *majorized* by y.)

7. Prove that the product of two doubly stochastic matrices is a doubly stochastic matrix.

8. Let A be a doubly stochastic matrix of order n. Let A' be a matrix of order $n - 1$ obtained by deleting the row and column of a positive element of A. Prove that $per(A) > 0$.

References

G. Birkhoff[1946], Tres observaciones sobre el algebra lineal, *Univ. Nac. Tucumán Rev.* Ser. A, pp. 147–151.

P. Hall[1935], On representatives of subsets, *J. London Math. Soc.*, 10, pp. 26–30.

D. König[1936], *Theorie der endlichen und unendlichen Graphen*, Leipzig, reprinted by Chelsea[1960], New York.

L. Mirsky[1971], *Transversal Theory*, Academic Press, New York.

H.J. Ryser[1963], *Combinatorial Mathematics*, Carus Mathematical Monograph No. 14, Math. Assoc. of Amer., Washington, D.C.

1.3   Set Intersections

We return to the m not necessarily distinct subsets $X_1, X_2, \ldots, X_m$ of an n-set X. Up to now we have discussed in some detail the formal structure of the $(0,1)$-matrix A of size m by n which is the incidence matrix for this configuration of subsets. In what follows the algebraic properties of the matrix A will begin to play a much more dominant role.

We are now concerned with the cardinalities of the set intersections $X_i \cap X_j$, and in order to study this concept we multiply the above matrix A by its transpose. We thereby obtain the matrix equation

$$AA^T = S. \tag{1.4}$$

The matrix S of (1.4) is a symmetric matrix of order m with nonnegative integral elements. The element s_{ij} in the (i, j) position of S records the cardinality of the set intersection $X_i \cap X_j$, namely,

$$s_{ij} = |X_i \cap X_j|, \quad (i, j = 1, 2, \ldots, m).$$

The main diagonal elements of S display the cardinalities of the m subsets $X_1, X_2, \ldots, X_m$. It should be noted that all of this information on the cardinalities of the set intersections is exhibited in (1.4) in an exceedingly compact form.

We mention next two variants of (1.4). We may reverse the order of multiplication of the matrix A and its transpose and this yields the matrix equation

$$A^T A = T. \tag{1.5}$$

The matrix T of (1.5) is a symmetric matrix of order n with nonnegative integral elements. The element t_{ij} in the (i, j) position of T records the number of times that the elements x_i and x_j occur among the subsets $X_1, X_2, \ldots, X_m$. The main diagonal elements of T display the totality of the occurrences of each of the n elements among the m subsets. The matrix T may also be regarded as recording the cardinalities of the set intersections of the dual configuration.

Our second variant of (1.4) involves the complement C of the incidence matrix A. We may multiply A by the transpose of C and this yields the matrix equation

$$AC^T = W. \tag{1.6}$$

This matrix equation differs noticeably from the two preceding equations. The matrix W need no longer be symmetric. The element w_{ij} in the (i, j) position of W records the cardinality of the set difference $X_i - X_j$. (The set difference $X_i - X_j$ is the set of all elements in X_i but not in X_j.) The matrix W has 0's in the m main diagonal positions.

We recall that for a matrix A with real elements we have

$$\operatorname{rank}(A) = \operatorname{rank}(AA^T).$$

Hence the matrices A and S of (1.4) satisfy

$$\operatorname{rank}(S) = \operatorname{rank}(A) \leq m, n. \tag{1.7}$$

Thus it follows from (1.7) that if S is nonsingular, then we must have

$$m \leq n. \tag{1.8}$$

The inequality (1.8) is of interest because it tells us that the algebraic requirement of the nonsingularity of S automatically imposes a constraint between the two integral parameters m and n. In many investigations the extremal configurations with $m = n$ are especially significant. For the dual configuration it follows that if T is nonsingular, then we must have $n \leq m$.

We make no attempt now to study the matrix equation (1.4) and its variants (1.5) and (1.6) in their full generality. We look at a very special case of (1.4) and show that this already leads us to important unanswered questions.

Let t be a positive integer and suppose that A is a $(0,1)$-matrix of size m by n that satisfies the matrix equation

$$AA^T = tI + J. \tag{1.9}$$

Thus in (1.9) we have selected our symmetric matrix S of (1.4) in a particularly simple form, namely with $t + 1$ in the m main diagonal positions and with 1's in all other positions. In order to evaluate the determinant of $tI + J$ we first subtract column 1 from all other columns and we then add the last $m - 1$ rows to the first row. This tells us that

$$\det(tI + J) = (t + m)t^{m-1} \neq 0.$$

Thus the matrix $tI + J$ is nonsingular, and by (1.8) we may conclude that $m \leq n$.

Now suppose that $m = n$. We show that in this case the incidence matrix A possesses a number of remarkable symmetries. Since the matrix A is square of order n we may apply the multiplicative law of determinants to the matrix equation (1.9). Thus we have

$$\det(AA^T) = \det(A)\det(A^T) = (\det(A))^2 = (t + n)t^{n-1}$$

and

$$\det(A) = \pm(t + n)^{1/2}t^{(n-1)/2}. \tag{1.10}$$

Since A is a $(0,1)$-matrix it follows that the expression on the right side of (1.10) is of necessity an integer. It also follows from (1.9) that all of the row sums of A are equal to $t + 1$. Thus we may write

$$AJ = (t + 1)J. \tag{1.11}$$

But A is a nonsingular matrix and hence the inverse of A satisfies

$$A^{-1}J = (t + 1)^{-1}J.$$

Moreover, it follows from (1.9) that

$$AA^T J = tJ + J^2 = (t + n)J$$

and hence

$$A^T J = (t + 1)^{-1}(t + n)J.$$

We next take transposes of both sides of this equation and obtain

$$JA = (t + 1)^{-1}(t + n)J. \tag{1.12}$$

The multiplication of (1.12) by J implies

$$JAJ = n(t + 1)^{-1}(t + n)J.$$

But from (1.11) we also have

$$JAJ = n(t+1)J,$$

whence it follows that

$$n = t^2 + t + 1. \tag{1.13}$$

This additional relation between n and t allows us to write (1.10) in the form

$$\det(A) = \pm(t+1)t^{t(t+1)/2},$$

and we now see that our formula for $\det(A)$ is, indeed, an integer. We may substitute (1.13) into (1.12) and with (1.11) obtain

$$AJ = JA = (t+1)J. \tag{1.14}$$

The equations of (1.14) tell us that all of the line sums of A are equal to $t + 1$.

We next investigate the matrix product $A^T A$ and note that

$$A^T A = A^{-1}(AA^T)A = A^{-1}(tI + J)A = tI + A^{-1}JA = tI + J.$$

A matrix of order n with real elements is called *normal* provided that it commutes under multiplication with its transpose. It follows that our matrix A is normal and satisfies

$$AA^T = A^T A = tI + J. \tag{1.15}$$

We may also readily verify that the complement C of A satisfies

$$AC^T = C^T A = t(J - I)$$

and

$$CC^T = C^T C = tI + t(t-1)J.$$

We now discuss some specific solutions of the matrix equation (1.15). We have shown that the order n of A satisfies (1.13) so that there is only a single integer parameter t involved. For the case in which $t = 1$ it follows readily that all solutions of (1.15) are given by the (0,1)-matrices of order 3 with all line sums equal to 2. These six matrices yield a single configuration in the sense of isomorphism.

The configuration associated with a solution of (1.15) for $t > 1$ is called a *finite projective plane* of *order t*. We exhibit the incidence matrix for the projective plane of order 2:

$$A = \begin{bmatrix} 1 & 1 & 1 & 0 & 0 & 0 & 0 \\ 1 & 0 & 0 & 1 & 1 & 0 & 0 \\ 1 & 0 & 0 & 0 & 0 & 1 & 1 \\ 0 & 1 & 0 & 1 & 0 & 1 & 0 \\ 0 & 1 & 0 & 0 & 1 & 0 & 1 \\ 0 & 0 & 1 & 1 & 0 & 0 & 1 \\ 0 & 0 & 1 & 0 & 1 & 1 & 0 \end{bmatrix}. \tag{1.16}$$

This is the "smallest" finite projective plane, and it is easy to verify that the projective plane of order 2 is unique in the sense of isomorphism. We remark in passing that the incidence matrix A of (1.16) possesses a most unusual property:

$$\mathrm{per}(A) = |\det(A)| = 24.$$

Thus all of the 24 permutations that contribute to $\det(A)$ are of the same sign.

Finite projective planes have been constructed for all orders t that are equal to the power of a prime number. No planes have as yet been constructed for any other orders, but they are known to be impossible for infinitely many values of t. For a long time the smallest undecided case was $t = 10$. Notice that the associated incidence structure is already of order 111. Using sophisticated computer calculations, Lam, Thiel and Swierzc[1989] have recently concluded that there is no finite projective plane of order 10. The smallest order for which nonisomorphic planes exist is $t = 9$.

One of the major unsolved problems in combinatorics is the determination of the precise range of values of t for which projective planes of order t exist. The determination of the number of nonisomorphic solutions for a general t appears to be well beyond the range of present day techniques. These extremal configurations are of the utmost importance and have many ramifications. They and their generalizations will be studied in some detail in the sequel to this book, *Combinatorial Matrix Classes*.

We next consider a finite projective plane whose associated incidence matrix is symmetric. The proof of the following theorem illustrates the effective use of matrix algebra techniques.

Theorem 1.3.1. *Let a finite projective plane* Π *be such that its associated incidence matrix A is symmetric. Suppose further that the order t of* Π *is not equal to an integral square. Then the incidence matrix A of* Π *contains exactly $t + 1$ 1's on its main diagonal.*

Proof. We first recall the following fundamental property concerning the eigenvalues (characteristic roots) of a matrix. Let A be a matrix of order n with elements in a field F and let the n eigenvalues of A be $\lambda_1, \lambda_2, \ldots, \lambda_n$. Let $f(A)$ be an arbitrary polynomial in the matrix A. Then the n eigenvalues of $f(A)$ are $f(\lambda_1), f(\lambda_2), \ldots, f(\lambda_n)$.

Since the incidence matrix A of Π is symmetric it follows that we may write (1.15) in the form

$$A^2 = tI + J. \tag{1.17}$$

The characteristic polynomial $f(\lambda)$ of $tI + J$ equals

$$f(\lambda) = \det(\lambda I - (tI + J)) = (\lambda - (t+1)^2)(\lambda - t)^{t^2+t}. \tag{1.18}$$

The calculation of $f(\lambda)$ in (1.18) is much the same as the one carried out earlier for $\det(tI + J)$. Thus we see that the $n = t^2 + t + 1$ eigenvalues of $tI + J$ are $(t+1)^2$ of multiplicity 1 and t of multiplicity $t^2 + t$. By (1.17) and the property concerning eigenvalues quoted at the outset of the proof it follows that the n eigenvalues of A are either $t+1$ or else $-(t+1)$ of multiplicity 1, and $\pm\sqrt{t}$ of appropriate multiplicities. Let u denote the column vector of n 1's. The matrix A has all its row sums equal to $t+1$ so that

$$Au = (t+1)u. \tag{1.19}$$

Equation (1.19) tells us that u is an eigenvector of A with associated eigenvalue $t+1$, and thus $-(t+1)$ does not arise as an eigenvalue of A.

The *trace* of a matrix of order n is the sum of the n main diagonal elements of the matrix and this in turn is equal to the sum of the n eigenvalues of the matrix. Thus there exists an integer e determined by the multiplicities of the eigenvalues $\pm\sqrt{t}$ of our incidence matrix A such that we may write

$$\mathrm{tr}\,(A) = a_{11} + a_{22} + \cdots + a_{nn} = \lambda_1 + \lambda_2 + \cdots + \lambda_n = t + 1 + e\sqrt{t}.$$

We know that A is a (0,1)-matrix so that $\mathrm{tr}\,(A)$ is an integer. But now using for the first time our hypothesis that t is not equal to an integral square it follows that we must have $e = 0$. $\qquad\square$

We note that the incidence matrix A of (1.16) for the projective plane of order 2 is symmetric. Consequently we now see that it is no accident that exactly three 1's appear on its main diagonal.

Exercises

1. Show that the determinant of the matrix $tI + aJ$ of order n equals $t^{n-1}(t + an)$.
2. Show that the n eigenvalues of the matrix $tI + aJ$ of order n are t with multiplicity $n-1$ and $t + an$.

3. Let A be an m by n (0,1)-matrix which satisfies the matrix equation $AA^T = tI + aJ$ where $t \neq 0$. Prove that $n \geq m$.
4. Let A be a (0,1)-matrix of order n which satisfies the matrix equation $AA^T = tI + aJ$. Generalize the argument given in the text for $a = 1$ to prove that A is a normal matrix.
5. Verify that the projective plane of order 2 is unique in the sense of isomorphism.
6. Verify that the incidence matrix A of the projective plane of order 2 satisfies $\text{per}(A) = |\det(A)| = 24$.
7. Determine a formula for the permanent of the matrix $tI + aJ$ of order n in terms of derangement numbers D_k. (D_k is the number of permutations of $\{1, 2, \ldots, k\}$ which have no fixed point.)
8. Let S denote a nonzero symmetric matrix of order $m \geq 2$ with nonnegative integral elements and with 0's in all of the main diagonal positions. Prove that there exists a diagonal matrix D of order m, an integer n and a (0,1)-matrix A of size m by n such that $AA^T = D + S$. Indeed show that a matrix A can be found with all column sums equal to 2.

References

C.W.H. Lam, L.H. Thiel and S. Swierzc[1989], The nonexistence of finite projective planes of order 10, *Canad. J. Math.*, XLI, pp. 1117–1123.

H.J. Ryser[1963], *Combinatorial Mathematics*, Carus Mathematical Monograph No. 14, Math. Assoc. of Amer., Washington, D.C.

1.4 Applications

We now apply the terminology and concepts of the preceding sections to prove several elementary theorems. The results are appealing in their simplicity and give us additional insight into the structure of (0,1)-matrices. We recall that a submatrix of order m of a matrix A of order n is called *principal* provided that the submatrix is obtained from A by deleting $n - m$ rows and $n - m$ columns of A with both sets of deleted rows and columns numbered identically $i_1, i_2, \ldots, i_{n-m}$. This definition of principal submatrix is equivalent to the assertion that the submatrix may be placed in the upper left corner of A by simultaneous permutations of the lines of A.

Theorem 1.4.1. *Let A be a $(0, 1)$-matrix of order n and suppose that A contains no column of 0's. Then A contains a principal submatrix which is a permutation matrix.*

Proof. The proof is by induction on n. The result is certainly valid in case $n = 1$ so that we may assume that $n > 1$. Let A contain e columns with column sums equal to 1 and $n - e$ columns with column sums greater than 1. We simultaneously permute the lines of A so that the e columns with column sum equal to 1 are the initial columns of the permuted matrix. We designate the permuted matrix by A' and note that it suffices to prove the theorem for A'. Let A_1 denote the principal submatrix of order e in the upper left corner of A' and let A_2 denote the principal submatrix of order $n - e$ in the lower right corner of A'.

In the event that A_1 is empty we delete a row of A_2 and its corresponding column. We then apply the induction hypothesis to the submatrix of order $n - 1$ and the result follows. Now suppose that A_1 is not empty and that A_1 contains a row of 0's. We now delete this row in A' and its corresponding column and once again apply the induction hypothesis to the submatrix of order $n - 1$. There remains the alternative case in which A_1 contains no row of 0's. Then A_1 has all of its row sums greater than or equal to 1 and all of its column sums less than or equal to 1. This state of affairs now implies that A_1 has all of its line sums equal to 1. Thus A_1 itself is the required principal submatrix of A'.   □

We now use the preceding theorem to characterize the $(0,1)$-matrices of order n whose permanents are equal to 1 (Brualdi[1966]).

Theorem 1.4.2. *Let A be a $(0, 1)$-matrix of order n. Then $\mathrm{per}(A) = 1$ if and only if the lines of A may be permuted to yield a triangular matrix with 1's in the n main diagonal positions and with 0's above the main diagonal.*

Proof. The proof is immediate in case A is permutable to triangular form. We use induction on n for the reverse implication. The result is obvious for $n = 1$. Since $\mathrm{per}(A) = 1$ we may permute the lines of A so that n 1's appear on the main diagonal of the matrix. We designate the permuted matrix by A' and suppose that A' has all of its row sums greater than 1. Then the transpose of the matrix $A' - I$ satisfies the requirements of Theorem 1.4.1 and hence contains a principal submatrix which is a permutation matrix. But then it follows that $\mathrm{per}(A) = \mathrm{per}(A') > 1$ and this is a contradiction. Hence A contains a row with a single 1. Thus we may permute the lines of A so that row 1 of the matrix contains a 1 in the $(1,1)$ position and 0's elsewhere. We now delete the first row and column of this matrix and apply the induction hypothesis to this submatrix of order $n - 1$.   □

A *triangle* of a $(0,1)$-matrix A is a submatrix of A of order 3 such that all of the line sums of the submatrix are equal to 2. The following theorem of Ryser[1969] deals with the set intersections of configurations whose incidence matrices contain no triangles.

Theorem 1.4.3. *Let A be a $(0, 1)$-matrix of size m by n. Suppose that A contains no triangles and that every element of AA^T is positive. Then A contains a column of m 1's.*

Proof. The proof is by induction on m. The result is valid for both $m = 1$ and $m = 2$ so that we may assume that $m \geq 3$. We delete row 1 of A and apply the induction hypothesis to the submatrix of A consisting of the last $m - 1$ rows of A. This submatrix contains a column of $m - 1$ 1's. Then either A contains a column of m 1's and we are done, or else A contains a column with a 0 in the first position and with 1's in the remaining $m - 1$ positions. We repeat the argument on A with row 2 of A deleted. Then either A contains a column of m 1's and we are done, or else A contains a column with a 0 in the second position and with 1's in the remaining $m - 1$ positions. We finally repeat the argument a third time on A with row 3 of A deleted. But now A cannot contain a column with a 0 in the third position and with 1's in the remaining positions because such a column yields a triangle within A. Hence the matrix A contains a column of m 1's as desired.   □

An extensive literature in the combinatorial geometry of convex sets is concerned with "Helly type" theorems (Hadwiger et al.[1964]). The following elementary proposition affords a good illustration of a Helly type theorem. *Let there be given a finite number of closed intervals on the real line with the property that every pair of the intervals has a point in common. Then all of the intervals have a point in common.*

We show that the above proposition is actually a special case of Theorem 1.4.3. Let the closed intervals be labeled $X_1, X_2, \ldots, X_m$ and let the endpoints of these intervals occur at the following points on the real line

$$e_1 < e_2 < \cdots < e_n.$$

We now form the incidence matrix A of size m by n of intervals versus endpoints. Thus we set $a_{ij} = 1$ if the point e_j is contained in the interval X_i and we set $a_{ij} = 0$ in the contrary case. This incidence matrix has a very special form, namely, the 1's in each row occur consecutively. Now the 1's in every submatrix also occur consecutively in each of the rows of the submatrix and hence A contains no triangles. Furthermore, the pairwise intersection property of the intervals implies that every element of AA^T is positive. But then Theorem 1.4.3 asserts that the matrix A contains a column of m 1's and this means that all of the intervals have a point in common.

We digress and consider in somewhat oversimplified form a problem from archaeology (Kendall[1969] and Shuchat[1984]). Suppose that we have a set of *graves* $G_1, G_2, \ldots, G_m$ and a set of *artifacts* (or aspects of artifacts) $a_1, a_2, \ldots, a_n$ collected from these graves. We form the incidence matrix of size m by n of graves versus artifacts in the usual way. Suppose that it is possible for us to permute the rows of A so that the 1's in each column occur consecutively. Then such a permutation of the rows of A determines a chronology of the graves and this in turn assigns a sequence date to each artifact. An incidence matrix in which the 1's in each column occur consecutively is called a *Petrie matrix* in honor of Flinders Petrie, a noted English Egyptologist. We have encountered Petrie matrices (or their transposes) in our discussion of the Helly type theorem as well as in the rectangle partitioning problem exemplified by Figure 1.1.

Our next theorem deals directly with set intersections and yields a considerable refinement of the inequality (1.8) for configurations related to finite projective planes.

Theorem 1.4.4.   *Let A be a $(0,1)$-matrix of size $m = t^2 + t + 1$ by n. Suppose that A contains no column of 0's and that A satisfies the matrix equation*

$$AA^T = tI + J \ (t \geq 2). \tag{1.20}$$

Then the only possible values of n occur for $n = t^2 + t + 1$ and for $n = t^3 + t^2 + t + 1$. The first case yields a projective plane of order t and the second case yields the unique configuration in which A contains a column of 1's.

Proof. We first note that the assumption that A contain no column of 0's is a natural one because such columns can be adjoined to A without affecting the general form of the matrix equation (1.20).

We suppose next that A contains a column of 1's. Then it follows from (1.20) that all of the remaining column sums of A are equal to 1 and hence A has a totality of

$$n = t(t^2 + t + 1) + 1 = t^3 + t^2 + t + 1$$

columns. The matrix A is unique apart from column permutations.

We now deal with the case in which A does not contain a column of 1's. We denote the sum of column 1 of A by s. We permute the rows of A so that the s 1's in column 1 of A occupy the initial positions in column 1, and we then permute the remaining columns of A so that the $t + 1$ 1's in row 1 occupy the initial positions in row 1. We designate the resulting matrix by A'. Then by (1.20) the first $t + 1$ columns of A' contain exactly one 1 in each of rows $2, 3, \ldots, m$. Hence the total number of 1's in the first $t + 1$ columns of A' is equal to

$$(t + 1) + (t^2 + t) = (t + 1)^2.$$

Now by construction row $s + 1$ of A' has a 0 in the initial position. But by (1.20) row $s + 1$ of A' has inner product 1 with each of rows $1, 2, \ldots, s$ of A'. Since the s 1's in column 1 of A' occur in the initial positions, it follows from (1.20) that row $s + 1$ of A' contains at least s 1's. But row $s + 1$ contains exactly $t + 1$ 1's and hence

$$s \leq t + 1. \tag{1.21}$$

The argument applied to column 1 of A holds for an arbitrary column of A, and hence every column of A satisfies (1.21). We have noted that the total number of 1's in the first $t + 1$ columns of A' is equal to $(t+1)^2$, and this in conjunction with (1.21) tells us that $s = t + 1$. But then all of the column sums of A are equal to $t + 1$, and hence all of the line sums of A are equal to $t + 1$. This means that A is a square and $m = n$.   □

Our concluding theorem in this chapter involves an application of (0,1)-matrices to number theory. We study the following integral matrix B of order n:

$$B = [b_{ij}] = [(i,j)], \quad (i, j = 1, 2, \dots, n) \tag{1.22}$$

where (i, j) denotes the positive greatest common divisor of the integers i and j.

Let m be a positive integer and let $\phi(m)$ denote the Euler ϕ-function of m. We recall that $\phi(m)$ is defined as the number of positive integers less than or equal to m and relatively prime to m. We also recall that

$$m = \sum_{d \mid m} \phi(d), \tag{1.23}$$

where the summation extends over all of the positive divisors d of m.

We now prove a classical theorem of Smith[1876] using the techniques of Frobenius[1879].

Theorem 1.4.5.   *The determinant of the matrix B of (1.22) satisfies*

$$\det(B) = \prod_{i=1}^{n} \phi(i). \tag{1.24}$$

Proof. Let $A = [a_{ij}]$ be the (0,1)-matrix of order n defined by the relationships $a_{ij} = 1$ if j divides i and $a_{ij} = 0$ if j does not divide i $(i, j = 1, 2, \dots, n)$. We define the diagonal matrix

$$\Phi = \text{diag}[\phi(1), \phi(2), \dots, \phi(n)]$$

of order n whose main diagonal elements are $\phi(1), \phi(2), \dots, \phi(n)$. Then

$$A\Phi A^T = [a_{ij}]\Phi[a_{ji}] = [a_{ij}\phi(j)][a_{ji}] = \left[\sum_{t=1}^{n} a_{it}\phi(t)a_{jt} \right].$$

The definition of the (0,1)-matrix A implies that

$$\sum_{t=1}^{n} a_{it}\phi(t)a_{jt} = \sum \phi(d_{ij}),$$

where d_{ij} ranges over all of the positive common divisors of i and j. But then by (1.23) we have

$$\sum \phi(d_{ij}) = (i,j),$$

whence

$$A\Phi A^T = B.$$

The (0,1)-matrix A of order n is triangular with 1's in the n main diagonal positions. Since the determinant function is multiplicative it follows that

$$\det(B) = \det(\Phi).$$

□

Exercises

1. Let A be a (0,1)-matrix of order n and suppose that $\mathrm{per}(A) = 2$. Show that there exists an integer $k \geq 2$ and a square submatrix B of A whose lines can be permuted to obtain a (0,1)-matrix with 1's exactly in the positions $(1,1), (2,2), \ldots, (k,k), (1,2), \ldots, (k-1,k), (k,1)$.
2. Deduce that the matrix B of Smith in (1.22) is a positive definite matrix.
3. Let $X = \{x_1, x_2, \ldots, x_n\}$ be a set of n distinct positive integers. Let $A = [a_{ij}]$ be the *greatest common divisor matrix* for X defined by $a_{ij} = (x_i, x_j), (i, j = 1, 2, \ldots, n)$. If X is *factor closed* in the sense that each positive integral divisor of an element in X is also in X, then generalize the argument in the proof of Theorem 1.4.5 to evaluate the determinant of A. Prove that the matrix A is positive definite for all X (Beslin and Ligh[1989]).

References

S. Beslin and S. Ligh[1989], Greatest common divisor matrices, *Linear Alg. Applics.*, 118, pp. 69–76.

R.A. Brualdi[1966], Permanent of the direct product of matrices, *Pac. J. Math.*, 16, pp. 471–482.

G. Frobenius[1879], Theorie der linearen Formen mit ganzen Coefficienten, *J. für reine und angew. Math.*, 86, pp. 146–208.

H. Hadwiger, H. Debrunner and V. Klee[1964], *Combinatorial Geometry in the Plane*, Holt, Rinehart and Winston, New York.

D.G. Kendall[1969], Incidence matrices, interval graphs and seriation in archaeology, *Pacific J. Math.*, 28, pp. 565–570.

H.J. Ryser[1969], Combinatorial configurations, *SIAM J. Appl. Math.*, 17, pp. 593–602.

A. Schuchat[1984], Matrix and network models in archaeology, *Math. Magazine*, 57, pp. 3–14.

H.J.S. Smith[1876], On the value of a certain arithmetical determinant, *Proc. London Math. Soc.*, 7, pp. 208–212.

2

Matrices and Graphs

2.1 Basic Concepts

A *graph G* (*simple graph*) consists of a finite set $V = \{a, b, c, \ldots\}$ of elements called *vertices* (*points*) together with a prescribed set E of *unordered* pairs of *distinct* vertices of V. (The set E is necessarily finite.) The number n of elements in the finite set V is called the *order* of the graph G. Every unordered pair α of vertices a and b in E is called an *edge* (*line*) of the graph G, written

$$\alpha = \{a, b\} = \{b, a\}.$$

We call a and b the *endpoints* of α. Two vertices on the same edge or two distinct edges with a common vertex are *adjacent*. Also, an edge and a vertex are *incident* with one another if the vertex is contained in the edge. Those vertices incident with no edge are *isolated*. A *complete graph* is one in which all possible pairs of vertices are edges. Let G be a graph and let K be the complete graph with the same vertex set V. Then the *complement* $\overline{G}$ of G is the graph with vertex set V and with edge set equal to the set of edges of K minus those of G.

A *subgraph* of a graph G consists of a subset V' of V and a subset E' of E that themselves form a graph. If E' contains all edges of G both of whose endpoints belong to V', then the subgraph is called an *induced subgraph* and is denoted by $G(V')$. A *spanning subgraph* of G has the same vertex set as G. Two graphs G and G' are *isomorphic* provided there exists a 1-1 correspondence between their vertex sets that preserves adjacency. Two complete graphs with the same order are isomorphic, and we denote a complete graph of order n by K_n.

If the definition of a graph is altered to allow a pair of vertices to form more than one distinct edge, then the structure is called a *multigraph*. Its edges are called *multiedges* (*multilines*) and the number of distinct edges of the form $\{a, b\}$ is called the *multiplicity* $m\{a, b\}$ of the edge $\{a, b\}$. The further generalization by allowing *loops*, edges of the form $\{a, a\}$ making a vertex adjacent to itself, results in a *general graph*. For both multigraphs and general graphs we require that the edge sets be finite. Terms such as *order*, *endpoints*, *adjacent*, *incident*, *isolated*, etc. carry over directly to multigraphs and general graphs.

Let G be a multigraph. Then the *degree* (*valency*) of a vertex in G is the number of edges incident with the vertex. Since each edge of G has two distinct endpoints, the sum of the degrees of the vertices of G is twice the number of its edges. The graph G is *regular* if all vertices have the same degree. If there are precisely k edges incident with each vertex of a graph, then we say that the graph is *regular of degree* k. A regular graph of degree 3 is called *cubic*.

One may ask for the number of graphs of a specified order n. This number has been determined in a certain sense. But the answer is far from elementary and we refer the reader to Harary and Palmer[1973] for a discussion of a variety of problems dealing with graphical enumeration.

Exercises

1. Prove there are as many graphs of order n with k edges as there are with $\binom{n}{2} - k$ edges. Determine the number of graphs of order at most 5.
2. Prove that a graph always has two distinct vertices with the same degree. Show by example that this need not hold for multigraphs.
3. Prove that a cubic graph has an even number of vertices.

References

C. Berge[1976], *Graphs and Hypergraphs*, North-Holland, Amsterdam.

N. Biggs[1974], *Algebraic Graph Theory*, Cambridge Tracts in Mathematics No. 67, Cambridge University Press, Cambridge.

B. Bollobás[1979], *Graph Theory*, Springer-Verlag, New York.

J.A. Bondy and U.S.R. Murty[1976], *Graph Theory with Applications*, North-Holland, New York.

F. Harary[1969], *Graph Theory*, Addison-Wesley, Reading, Mass.

F. Harary and E.M. Palmer[1973], *Graphical Enumeration*, Academic Press, New York.

W.T. Tutte[1984], *Graph Theory*, Encyclopedia of Mathematics and Its Applications, Vol. 21, Addison-Wesley, Reading, Mass.

R.J. Wilson[1972], *Introduction to Graph Theory*, Academic Press, New York.

2.2 The Adjacency Matrix of a Graph

Let G denote a general graph of order n with vertex set

$$V = \{a_1, a_2, \ldots, a_n\}.$$

We let a_{ij} equal the multiplicity $m\{a_i, a_j\}$ of the edges of the form $\{a_i, a_j\}$. This means, of course, that $a_{ij} = 0$ if there are no edges of the form $\{a_i, a_j\}$. Also, $m\{a_i, a_i\}$ equals the number of loops at vertex a_i. The resulting matrix

$$A = [a_{ij}], \quad (i, j, = 1, 2, \ldots, n)$$

of order n is called the *adjacency matrix* of G. The matrix A characterizes G.

We note that A is a symmetric matrix with nonnegative integral elements. The trace of A denotes the number of loops in G. If G is a multigraph, then the trace of A is zero and the sum of line i of A equals the degree of vertex a_i. If G is a graph, then A is a symmetric (0,1)-matrix of trace zero.

The concept of graph isomorphism has a direct interpretation in terms of the adjacency matrix of the graph. Thus let G and G' denote two general graphs of order n and let the adjacency matrices of these graphs be denoted by A and A', respectively. Then the general graphs G and G' are isomorphic if and only if A is transformable into A' by simultaneous permutations of the lines of A. Thus G and G' are isomorphic if and only if there exists a permutation matrix P of order n such that

$$PAP^T = A'.$$

Let G be a general graph. A sequence of m successively adjacent edges

$$\{a_0, a_1\}, \{a_1, a_2\}, \ldots, \{a_{m-1}, a_m\}, (m > 0)$$

is called a *walk* of *length* m, and is also denoted by

$$a_0 \to a_1 \to a_2 \to \cdots \to a_{m-1} \to a_m$$

and by

$$a_0, a_1, a_2, \ldots, a_{m-1}, a_m.$$

The vertices a_0 and a_m are the *endpoints* of the walk. The walk is *closed* or *open* according as $a_0 = a_m$ or $a_0 \neq a_m$. A walk with distinct edges is called a *trail*. A walk with distinct edges and in addition distinct vertices (except, possibly, $a_0 = a_m$) is called a *chain*. A closed chain is called a *cycle*. Notice that in a graph a cycle must contain at least 3 edges. But in a general graph a loop or a pair of multiple edges form a cycle.

Let us now form

$$A^2 = \left[\sum_{t=1}^{n} a_{it} a_{tj} \right], \quad (i, j = 1, 2, \ldots, n). \tag{2.1}$$

Then (2.1) implies that the element in the (i, j) position of A^2 equals the number of walks of length 2 with a_i and a_j as endpoints. In general, the element in the (i, j) position of A^k equals the number of walks of length k with a_i and a_j as endpoints. The numbers for closed walks appear on the main diagonal of A^k.

Let G be the complete graph K_n of order n. We determine the number of walks of length k in K_n with a_i and a_j as endpoints. The adjacency matrix of K_n is

$$A = J - I.$$

We know that $J^e = n^{e-1}J$ so that

$$A^k = \left[n^{k-1} - \binom{k}{1} n^{k-2} + \binom{k}{2} n^{k-3} - \cdots + (-1)^{k-1} \binom{k}{k-1} \right] J + (-1)^k I.$$

But

$$(n-1)^k = n^k - \binom{k}{1} n^{k-1} + \binom{k}{2} n^{k-2} - \cdots + (-1)^{k-1} \binom{k}{k-1} n + (-1)^k$$

and hence we have

$$A^k = \left(\frac{(n-1)^k - (-1)^k}{n} \right) J + (-1)^k I.$$

We return to the general graph G and its adjacency matrix A. The polynomial

$$f(\lambda) = \det(\lambda I - A)$$

is called the *characteristic polynomial* of G. The collection of the n eigenvalues of A is called the *spectrum* of G. Since A is symmetric the spectrum of G consists of n real numbers.

Suppose that G and G' are isomorphic general graphs. Then we have noted that there exists a permutation matrix P such that the adjacency matrices A and A' of G and G', respectively, satisfy

$$PAP^T = A'.$$

But the transpose of a permutation matrix is equal to its inverse. Thus A and A' are similar matrices and hence G and G' have the same spectrum. Two nonisomorphic general graphs G and G' with the same spectrum are called *cospectral*. We exhibit in Figure 2.1 two pairs of cospectral graphs of orders 5 and 6 with characteristic polynomials $f(\lambda) = (\lambda - 2)(\lambda + 2)\lambda^3$ and $f(\lambda) = (\lambda - 1)(\lambda + 1)^2(\lambda^3 - \lambda^2 - 5\lambda + 1)$, respectively.

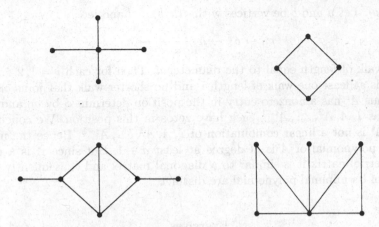

Figure 2.1.   Two pairs of cospectral graphs.

A general graph G is *connected* provided that every pair of vertices a and b is joined by a walk with a and b as endpoints. A vertex is regarded as trivially connected to itself. Otherwise, the general graph is *disconnected*. Connectivity between vertices is reflexive, symmetric, and transitive. Hence connectivity defines an equivalence relation on the vertices of G and yields a partition

$$V_1 \cup V_2 \cup \cdots \cup V_t,$$

of the vertices of G. The induced subgraphs $G(V_1), G(V_2), \ldots, G(V_t)$ of G formed by taking the vertices in an equivalence class and the edges incident to them are called the *connected components* of G. For most problems concerning G it suffices to study only the connected components of G.

Connectivity has a direct interpretation in terms of the adjacency matrix A of G. Thus we may simultaneously permute the lines of A so that A is transformed into a direct sum of the form

$$A_1 \oplus A_2 \oplus \cdots \oplus A_t,$$

where A_i is the adjacency matrix of the connected component $G(V_i), (i = 1, 2, \ldots, t)$.

Let G be a connected general graph. The length of the shortest walk between two vertices a and b is the *distance* $d(a, b)$ between a and b in G. A vertex is regarded as distance 0 from itself. The maximum value of the distance function over all pairs of vertices is called the *diameter* of G.

Theorem 2.2.1.   *Let G be a connected general graph of diameter d. Then G has at least $d + 1$ distinct eigenvalues in its spectrum.*

Proof. Let a and b be vertices with $d(a, b) = d$ and let

$$a_0 \to a_1 \to a_2 \to \cdots \to a_d$$

be a walk of length equal to the diameter d. Then for each $i = 1, 2, \ldots, d$ there is at least one walk of length i and no shorter walk that joins a_0 to a_i. Thus A^i has a nonzero entry in the position determined by a_0 and a_i, whereas $I, A, A^2, \ldots, A^{i-1}$ each have zeros in this position. We conclude that A^i is not a linear combination of $I, A, A^2, \ldots, A^{i-1}$. Hence the minimum polynomial of A is of degree at least $d + 1$. But since A is a real symmetric matrix it is similar to a diagonal matrix and consequently the zeros of its minimal polynomial are distinct. $\square$

Exercises

1. Prove that the complement of a disconnected graph is a connected graph.
2. Determine the spectrum of the complete graph K_n of order n.
3. Show that k is an eigenvalue of a regular graph of degree k.
4. Let G be a graph of order n. Suppose that G is regular of degree k and let $\lambda_1 = k, \lambda_2, \ldots, \lambda_n$ be the spectrum of G. Prove that the spectrum of the complement of G is $n - 1 - k, -1 - \lambda_2, \ldots, -1 - \lambda_n$.
5. Let $f(\lambda) = \lambda^n + c_1 \lambda^{n-1} + c_2 \lambda^{n-2} + \cdots + c_n$ be the characteristic polynomial of a graph G of order n. Prove that c_1 equals 0, c_2 equals -1 times the number of edges of G, and c_3 equals -2 times the number of cycles of length 3 of G (a cycle of length 3 in a graph is sometimes called a *triangle*).
6. Let $K_{1,n-1}$ be the graph of order n whose vertices have degrees $n - 1, 1, \ldots, 1$, respectively. ($K_{1,n-1}$ is the *star* of order n.) Prove that the spectrum of $K_{1,n-1}$ is $\pm\sqrt{n - 1}, 0, \ldots, 0$.
7. Prove that there does not exist a connected graph which is cospectral with the star $K_{1,n-1}$.
8. Let G be a connected graph of order n which is regular of degree 2. The edges of G thus form a cycle of length n, and G is sometimes called a *cycle graph* of order n. Determine the spectrum of G.

References

N. Biggs[1974], *Algebraic Graph Theory*, Cambridge Tracts in Mathematics No. 67, Cambridge University Press, Cambridge.

D.M. Cvetković, M. Doob and H. Sachs[1982], *Spectra of Graphs – Theory and Application*, 2d ed., Deutscher Verlag der Wissenschaften, Berlin, Academic Press, New York.

D. Cvetković, M. Doob, I. Gutman and A. Torĝasev[1988], *Recent Results in the Theory of Graph Spectra*, Annals of Discrete Mathematics No. 36, Elsevier Science Publishers, New York.

W. Haemers[1979], *Eigenvalue Techniques in Design and Graph Theory*, Mathematisch Centrum, Amsterdam.

A.J. Schwenk and R.J. Wilson[1978], On the eigenvalues of a graph, *Selected Topics in Graph Theory* (L.W. Beineke and R.J. Wilson, eds.), Academic Press, New York, pp. 307–336.

2.3 The Incidence Matrix of a Graph

Let G be a general graph of order n with vertices $a_1, a_2, \ldots, a_n$ and edges $\alpha_1, \alpha_2, \ldots, \alpha_m$. We set $a_{ij} = 1$ if vertex a_j is on edge α_i and we set $a_{ij} = 0$ otherwise. The resulting $(0,1)$-matrix

$$A = [a_{ij}], \quad (i = 1, 2, \ldots, m; j = 1, 2, \ldots, n)$$

of size m by n is called the *incidence matrix* of G. The matrix A is in fact the conventional incidence matrix in which the edges are regarded as subsets of vertices. Each row of A contains at least one 1 and not more than two 1's. The rows with a single 1 in A correspond to the edges in G that are loops. Identical rows in A correspond to multiple edges in G.

The simple row structure of the matrix A is misleading because A describes the full complexity of the general graph G. For example, there is no computationally effective procedure known for the determination of the minimal number of columns in A with the property that these columns of A collectively contain at least one 1 in each of the m rows of A. In terms of G this quantity is the minimal number of vertices in G that touch all edges.

The incidence matrix and the adjacency matrix of a multigraph are related in the following way.

Theorem 2.3.1.   *Let G be a multigraph of order n. Let A be the incidence matrix of G and let B be the adjacency matrix of G. Then*

$$A^T A = D + B,$$

where D is a diagonal matrix of order n whose diagonal entry d_i is the degree of the vertex a_i of G, $(i = 1, 2, \ldots, n)$.

Proof. The inner product of columns i and j of A $(i \neq j)$ equals the multiplicity $m\{a_i, a_j\}$ of the edge $\{a_i, a_j\}$. The inner product of column i of A with itself equals the degree of the vertex a_i.   □

Now let G denote a graph of order n with vertices $a_1, a_2, \ldots, a_n$ and edges $\alpha_1, \alpha_2, \ldots, \alpha_m$. We assign to each of the edges of G one of the two possible orientations and thereby transform G into a graph in which each of the edges of G is assigned a direction. We set $a_{ij} = 1$ if a_j is the "initial" vertex of α_i, we set $a_{ij} = -1$ if a_j is the "terminal" vertex of α_i and we set $a_{ij} = 0$ if a_j is not an endpoint of α_i. The resulting $(0, 1, -1)$-matrix

$$A = [a_{ij}], \quad (i = 1, 2, \ldots, m; j = 1, 2, \ldots, n)$$

of size m by n is called the *oriented incidence matrix* of G. Each row of A contains exactly two nonzero entries, one of which is 1 and the other -1.

We note that in the notation of Theorem 2.3.1 the oriented incidence matrix satisfies

$$A^T A = D - B. \tag{2.2}$$

The matrix $A^T A$ in (2.2) is called the *Laplacian matrix* (also called the *admittance matrix*) of G. It follows from (2.2) that the Laplacian matrix is independent of the particular orientation assigned to G. The Laplacian matrix will be discussed further in section 2.5.

The oriented incidence matrix is used to determine the number of connected components of G.

Theorem 2.3.2. *Let G be a graph of order n and let t denote the number of connected components of G. Then the oriented incidence matrix A of G has rank $n - t$. In fact, each matrix obtained from A by deleting t columns, one corresponding to a vertex of each component, has rank $n - t$. A submatrix A' of A of order $n-1$ has rank $n-t$ if and only if the spanning subgraph G' of G whose edges are those corresponding to the rows of A' has t connected components.*

Proof. Let the connected components of G be denoted by

$$G(V_1), G(V_2), \ldots, G(V_t).$$

Then we may label the vertices and edges of G so that the oriented incidence matrix A is a direct sum of the form

$$A_1 \oplus A_2 \oplus \cdots \oplus A_t,$$

where A_i displays the vertices and edges in $G(V_i)$, $(i = 1, 2, \ldots, t)$. Let $G(V_i)$ contain n_i vertices. We prove that the rank of A_i equals $n_i - 1$. The conclusion then follows by addition.

Let β_j denote the column of A_i corresponding to the vertex a_j of $G(V_i)$. Since each row of A_i contains exactly one 1 and one -1, it follows that the sum of the columns of A_i is the zero vector. Hence the rank of A_i is at most $n_i - 1$. Suppose then that we have a linear relation $\sum b_j \beta_j = 0$, where the summation is over all columns of A_i and not all the coefficients are zero. Let us suppose that column β_k has $b_k \neq 0$. This column has nonzero entries in those rows corresponding to the edges incident with a_k. For each such row there is just one other column β_l with a nonzero entry in that row. In order for the dependency to hold we must have $b_k = b_l$. Hence if $b_k \neq 0$, then $b_l = b_k$ for all vertices a_l adjacent to a_k. Since $G(V_i)$ is connected it follows that all of the coefficients b_j are equal and the linear relation is merely a multiple of our earlier relation $\sum \beta_j = 0$. Hence the rank of $G(V_i)$ is $n_i - 1$, and deleting any column of A_i results in a matrix of rank $n_i - 1$. Finally we observe that the last conclusion of the theorem follows by applying the earlier conclusions to G'. □

A matrix A with integral elements is *totally unimodular* if every square submatrix of A has determinant $0, 1$, or -1. It follows at once that a totally unimodular matrix is a $(0, 1, -1)$-matrix.

The following theorem is due to Hoffman and Kruskal[1956].

Theorem 2.3.3.   *Let A be an m by n matrix whose rows are partitioned into two disjoint sets B and C and suppose that the following four properties hold:*

(i)   Every entry of A is $0, 1$, or -1.

(ii)   Every column of A contains at most two nonzero entries.

(iii)   If two nonzero entries in a column of A have the same sign, then the row of one is in B and the row of the other is in C.

(iv)   If two nonzero entries in a column of A have opposite signs, then the rows of both are in B or in C.

Then the matrix A is totally unimodular.

Proof. An arbitrary submatrix of A also satisfies the hypothesis of the theorem. Hence it suffices to prove that an arbitrary square matrix A satisfying the hypotheses of the theorem has $\det(A)$ equal to $0, 1$, or -1. The proof is by induction on n. For $n = 1$ the theorem follows trivially from (i). Suppose that every column of A has two nonzero entries. Then the sum of the rows in B equals the sum of the rows in C and $\det(A) = 0$. This assertion is also valid in case $B = \emptyset$ or $C = \emptyset$. Also if some column of A has all 0's, then $\det(A) = 0$. Hence we are left with the case in which some column of A has exactly one nonzero entry. We expand $\det(A)$ by this column and apply the induction hypothesis.   □

The preceding result implies the following theorem of Poincaré[1901].

Corollary 2.3.4.   *The oriented incidence matrix A of a graph G is totally unimodular.*

Proof. We apply Theorem 2.3.3 to the matrix A^T with $C = \emptyset$.   □

A square $(0, 1, -1)$-matrix is *Eulerian* provided all line sums are even integers. Camion[1965] (see also Padberg[1976]) has established the following theorem giving a necessary and sufficient condition for total unimodularity which we state without proof.

Theorem 2.3.5.   *A $(0, 1, -1)$-matrix A of size m by n is totally unimodular if and only if the sum of the elements in each Eulerian submatrix is a multiple of 4.*

Totally unimodular matrices are intimately related to a special class of *matroids* (see White[1986]) called *unimodular matroids* (also called *regular matroids*). These are matroids which can be coordinatized by a matrix A over the rational field for which A is totally unimodular. Unimodular matroids were characterized by Tutte[1958] (see also Gerards[1989]) in a very striking theorem. Another striking characterization of unimodular matroids was obtained by Seymour[1980]. The characterizations of Tutte and of Seymour are in terms of the linear dependence structure of the columns of the matrix A.

A *tree* is a connected graph that contains no cycle. We will assume a familiarity with a few of the most elementary properties of trees (Brualdi [1977] or Wilson[1972]). Let T be a graph of order n. Then the following statements are equivalent: (1) T is a tree; (2) T contains no cycles and has exactly $n - 1$ edges; (3) T is connected and has exactly $n - 1$ edges; (4) each pair of distinct vertices of T is joined by exactly one-chain.

Let T be a tree of order n with vertices $a_1, a_2, \ldots, a_n$ and edges $\alpha_1, \alpha_2, \ldots, \alpha_{n-1}$. We suppose that the edges of T have been oriented. Let (s_i, t_i), $(i = 1, 2, \ldots, l)$ be l ordered pairs of vertices of T. We set $m_{ij} = 1$ if the unique chain γ in T joining s_i and t_i uses the edge α_j in its assigned direction, we set $m_{ij} = -1$ if the chain γ uses the edge α_j in the direction opposite to its assigned direction, and we set $m_{ij} = 0$ if γ does not use the edge α_j. The resulting $(0, 1, -1)$-matrix

$$M = [m_{ij}], \quad (i = 1, 2, \ldots, l; \ j = 1, 2, \ldots, n - 1)$$

of size l by $n - 1$ is called a *network matrix* (Tutte[1965]). If we delete the column of the network matrix A corresponding to the arc α_k of G, the resulting matrix is a network matrix for the tree obtained from T by contracting the edge α_k, that is by deleting the arc α_k and identifying its two endpoints. It follows that submatrices of network matrices are also network matrices.

Theorem 2.3.6.   *A network matrix M corresponding to the oriented tree T is a totally unimodular matrix.*

Proof. We continue with the notation in the preceding paragraph. Let G be the graph with vertices $a_1, a_2, \ldots, a_n$ and edges $\{s_i, t_i\}, (i = 1, 2, \ldots, l)$. We orient each edge $\{s_i, t_j\}$ from s_i to t_j. Let A be the l by n oriented incidence matrix of G, and let B be the $n - 1$ by n oriented incidence matrix of T. Let A' and B' result from A and B, respectively, by deleting the last column (the column corresponding to vertex a_n in each case). From the definitions of the matrices involved we obtain the relation $MB = A$. Hence $MB' = A'$, and since by Theorem 2.3.2 B' is invertible, the relation $M = A'B'^{-1}$ holds. By Theorem 2.3.3 the matrix

$$\begin{bmatrix} I_{n-1} \\ B' \\ A' \end{bmatrix}$$

is totally unimodular. It follows that the matrix

$$\begin{bmatrix} I_{n-1} \\ B' \\ A' \end{bmatrix} B'^{-1} = \begin{bmatrix} B'^{-1} \\ I_{n-1} \\ A'B'^{-1} \end{bmatrix}$$

has the property that all of its submatrices of order $n-1$ have determinants equal to one of $0, 1$ and -1. This implies that the matrix $M = A'B'^{-1}$ is totally unimodular.   □

Seymour's characterization of unimodular matroids can be restated in matrix terms. In this characterization the totally unimodular matrices

$$\begin{bmatrix} -1 & 1 & 0 & 0 & 1 \\ 1 & -1 & 1 & 0 & 0 \\ 0 & 1 & -1 & 1 & 0 \\ 0 & 0 & 1 & -1 & 1 \\ 1 & 0 & 0 & 1 & -1 \end{bmatrix}, \tag{2.3}$$

and

$$\begin{bmatrix} 1 & 1 & 0 & 0 & 1 \\ 0 & 1 & 1 & 0 & 1 \\ 0 & 0 & 1 & 1 & 1 \\ 1 & 0 & 0 & 1 & 1 \\ 1 & 1 & 1 & 1 & 1 \end{bmatrix} \tag{2.4}$$

have an exceptional role.

The following theorem of Seymour[1982] asserts that a totally unimodular matrix which is not a network matrix, the transpose of a network matrix, or one of the two exceptional matrices above admits a "diagonal decomposition" into smaller totally unimodular matrices.

Theorem 2.3.7.   *Let A be a totally unimodular matrix. Then one of the following properties holds:*

   (i) A is a network matrix or the transpose of a network matrix;
   (ii) A can be obtained from one of the two exceptional matrices above by line permutations and by the multiplication of some of their lines by -1*;*
   (iii) The lines of A can be permuted to obtain a matrix

$$\begin{bmatrix} B_{11} & B_{12} \\ B_{21} & B_{22} \end{bmatrix},$$

   where B_{11} and B_{22} are totally unimodular matrices, and either
   (a) $\operatorname{rank}(B_{12}) + \operatorname{rank}(B_{21}) = 0$
       and B_{11} and B_{22} each have at least one line, or

(b) rank (B_{12}) + rank (B_{21}) = 1

and B_{11} and B_{22} *each have at least two lines, or*

(c) rank (B_{12}) + rank (B_{21}) = 2

and B_{11} and B_{22} *each have at least six lines.*

The above theorem implies that it is possible to construct the entire class of totally unimodular network matrices from the class of network matrices and the two exceptional matrices (2.3) and (2.4). Neither of these exceptional matrices is a network matrix or the transpose of a network matrix. If the matrix A satisfies *(iii)* above, then it does not necessarily follow that A is totally unimodular. However, a result of Brylawski[1975] gives a list of conditions such that a matrix A which satisfies *(iii)* is totally unimodular and such that every totally unimodular matrix can be constructed in the manner of *(iii)* starting from the network matrices, the transposes of the network matrices and the two exceptional matrices. These conditions are difficult to state and we refer the reader to the original paper by Brylawski. A consequence of Seymour's characterizations of unimodular matroids and totally unimodular matrices is the existence of an algorithm to determine whether a $(0, 1, -1)$-matrix is totally unimodular whose number of steps is bounded by a polynomial function in the number of lines of the matrix (see Schrijver[1986]).

Exercises

1. Verify that the matrices (2.3) and (2.4) are totally unimodular.
2. Prove that each nonsingular submatrix of a totally unimodular matrix has an integral inverse.
3. Let A be a totally unimodular matrix of size m by n. Let b be a matrix of size m by 1 each of whose elements is an integer. Prove that the consistent equation $Ax = b$ has an integral solution.
4. Let A be a totally unimodular matrix and let B be a nonsingular submatrix of A of order k. Prove that for each nonzero $(0, 1, -1)$-vector y of size k, the greatest common divisor of the elements of the vector yB equals 1. [Indeed Chandrasekaran has shown that this property characterizes totally unimodular matrices (see Schrijver[1986]).]
5. Let A be a nonsingular $(0, 1, -1)$-matrix and suppose that $|\det(A)| \neq 1$. Prove that A has a square submatrix B with $|\det(B)| = 2$.

References

N. Biggs[1974], *Algebraic Graph Theory*, Cambridge Tracts in Mathematics No. 67, Cambridge University Press, Cambridge.

R.A. Brualdi[1977], *Introductory Combinatorics*, Elsevier Science Publishers, New York.

T. Brylawski[1975], Modular constructions for combinatorial geometries, *Trans. Amer. Math. Soc.*, 203, pp. 1–44.

P. Camion[1965], Characterization of totally unimodular matrices, *Proc. Amer. Math. Soc.*, 16, pp. 1068–1073.

A.M.H. Gerards[1989], A short proof of Tutte's characterization of totally unimodular matroids, *Linear Alg. Applics.*, 114/115, pp. 207–212.

A.J. Hoffman and J.B. Kruskal[1956], Integral boundary points of convex polyhedra, *Annals of Math. Studies* No. 38, Princeton University Press, Princeton,
pp. 223–246.

M.W. Padberg[1976], A note on the total unimodularity of matrices, *Discrete
Math.*, 14, pp. 273–278.

H. Poincaré[1901], Second complément à l'analysis situs, *Proc. London Math.
Soc.*, 32, pp. 277–308.

A. Schrijver[1986], *Theory of Linear and Integer Programming*, Wiley, New York.

P.D. Seymour[1980], Decomposition of regular matroids, *J. Combin. Theory, Ser.
B*, 28, pp. 305–359.

[1982], Applications of the regular matroid decomposition, *Colloquia Math.
Soc. János Bolyai*, No. 40 Matroid Theory, Szeged (Hungary), pp. 345–357.

W.T. Tutte[1958], A homotopy theorem for matroids, I and II, *Trans. Amer.
Math. Soc.*, 88, pp. 144–174.

N. White (ed.)[1986], *Theory of Matroids*, Encyclopedia of Maths. and Its Applics. Cambridge University Press, Cambridge.

[1987], Unimodular matroids, *Combinatorial Geometries*, ed. N. White, Encyclopedia of Maths. and Its Applics., Cambridge University Press, Cambridge.

R.J. Wilson[1972], *Introduction to Graph Theory*, Academic Press, New York.

2.4 Line Graphs

Let G denote a graph of order n on m edges. The *line graph* $L(G)$ of G
is the graph whose vertices are the edges of G and two vertices of $L(G)$
are adjacent if and only if the corresponding edges of G have a vertex in
common. The line graph $L(G)$ is of order m.

Theorem 2.4.1. *Let A be the incidence matrix of a graph G on m edges
and let B_L be the adjacency matrix of the line graph $L(G)$. Then*

$$AA^T = 2I_m + B_L. \tag{2.5}$$

Proof. For $i \neq j$ the entry in the (i,j) position of B_L is 1 if the edges
α_i and α_j of G have a vertex in common and 0 otherwise. But the same
conclusion holds for the entry in the (i,j) position of AA^T. The main
diagonal elements are as indicated. $\square$

Theorem 2.4.1 implies a severe restriction on the spectrum of a line graph.

Theorem 2.4.2. *If λ is an eigenvalue of the line graph $L(G)$, then $\lambda \geq -2$.
If G has more edges than vertices, then $\lambda = -2$ is an eigenvalue of $L(G)$.*

Proof. The symmetric matrix AA^T is positive semidefinite and hence its
eigenvalues are nonnegative. But if α is an eigenvector of B_L associated
with the eigenvalue λ, then (2.5) implies

$$AA^T\alpha = (2 + \lambda)\alpha$$

so that $\lambda \geq -2$. If G has more edges than vertices, then AA^T is singular and hence 0 is an eigenvalue of AA^T.   □

If the graph G is regular of degree k, then the number of its edges is $m = nk/2$ and the line graph $L(G)$ is regular of degree $2(k-1)$. The following theorem of Sachs[1967] shows that in this case the characteristic polynomials of G and $L(G)$ are related in an elementary way.

Theorem 2.4.3.   *Let G be a graph of order n which is regular of degree k on m edges. Let $f(\lambda)$ and $g(\lambda)$ be the characteristic polynomials of G and $L(G)$, respectively. Then*

$$g(\lambda) = (\lambda + 2)^{m-n} f(\lambda + 2 - k).$$

Proof. We recall that two matrix products of the form XY and YX have the same collection of eigenvalues apart from zero eigenvalues. Thus the incidence matrix A of G satisfies

$$\det(\lambda I_m - AA^T) = \lambda^{m-n} \det(\lambda I_n - A^T A). \qquad (2.6)$$

We let B and B_L denote the adjacency matrices of G and $L(G)$, respectively. Then using Theorems 2.3.1 and 2.4.1, we obtain

$$
\begin{aligned}
\det(\lambda I_m - B_L) &= \det((\lambda + 2)I_m - AA^T) \\
&= (\lambda + 2)^{m-n} \det((\lambda + 2)I_n - A^T A) \\
&= (\lambda + 2)^{m-n} \det((\lambda + 2 - k)I_n - B). \qquad \square
\end{aligned}
$$

The study of the spectral properties of line graphs was initiated by A.J. Hoffman and extensively investigated by him and his associates over a period of many years. The condition $\lambda \geq -2$ of Theorem 2.4.2 imposes severe restrictions on the spectrum of $L(G)$. But graphs other than line graphs exist that also satisfy this requirement. Generalized line graphs were introduced by Hoffman[1970,1977] as a class of graphs more general than line graphs that satisfy $\lambda \geq -2$. We briefly discuss these contributions.

Let t be a nonnegative integer. The *cocktail party graph* $CP(t)$ of order $2t$ is the graph with vertices $b_1, b_2, \ldots, b_{2t}$ in which each pair of distinct vertices form an edge with the exception of the pairs $\{b_1, b_2\}, \{b_3, b_4\}, \ldots, \{b_{2t-1}, b_{2t}\}$. The cocktail party graph $CP(t)$ can be obtained from the complete graph K_{2t} of order $2t$ by deleting t edges no two of which are adjacent. If $t = 0$, then the cocktail party graph is a graph with no vertices. Now let G be a graph of order n with vertices $a_1, a_2, \ldots, a_n$, and let $k_1, k_2, \ldots, k_n$ be an n-tuple of nonnegative integers. The *generalized line graph* $L(G; k_1, k_2, \ldots, k_n)$ is the graph of order $n + 2(k_1 + k_2 + \cdots + k_n)$ defined as follows. We begin with the line graph $L(G)$ and disjoint cocktail party graphs $CP(k_1), CP(k_2), \ldots, CP(k_n)$. Each vertex of $L(G)$ is an edge $\alpha = \{a_i, a_j\}$ of G and we put edges between α and each vertex of $CP(k_i)$ and of $CP(k_j)$. If λ is an eigenvalue of a generalized line graph, then $\lambda \geq -2$.

Theorem 2.4.4.   *If G is a connected graph of order greater than 36 for which each eigenvalue λ satisfies the condition $\lambda \geq -2$, then G is a generalized line graph.*

The following theorem of Hoffman and Ray-Chaudhuri characterizes regular graphs satisfying $\lambda \geq -2$.

Theorem 2.4.5.   *If G is a regular connected graph of order greater than 28 for which each eigenvalue λ satisfies the condition $\lambda \geq -2$, then G is either a line graph or a cocktail party graph.*

Finally, we mention the important paper by Cameron, Goethals, Seidel and Shult [1976] in which the above two theorems are obtained by appealing to the classical root systems.

Exercises

1. Determine the spectrum of the line graph $L(K_n)$.
2. The complement of the line graph of K_5 is a cubic graph of order 10 and is known as the *Petersen graph*. Determine the spectrum of the Petersen graph.
3. Find an example of a graph of order 4 which is not isomorphic to the line graph of any graph. Deduce that the Petersen graph is not isomorphic to a line graph.
4. Show that the spectrum of the cocktail party graph $CP(t)$ of order $2t$ is $2t-2$, 0 (with multiplicity t), and -2 (with multiplicity $t-1$).
5. Let B be the adjacency matrix of a generalized line graph G. Determine a matrix N such that $NN^T = 2I + B$, and then deduce that $\lambda \geq -2$ for each eigenvalue λ of G.
6. Let A be the incidence matrix of a tree of order n. Prove that the rank of A equals $n - 1$.
7. Let G be a connected graph of order n on n edges and let A be the incidence matrix of G. The graph G has a unique cycle γ. Prove that A has rank n if γ has odd length and that A has rank $n - 1$ if γ has even length. Deduce that the incidence matrix of a connected graph of order n has rank n if it has an odd length cycle and has rank $n - 1$ otherwise.
8. Let G be a connected graph of order n. Prove that the multiplicity of 0 as an eigenvalue of the line graph $L(G)$ equals $m - n$ if G has an odd length cycle and equals $m - n + 1$ otherwise.

References

N. Biggs[1974], *Algebraic Graph Theory*, Cambridge Tracts in Mathematics No. 67, Cambridge University Press, Cambridge.

P.J. Cameron, J.M. Goethals, J.J. Seidel and E.E. Shult[1976], Line graphs, root systems, and elliptic geometry, *J. Algebra*, 43, pp. 305–327.

D. Cvetković, M. Doob, I. Gutman and A. Torğasev[1988], *Recent Results in the Theory of Graph Spectra*, Annals of Discrete Mathematics No. 36, Elsevier Science Publishers, New York.

A.J. Hoffman[1970], $-1 - \sqrt{2}$? *Combinatorial Structures and Their Applications*, Gordon and Breach, New York, pp. 173–176.

[1977], On graphs whose least eigenvalue exceeds $-1 - \sqrt{2}$, *Linear Alg. Applics.*, 16, pp. 153–165.

A.J. Hoffman and D.K. Ray-Chaudhuri, On a spectral characterization of regular line graphs, unpublished manuscript.

H. Sachs[1967], Über Teiler, Faktoren und charakteristische Polynome von Graphen II, *Wiss. Z. Techn. Hochsch. Ilmenau*, 13, pp. 405–412.

2.5 The Laplacian Matrix of a Graph

Let G denote a graph of order n with vertices $a_1, a_2, \ldots, a_n$ and edges $\alpha_1, \alpha_2, \ldots, \alpha_m$. Let A be the oriented incidence matrix of G of size m by n, and let B be the adjacency matrix of G. Recall that the Laplacian matrix of G is the matrix of order n

$$F = A^T A = D - B$$

where D is a diagonal matrix of order n whose diagonal entry d_i is the degree of the vertex a_i of G, $(i = 1, 2, \ldots, n)$. By Theorem 2.3.2 the matrix A, and hence the Laplacian matrix F has rank at most equal to $n - 1$. Thus the matrix F is a singular matrix. A *spanning tree* T of G is a spanning subgraph of G which forms a tree. Every connected graph contains a spanning tree. Let U be a subset of the edges of G. Then we denote by $\langle U \rangle$ the subgraph of G consisting of the edges of U and all vertices of G incident with at least one edge of U. The following lemma is an immediate consequence of Theorem 2.3.2.

Lemma 2.5.1.   *Let U be an $(n-1)$-subset of edges of the connected graph G of order n. Let A_U denote a submatrix of order $n - 1$ of the oriented incidence matrix A of G consisting of the intersection of the $n - 1$ rows of A corresponding to the edges of U and any set of $n - 1$ columns of A. Then A_U is nonsingular if and only if $\langle U \rangle$ is a spanning tree of G.*

The *complexity* of a graph G of order n is the number of spanning trees of G. We denote the complexity of G by $c(G)$. In case G is disconnected we have $c(G) = 0$.

Lemma 2.5.2.   *Let A be the oriented incidence matrix of a graph G of order n. Then the adjugate of the Laplacian matrix*

$$F = A^T A = D - B$$

is a multiple of J.

Proof. If G is disconnected, then $\operatorname{rank}(F) = \operatorname{rank}(A) < n - 1$ and hence $\operatorname{adj}(F) = O$.

If G is connected, then $\mathrm{rank}(F) = n - 1$. But since

$$F\,\mathrm{adj}(F) = \det(F)I = O$$

it follows that each column of $\mathrm{adj}(F)$ is in the kernel of F. But this kernel is a one-dimensional space spanned by the vector $e_n = (1, 1, \ldots, 1)^T$. Hence each column of $\mathrm{adj}(F)$ is a multiple of u. But $F = A^T A$ is symmetric and this implies that $\mathrm{adj}(F)$ is also symmetric. Hence it follows that $\mathrm{adj}(A)$ is a multiple of J. $\square$

We now obtain a classical formula.

Theorem 2.5.3.   *In the above notation we have*

$$\mathrm{adj}(F) = c(G)J.$$

Proof. By Lemma 2.5.2 we need only show that one cofactor of F is equal to $c(G)$. Let A_0 denote the matrix obtained from A by removing the last column of A. It follows that $\det(A_0^T A_0)$ is a cofactor of F. Now let A_U denote a submatrix of order $n - 1$ of A_0 whose rows correspond to the edges in an $(n - 1)$-subset U of the edges of G. Then by the Binet-Cauchy theorem we have

$$\det(A_0^T A_0) = \sum \det(A_U^T)\det(A_U),$$

where the summation is over all possible choices of U. By Lemma 2.5.1 we have that A_U is nonsingular if and only if $\langle U \rangle$ is a spanning tree of G, and in this case by Corollary 2.3.4 we have $\det(A_U) = \pm 1$. But $\det(A_U) = \det(A_U^T)$ so that $\det(A_0^T A_0) = c(G)$ and the conclusion follows. $\square$

For the complete graph K_n of order n we have $F = nI - J$, and an easy calculation (cf. Exercise 1, Sec. 1.3) yields the famous Cayley formula [1889]

$$c(K_n) = n^{n-2}$$

for the number of labeled trees of order n.

Theorem 2.5.3 may be formulated in an even more elegant form (Temperley [1964]).

Theorem 2.5.4.   *The complexity of a graph G of order n is given by the formula*

$$c(G) = n^{-2}\det(F + J).$$

Proof. We have $J^2 = nJ$ and $FJ = O$ so that

$$(F + J)(nI - J) = nF.$$

We take the adjugate of both sides and use Theorem 2.5.3 to obtain

$$n^{n-2} J \operatorname{adj}(F + J) = n^{n-1} c(G) J.$$

We now multiply by $F + J$ and this gives

$$\det(F + J) J = n^2 c(G) J,$$

as desired.   □

Now let $(x_1, x_2, \ldots, x_n)^T$ be a real n-vector. Then

$$x^T F x = x^T A^T A x = \sum_{\alpha_t = \{a_i, a_j\}} (x_i - x_j)^2 \qquad (2.7)$$

where the summation is over all m edges $\alpha_t = \{a_i, a_j\}$ of G. The matrix F is a positive semidefinite symmetric matrix. Moreover, 0 is an eigenvalue of F with corresponding eigenvector $e_n = (1, 1, \ldots, 1)^T$. Let $\mu = \mu(G)$ denote the second smallest eigenvalue of F. In case $n = 1$ we define μ to be 0. We have from Theorem 2.3.2 that $\mu \geq 0$ with equality if and only if G is a disconnected graph. Fiedler[1973] defined μ to be the *algebraic connectivity* of the graph G. The algebraic connectivity of the complete graph K_n of order n is n (cf. Exercise 2, Sec. 2.2).

Let U be the set of all real n-tuples x such that $x^T x = 1$ and $x^T e_n = 0$. From the theory of symmetric matrices (see, e.g., Horn and Johnson[1987]) we obtain the characterization

$$\mu = \min\{x^T F x \mid x \in U\} \qquad (2.8)$$

for the algebraic connectivity of the graph G. It follows from equations (2.7) and (2.8) that if G' is a spanning subgraph of G then $\mu(G') \leq \mu(G)$. Thus for graphs with the same set of vertices the algebraic connectivity is a nondecreasing function of the edges.

Theorem 2.5.5. *Let G_t be a graph of order n which is obtained from the graph G by removing the vertex a_t and all edges incident with a_t. Then*

$$\mu(G_t) \geq \mu(G) - 1.$$

Proof. Let F_t be the Laplacian matrix of G_t. First suppose that a_t is adjacent to all other vertices in G. Then the matrix $F_t + I$ is a principal submatrix of order $n-1$ of the Laplacian matrix F of G. By the interlacing inequalities for the eigenvalues of symmetric matrices we have

$$\mu(G_t) + 1 \geq \mu(G).$$

In the general case we use the fact that the algebraic connectivity is a nondecreasing function of the edges.   □

There are two standard ways to measure the extent to which a graph is connected. The *vertex connectivity* of the graph G is the smallest number of vertices whose removal from G, along with each edge incident with at least one of the removed vertices, leaves either a disconnected graph or a graph with a single vertex. Thus if G is not a complete graph, the vertex connectivity equals $n - p$ where p is the largest order of a disconnected induced subgraph of G. The *edge connectivity* of G is the smallest number of edges whose removal from G leaves a disconnected graph or a graph with one vertex. The edge connectivity is no greater than the minimum degree of the vertices of G. The vertex and edge connectivities of the complete graph K_n equal $n - 1$.

Theorem 2.5.6.   *Let G be a graph of order n which is not complete. The algebraic connectivity $\mu(G)$, the vertex connectivity $v(G)$ and the edge connectivity $e(G)$ of G satisfy*

$$\mu(G) \leq v(G) \leq e(G).$$

Proof. Since G is not complete there is a disconnected graph G^* which results from G by the removal of $v(G)$ vertices. Then $\mu(G^*) = 0$ and by repeated application of Theorem 2.5.5 we get $\mu(G^*) \geq \mu(G) - v(G)$. Now let $k = e(G)$ and let $\alpha_{i_1}, \alpha_{i_2}, \ldots, \alpha_{i_k}$ be a set of k edges whose removal from G results in a disconnected graph. This disconnected graph has exactly two connected components G_1 and G_2, and each of the removed edges joins a vertex of G_1 to a vertex of G_2. Let x_j be the vertex of α_{i_j} which belongs to $G_1, (j = 1, 2, \ldots, k)$. Notice that the vertices $x_1, x_2, \ldots, x_k$ are not necessarily distinct. If the removal of the vertices $x_1, x_2, \ldots, x_k$ disconnects G, then $v(G) \leq k = e(G)$. Otherwise $x_1, x_2, \ldots, x_k$ are the only vertices of G_1 and it follows that each vertex x_i has degree at most k. Hence each vertex x_i has degree exactly k. We now delete all the vertices adjacent to x_i and disconnect the graph G. Hence $v(G) \leq k = e(G)$.   □

We now assume that the graph G is connected and hence that the algebraic connectivity μ of G is positive. The following theorem of Fiedler[1975] shows that an eigenvector x of the Laplacian matrix corresponding to its eigenvalue μ contains easily accessible information about the graph G.

Theorem 2.5.7.   *Let G be a connected graph of order n with vertices $a_1, a_2, \ldots, a_n$. Let $x = (x_1, x_2, \ldots, x_n)^T$ be an eigenvector of the Laplacian matrix F of G corresponding to the eigenvalue μ. Let r be a nonnegative number, and define*

$$V_r = \{a_i | x_i + r \geq 0, 1 \leq i \leq n\}.$$

Then the induced subgraph $G(V_r)$ of G with vertex set V_r is connected.

Proof. We first prove the theorem under the assumption that $r = 0$. In this case $V_0 = \{a_i | x_i \geq 0, 1 \leq i \leq n\}$. Suppose to the contrary that the induced subgraph $G(V_0)$ is disconnected. We simultaneously permute the lines of F to obtain

$$\begin{bmatrix} F_1 & O & R \\ O & F_2 & S \\ R^T & S^T & F' \end{bmatrix}, \tag{2.9}$$

where F_1 corresponds to a connected component of $G(V_0)$ and F_2 corresponds to the remaining components of $G(V_0)$. We write

$$x = \begin{bmatrix} x_1 \\ x_2 \\ x' \end{bmatrix}$$

to conform to the partition of F in (2.9). The elements of the vectors x_1 and x_2 are nonnegative, and the elements of the vector x' are negative. The matrix

$$\begin{bmatrix} F_1 & O \\ O & F_2 \end{bmatrix} \tag{2.10}$$

is a principal submatrix of F. Since 0 is a simple eigenvalue of F, it follows from the interlacing inequalities for symmetric matrices that the multiplicity of 0 as an eigenvalue of the matrix (2.10) is at most 1. This implies that we may assume that the matrix F_1 is nonsingular. The equation $Fx = \mu x$ implies that

$$(F_1 - \mu I)x_1 = -Rx'. \tag{2.11}$$

The eigenvalues of $F_1 - \mu I$ are nonnegative and hence $F_1 - \mu I$ is a positive semidefinite matrix. The sign patterns of the matrices and vectors involved imply that the elements of the vector $-Rx'$ are nonpositive. Multiplying equation (2.11) on the left by x_1^T we obtain

$$x_1^T(F_1 - \mu I)x_1 = -x_1^T Rx' \leq 0.$$

Since the matrix $F_1 - \mu I$ is positive semidefinite, this implies that

$$(F_1 - \mu I)x_1 = O.$$

By (2.11) we also have $Rx' = O$. Since R has nonpositive entries and x' has negative entries, we have $R = O$, contradicting the hypothesis that G is a connected graph.

For general nonnegative numbers r we replace the vector x by $x + re_n$ and then proceed as above.   □

A very similar proof can be given for the following: Let r be a nonpositive number and define

$$V_r = \{a_i | x_i + r \leq 0, 1 \leq i \leq n\}.$$

Then the induced subgraph $G(V_r)$ of G is connected. More general results can be found in Fiedler[1975]. The algebraic connectivity of trees is studied in Grone and Merris[1987] and in Fiedler[1990]. A survey of the eigenvalues of the Laplacian matrix of graphs is given by Mohar[1988]. A more general survey of the Laplacian is given by Grone[1991].

Exercises

1. Determine the complexity of the Petersen graph.
2. Determine the algebraic connectivity of the star $K_{1,n-1}$ of order n.
3. Let G be a graph of order n and let d denote the smallest degree of a vertex of G. Prove that $\mu(G) \leq dn/(n-1)$ (Fiedler[1973]).
4. Let G be a connected graph of order n which is regular of degree k. Let the spectrum of G be $\lambda_1 = k, \lambda_2, \ldots, \lambda_n$. Use Theorem 2.5.4 to show that the complexity $c(G)$ satisfies

$$c(G) = \frac{1}{n} \prod_{i=2}^{n} (k - \lambda_i)$$

(Biggs[1974]).

References

N. Biggs[1974], *Algebraic Graph Theory*, Cambridge Tracts in Mathematics No. 67, Cambridge University Press, Cambridge.

A. Cayley[1889], A theorem on trees, *Quarterly J.Math.*, 23, pp. 376–378.

M. Fiedler[1973], Algebraic connectivity of graphs, *Czech. Math. J.*, 23, pp. 298–305.

[1975], A property of eigenvectors of nonnegative symmetric matrices and its application to graph theory, *Czech. Math. J.*, 25, pp. 619–633.

[1990], Absolute connectivity of trees, *Linear Multilin. Alg.*, 26, pp. 86–106.

R. Grone[1991], On the geometry and Laplacian of a graph, *Linear Alg. Applics.*, 150, pp. 167–178.

B. Grone and R. Merris[1987], Algebraic connectivity of trees, *Czech. Math. J.*, 37, pp. 660–670.

R. Horn and C.R. Johnson[1985], *Matrix Analysis*, Cambridge University Press, Cambridge.

B. Mohar[1988], The Laplacian spectrum of graphs, *Preprint Series Dept. Math. University E.K. Ljubljana*, 26, pp. 353–382.

H.N.V. Temperley[1964], On the mutual cancellation of cluster integrals in Mayer's fugacity series, *Proc. Phys. Soc.*, 83, pp. 3–16.

2.6 Matchings

A graph G is called *bipartite* provided that its vertices may be partitioned into two subsets X and Y such that every edge of G is of the form $\{a, b\}$ where a is in X and b is in Y. We call $\{X, Y\}$ a *bipartition* of G. A connected bipartite graph has a unique bipartition. In a bipartite graph the vertices may be colored red and blue in such a way that each edge of the graph has a red endpoint and a blue endpoint. Trees are simple instances of bipartite graphs.

Let A denote the adjacency matrix of a bipartite graph G with bipartition $\{X, Y\}$ where X is an m-set and Y is an n-set. Then we may write A in the following special form:

$$A = \begin{bmatrix} O & B \\ B^T & O \end{bmatrix}, \tag{2.12}$$

where B is a (0,1)-matrix of size m by n which specifies the adjacencies between the vertices of X and the vertices of Y. Without loss of generality we may select the notation so that $m \leq n$. The matrix B characterizes the bipartite graph G. We note the paradoxical nature of combinatorial representations. A general graph is characterized by a very special (0,1)-matrix called its incidence matrix, whereas a very special graph called a bipartite graph is characterized by an arbitrary (0,1)-matrix.

A *matching M* in the bipartite graph G is a subset of its edges no two of which are adjacent. A matching defines a one-to-one correspondence between a subset X' of X and a subset Y' of Y such that corresponding vertices of X' and Y' are joined by an edge of the matching. The cardinality $|M|$ of the matching M is the common cardinality of X' and Y'. A matching in G with cardinality t corresponds in the matrix B of (2.12) to a set of t 1's with no two of the 1's on the same line. In A it corresponds to a set of $2t$ symmetrically placed 1's with no two of the 1's on the same line. The cardinality of a matching cannot exceed m. The number, possibly zero, of matchings having cardinality m is given by $\operatorname{per}(B)$. The König theorem (Theorem 1.2.1) applied to the matrix B yields: *The maximum cardinality of a matching in the bipartite graph G equals the minimum cardinality of a set S of vertices such that each edge of G is incident with at least one vertex in S.*

Now let G be a general graph of order n with vertex set V. The concept of a *matching* may be carried over to the general graph G without any change in the definition. In investigating the largest cardinality of a matching we may assume without loss of generality that each edge of G has multiplicity 1. However, the possible presence of loops in G adds to the generality of the situation. Let A be the adjacency matrix of order n of G. Let M be a matching in G with cardinality t, and suppose that $p \geq 0$ of the edges of M are loops. The cardinality of the set of endpoints of the edges of M

is denoted by $\#(M)$. It follows that $\#(M) = 2t - p$. In the matrix A, M corresponds to a set of $2t-p$ symmetrically placed 1's with no two of the 1's on the same line. Each of the edges of M which is not a loop corresponds to two symmetrically placed 1's. Each of the loops in M corresponds to a single 1 on the main diagonal of A.

We now turn to the fundamental minimax theorem for matchings in general graphs which extends the König theorem for bipartite graphs. This theorem in its original form is due to Tutte[1947]. It was later extended by Berge[1958]. The modification presented here was mentioned in Brualdi[1976] and allows for the presence of loops in a matching. The proof we give is based on Anderson[1971], Brualdi[1971], Gallai[1963] and Mader[1973].

If S is a subset of the vertex set V of the general graph G of order n, we define $\mathcal{C}(G; S)$ to be the set of connected components of the induced subgraph $G(V-S)$ which have an odd number of vertices and no loops. The cardinality of $\mathcal{C}(G; S)$ is denoted by $p(G; S)$. We also define the function $f(G; S)$ by

$$f(G; S) = n - p(G; S) + |S|. \qquad (2.13)$$

Theorem 2.6.1.   *Let G be a general graph of order n whose vertex set is V. The maximum cardinality of the set of endpoints of a matching in G equals the minimum value of $f(G; S)$ over all subsets S of the vertex set V:*

$$\max\{\#(M) : M \text{ a matching in } G\} = \min\{f(G; S) : S \subseteq V\} \qquad (2.14)$$

Proof. Let M be a matching in G and let S be a subset of V. We first show that $\#(M) \leq f(G; S)$. Let $G(W)$ be a component of $G(V - S)$ which belongs to $\mathcal{C}(G; S)$. Then $|W|$ is odd and at most $(|W| - 1)/2$ edges in M are edges of $G(W)$. An edge of G which has one of its endpoints in W has its other endpoint in $W \cup S$ and this implies that there are at least $p(G; S) - |S|$ vertices of G which are not incident with any edge in M. Hence

$$\#(M) \leq n - (p(G; S) - |S|) = f(G; S). \qquad (2.15)$$

Since (2.15) holds for each matching M and each set S of vertices, the value of the expression on the left in (2.14) is at most equal to the value of the expression on the right.

We now denote the value of the expression on the right in (2.14) by m and prove by induction on m that G has a matching M with $\#(M) = m$. If $m = 0$, we may take M to be empty. Now let $m \geq 1$ and let S denote the collection of maximal subsets T of the vertex set V for which

$$n - p(G; T) + |T| = m. \qquad (2.16)$$

We choose T in S and first show that each component of $G(V - T)$ with an even number of vertices has a loop. Suppose to the contrary that $G(W)$ is a component of $G(V - T)$ such that $|W|$ is even and $G(W)$ has no loops. Let x be a vertex in W. Then $p(G; T \cup \{x\}) \geq p(G; T) + 1$ and hence by (2.16)

$$n - p(G; T \cup \{x\}) + |T \cup \{x\}| \leq m. \tag{2.17}$$

It follows that equality holds in (2.17) and we contradict the choice of T in S.

Now let $G(W)$ be a component of $G(V - T)$ which has at least one loop. We show that there is a matching M_W in $G(W)$ satisfying $\#(M_W) = |W|$. Let $\{x, x\}$ be a loop in $G(W)$ and consider the graph $G(W - \{x\})$. Suppose there is a subset U of $W - \{x\}$ satisfying $p(G(W - \{x\}); U) \geq |U| + 1$. Then

$$p(G(W); U \cup \{x\}) \geq |U \cup \{x\}| \tag{2.18}$$

and using (2.16) and (2.18) we obtain

$$n - p(G; T \cup U \cup \{x\}) + |T \cup U \cup \{x\}|$$
$$= n - p(G; T) + |T| + |U \cup \{x\}| - p(G(W); U \cup \{x\}) \leq m.$$

Again equality holds and we contradict the choice of T in S. Thus for all subsets U of $W - \{x\}$,

$$p(G(W - \{x\}); U) \leq |U|,$$

and hence

$$|W - \{x\}| - p(G(W - \{x\}); U) + |U| \geq |W - \{x\}|.$$

Since $|W - \{x\}| < m$, it follows from the induction hypothesis that $G(W - \{x\})$ has a matching M' with $\#(M') = |W - \{x\}|$. Then $M_W = M' \cup \{\{x, x\}\}$ is a matching of $G(W)$ with $\#(M_W) = |W|$.

We now deal exclusively with the components in $\mathcal{C}(G; T)$ and the edges between vertices of these components and the vertices of T. We distinguish two cases.

Case 1. Either $T \neq \emptyset$ or $\mathcal{C}(G; T)$ contains at least two components with more than one vertex.

Let $(G(W_i) : i \in I)$ be the components in $\mathcal{C}(G; T)$ satisfying $|W_i| > 1$, where we have indexed these components by some set I. The assumptions of this case imply that $|W_i| - 1 < m$ for all $i \in I$. Let j be any element in I and let z be any vertex in W_j. We next show that $G(W_j - \{z\})$ has a matching $M_j(z)$ satisfying $\#(M_j(z)) = |W_j| - 1$. If not, then by the induction hypothesis there exists $S \subseteq W_j - \{z\}$ such that

$$p(G(W_j - \{z\}); S) \geq |S| + 1.$$

The fact that $|W_j - \{z\}|$ is an even number now implies that

$$p(G(W_j - \{z\}) : S) \geq |S| + 2.$$

We then use (2.16) to calculate that

$$p(G; T \cup S \cup \{z\}) = p(G; T) - 1 + p(G(W_j - \{z\}); S)$$
$$\geq n + |T| - m - 1 + |S| + 2 = n + |S \cup T \cup \{z\}| - m.$$

Once again we have contradicted the choice of T in S, and we have established the existence of the matching $M_j(z)$ in $G(W_j - \{z\})$.

We now define a bipartite graph G^* with bipartition $\{S, I\}$. If $s \in S$ and $i \in I$, then $\{s, x\}$ is an edge of G^* if and only if there is an edge in G joining s and *some* vertex in W_i. We prove that G^* has a matching whose cardinality equals $|S|$. If not then by the König theorem there exist $X \subseteq S$ and $J \subseteq I$ such that $|J| < |X|$ and no edge in G^* has one endpoint in X and the other endpoint in $I - J$. We then conclude that

$$p(G; T - X) \geq p(G; T) - |J|.$$

Hence

$$n - p(G; T - X) + |T - X| \leq n - p(G; T) + |J| + |T| - |X| < m,$$

which contradicts the definition of m. Thus G^* has a matching whose cardinality equals $|S|$. This means that there exists a subset K of I with $|K| = |S|$ and vertices z_i in $W_i, (i \in K)$ such that G has a matching M' whose set of endpoints is $S \cup \{z_i : i \in K\}$. For each $i \in I - K$, let z_i be any chosen vertex in W_i. Then

$$M = M' \cup \cup_{i \in I} M_i(z_i) \cup \cup_W M_W,$$

where the last union is over those components $G(W)$ of $G(V - T)$ which contain a loop, is a matching in G with $\#(M) = m$.

Case 2. $T = \emptyset$ (that is, $S = \{\emptyset\}$) and $\mathcal{C}(G; \emptyset)$ has at most one component with more than one vertex.

In this case, $m = |V| - p(G; \emptyset)$ and $\mathcal{C}(\mathcal{G}; \emptyset)$ is the set of components of G with an odd number of vertices and no loops. In addition for all nonempty subsets S of V

$$m + 1 \leq |V| - p(G; S) + |S|.$$

Let $G(U)$ be a component of G such that the number of vertices of $G(U)$ is an odd number greater than 1 and $G(U)$ has no loops [by the assumptions of this case $G(U)$, if it exists, is unique]. We show that $G(U)$ has a matching M^* with $\#(M^*) = |U| - 1$. We choose an edge $\{x, y\}$ in $G(U)$. Let U' be obtained from U by removing the vertices x and y and let $S' \subseteq U'$. We let $S = S' \cup \{x, y\}$ and calculate

$$
\begin{aligned}
m + 1 &\leq |V| - p(G; S) + |S| \\
&= |V| - (p(G(U'); S') + p(G; \emptyset) - 1) + |S'| + 2 \\
&= (|V| - p(G; \emptyset) + 1) + |S'| - p(G(U'); S') + 2 \\
&= m + 1 + |S'| - p(G(U'); S') + 2.
\end{aligned}
$$

Hence

$$
p(G(U'); S') \leq |S'| + 2,
$$

and since $|U'|$ is odd,

$$
p(G(U'); S') \leq |S'| + 1.
$$

Thus for all subsets S' of U',

$$
|U'| - p(G(U'); S') + |S'| \geq |U'| - 1.
$$

We now apply the induction hypothesis and obtain a matching M' in $G(U')$ with $\#(M') = |U'| - 1$. Then $M^* = M' \cup \{\{x, y\}\}$ is a matching in $G(U)$ with $\#(M^*) = |U| - 1$. Now

$$
M = M^* \cup \cup_W M_W,
$$

where the last union is over those components $G(W)$ of G which have a loop, is a matching in G satisfying $\#(M) = m$.   □

Let G be a general graph of order n with vertex set V. A matching M with $\#(M) = n$ has the property that every vertex of G is an endpoint of an edge in M and is called a *perfect matching* or 1-*factor* of G. It follows from Theorem 2.6.1 that G has a perfect matching if and only if

$$
p(G; S) \leq |S|, \text{ for all } S \subseteq V.
$$

Now let A be a symmetric (0,1)-matrix of order n and let G be the general graph whose adjacency matrix is A. A perfect matching M in G corresponds to a set of n symmetrically placed 1's in A with no two of the 1's on the same line of A. Thus the above special case of Theorem 2.6.1 gives necessary and sufficient conditions for there to exist a symmetric permutation matrix P with $P \leq A$. We briefly describe these conditions in terms of the matrix A. Let S be a set of vertices of G with $|S| = k$. The adjacency matrix of the subgraph $G(V-S)$ is a principal submatrix A' of A of order $n-k$. The connected components of $G(V-S)$ correspond to certain "connected" principal submatrices $A_1, A_2, \ldots, A_t$ of A'. The condition (2.6) asserts that the number of submatrices $A_1, A_2, \ldots, A_t$ which are of odd order and have zero trace is at most k.

A $(0,1)$-matrix P of size m by n is a *subpermutation matrix of rank r* provided P has exactly r 1's, and no two 1's of P are on the same line of P. Let A be the adjacency matrix of a general graph G of order n. A subpermutation matrix P of rank r which is in addition symmetric corresponds to a matching in G of r edges. Hence Theorem 2.6.1 characterizes the maximum integer r such that A can be written in the form

$$A = P + X$$

where P is a symmetric subpermutation matrix of rank r and X is a non-negative matrix.

We next consider expressions of the form

$$A = P_1 + P_2 + \cdots + P_l \tag{2.19}$$

where $P_1, P_2, \ldots, P_l$ are symmetric permutation matrices of arbitrary ranks, and we seek to minimize the value of l in (2.19). Any l for which we have a decomposition of the form (2.19) is at least equal to the maximum line sum k of A. A theorem of Vizing[1964] concerning graphs asserts the existence of a decomposition (2.19) in which $l = k + 1$. We state this theorem in terms of matrices in a somewhat more general form to allow the presence of 1's on the main diagonal.

Theorem 2.6.2. *Let A be a symmetric $(0,1)$-matrix of order n, and let k be the maximum number of 1's in the off-diagonal positions of the lines of A. Then there exist symmetric subpermutation matrices $P_1, P_2, \ldots, P_{k+1}$ such that*

$$A = P_1 + P_2 + \cdots + P_{k+1}. \tag{2.20}$$

Proof. We first show that if the theorem is true in the case that the matrix A has zero trace, then it is true in general. Let A' be a symmetric $(0,1)$-matrix having at least one 1 on its main diagonal, and let k be the maximum number of 1's in the off-diagonal positions of the lines of A'. Let A be the matrix obtained from A by replacing the 1's on the main diagonal of A' with 0's. Suppose that there are subpermutation matrices $P_1, P_2, \ldots, P_{k+1}$ satisfying (2.20). Since the maximum number of 1's in a line of A is k, for each $i = 1, 2, \ldots, n$ at least one of the matrices $P_1, P_2, \ldots, P_{k+1}$ has only 0's in line i. It follows that we may replace certain 0's on the main diagonals of $P_1, P_2, \ldots, P_{k+1}$ and obtain symmetric subpermutation matrices $P'_1, P'_2, \ldots, P'_{k+1}$ satisfying $A' = P'_1 + P'_2 + \cdots + P'_{k+1}$.

We now prove the theorem under the added assumption that A has zero trace. Let G be the graph of order n whose adjacency matrix is A. The maximum degree of a vertex of G is k. Let σ be a function which assigns to each edge of G an integer from the set $\{1, 2, \ldots, k+1\}$. We think of σ as assigning a *color* to each edge of G from a set $\{1, 2, \ldots, k+1\}$ of $k+1$ colors. We call σ a $(k+1)$-*edge coloring* provided adjacent edges are always assigned different colors. Let F_i be the set of edges of G that are assigned color i by the edge coloring σ, and let P_i be the adjacency matrix of the spanning subgraph of G whose set of edges is F_i ($i = 1, 2, \ldots, k+1$). Then each P_i is a subpermutation matrix and $A = P_1 + P_2 + \cdots + P_{k+1}$. To complete the proof we show that a graph G has a $(k+1)$-edge coloring if k is the maximum degree of its vertices. The proof is by induction on the number of edges of G. Let $\alpha_1 = \{a, b_1\}$ be an edge of G. It suffices to show that if there is a $(k+1)$-edge coloring for G with the edge α_1 deleted, then there is a $(k+1)$-edge coloring for G.

Let σ be a $(k+1)$-edge coloring for the graph G' obtained by deleting the edge α_1 of G. There is a color t which is not assigned to any edge of G' which is incident with a. There is also a color t_1 which is not assigned to any edge incident with b_1. If $t = t_1$, then we may assign the color t to α_1 and thereby obtain a $(k+1)$-edge coloring of G. We now assume that there is an edge $\alpha_2 = \{a, b_2\}$ with color t_1. We remove the color t_1 from α_2 and assign the color t_1 to α_1. Let G_{t,t_1} be the subgraph of G consisting of those edges assigned colors t or t_1 and the vertices incident to these edges. The vertices a and b_1 belong to the same connected component C_1 of G_{t,t_1}. Suppose that b_2 is not a vertex of C_1. We then switch the colors t and t_1 on the edges of C_1. Now there is no edge of color t_1 incident to a or to b_2, and we may assign the color t_1 to the edge α_2 and obtain a $(k+1)$-edge coloring of G. We therefore assume that b_2 is a vertex of C_1. In particular, there is an edge with color t incident with b_2.

There is now a color t_2 different from t and t_1 which is not assigned to any edge incident with b_2. If the color t_2 were not assigned to some edge incident with a, we could assign t_2 to α_2 and obtain a $(k+1)$-edge coloring of G. We therefore proceed under the assumption that there is an edge $\alpha_3 = \{a, b_3\}$ which is assigned the color t_2. Arguing as above we may assume that a, b_2 and b_3 all belong to the same component of the subgraph G_{t,t_2} of G determined by the colors t and t_2. In particular, there is an edge with color t incident with b_3. We continue in this fashion and obtain a sequence of edges $\alpha_1 = \{a, b_1\}, \alpha_2 = \{a, b_2\}, \ldots, \alpha_k = \{a, b_k\}$ where after reassigning colors the edge α_i has color t_i ($i = 1, 2, \ldots, k-1$), edge α_k has no color assigned to it, and there is a color t_k different from colors t and t_{k-1} which is not assigned to any edge incident with b_k. We choose k to be the first integer such that $t_k = t_j$ for some $j < k-1$. The vertices a, b_j and b_{j+1} belong to the same connected component C_j of the subgraph G_{t,t_j} of G. Since there is no edge of color t at a and since there is no edge of color t_j at b_{j+1}, the component C_j consists of the vertices and edges of a chain

$$a \to b_j \to \cdots \to b_{j+1}.$$

Since $t_k = t_j$ and there is no edge incident with b_k which is assigned the color t_k, this chain cannot contain the vertex b_k. Thus b_k is a vertex of a connected component C^* of G_{t,t_j} different from C_j. We next switch the colors t and $t_j = t_k$ on the edges of C^*. Now there is no edge incident with b_k which is assigned the color t. We assign the color t to the edge $\alpha_k = \{a, b_k\}$ and thereby obtain a $(k + 1)$-coloring of G.   □

In the proof of Theorem 2.6.2 we have shown that a graph has a $(k+1)$-edge coloring if each of its vertices has degree at most equal to k. This conclusion need not hold for multigraphs. For example, the multigraph obtained by doubling each edge of the complete graph K_3 has six edges each pair of which are adjacent. Hence it has no 5-edge coloring. The smallest number t such that a multigraph G has a t-edge coloring is called the *chromatic index* of G. Thus Vizing's theorem asserts that the chromatic index of a graph for which the maximal degree of a vertex is k equals k or $k + 1$.

Vizing[1965] generalized Theorem 2.6.2 to include multigraphs. We state this theorem without proof in the language of matrices.

Theorem 2.6.3.   *Let A be a symmetric nonnegative integral matrix of order n. Let k be the maximum sum of the off-diagonal entries in the lines of A, and let m be the maximum element in all of A. Then there exist symmetric subpermutation matrices $P_1, P_2, \ldots, P_{k+m}$ such that*

$$A = P_1 + P_2 + \cdots + P_{k+m}.$$

A theorem of Shannon[1949] sometimes gives a better result than Theorem 2.6.3.

Theorem 2.6.4.   *Let A be a symmetric nonnegative integral matrix of order n. Let k be the maximum sum of the off-diagonal entries in the lines of A, and let l be the largest element on the main diagonal of A. Then there exist symmetric subpermutation matrices $P_1, P_2, \ldots, P_t$ with $t = k + \max\{\lceil k/2 \rceil, l\}$ such that*

$$A = P_1 + P_2 + \cdots + P_t.$$

Exercises

1. Prove the theorem of Petersen[1891]: Let G be a connected cubic graph such that each subgraph obtained from G by removing an edge is connected. Then G has a perfect matching.

2. Show that the chromatic index of the Petersen graph is 4.
3. Determine the chromatic index of the complete graph K_n, that is, the smallest number of symmetric subpermutation matrices into which the matrix $J_n - I_n$ can be decomposed.
4. Show that J_n can be decomposed into n symmetric permutation matrices.
5. Let G be a graph which is regular of degree k and suppose that k is a positive even integer. Prove that G can be decomposed into $k/2$ spanning subgraphs each of which is a regular graph of degree 2.

References

I. Anderson[1971], Perfect matchings of graphs, *J. Combin. Theory*, 10, pp. 183–186.

C. Berge[1958], Sur le couplage maximum d'un graphe, *C.R. Acad. Sciences* (Paris), 247, pp. 258–259.

R.A. Brualdi[1971], Matchings in arbitrary graphs, *Proc. Cambridge Phil. Soc.*, 69, pp. 401–407.

[1976], Combinatorial properties of symmetric non-negative matrices, *Teorie Combinatorie*, Toma II, Accademia Nazionale dei Lincei, Roma, pp. 99–120.

S. Fiorini and R.J. Wilson[1977], *Edge-colorings of graphs*, Pitman, London.

T. Gallai[1963], Neuer Beweis eines Tutte'schen Satzes, *Magyar Tud. Akad. Közl.*, 8, pp. 135–139.

W. Mader[1973], Grad und lokaler Zusammenhang in endlichen Graphen, *Math. Ann.*, 205, pp. 9–11.

C.E. Shannon[1949], A theorem on coloring the lines of a network, *J. Math. Phys.*, 28, pp. 148–151.

W.T. Tutte[1947], The factorization of linear graphs, *J. London Math. Soc.*, 22, pp. 107–111.

[1952], The factors of graphs, *Canadian J. Math.*, 4, pp. 314–328.

[1981], Graph factors, *Combinatorica*, 1, pp. 79–97.

V.G. Vizing[1964], On an estimate of the chromatic class of a p-graph (in Russian), *Diskret. Analiz.*, 3, pp. 25–30.

[1965], The chromatic class of a multigraph (in Russian), *Cybernetics*, 3, pp. 32–41.

3

Matrices and Digraphs

3.1 Basic Concepts

A *digraph* (*directed graph*) D consists of a finite set V of elements called *vertices* (*points*) together with a prescribed set E of *ordered* pairs of *not necessarily distinct* vertices of V. Every ordered pair α of vertices a and b in E is called an *arc* (*directed edge, directed line*) of the digraph D, written

$$\alpha = (a, b).$$

Notice that a digraph may contain both the arcs (a, b) and (b, a) as well as *loops* of the form (a, a). The generalization of a digraph by allowing multiple arcs results in a *directed general graph* (*general digraph*). Here the arcs sets are required to be finite. Most of the terminology in section 2.1 carries over without ambiguity to the directed case. The vertices a and b of an arc $\alpha = (a, b)$ are the *endpoints* of α, but now a is called the *initial vertex* and b is called the *terminal vertex* of α. The number of arcs issuing from a vertex is the *outdegree* of the vertex. The number of entering arcs is the *indegree* of the vertex. We agree that a loop at a vertex contributes one to the outdegree and also one to the indegree. If the outdegrees and indegrees equal a fixed integer k for every vertex of D, then D is *regular of degree k*.

Let D be a general digraph of order n whose set of vertices is $V = \{a_1, a_2, \ldots, a_n\}$. We let a_{ij} equal the multiplicity $m(a_i, a_j)$ of the arcs of the form (a_i, a_j). The resulting matrix

$$A = [a_{ij}], (i, j, = 1, 2, \ldots, n)$$

of order n is called the *adjacency matrix* of D. The entries of A are non-negative integers. But A need no longer be symmetric. In the event that A is symmetric, then the general digraph D is said to be *symmetric*.

The sum of row i of the adjacency matrix A is the outdegree of vertex a_i. The sum of column j of A is the indegree of vertex a_j. Notice how nicely loops behave in that they contribute one to both the outdegree and indegree. The assertion that D is regular of degree k is equivalent to the assertion that A has all of its line sums equal to k.

In the general digraph we now deal with *directed walks* of the form

$$(a_0, a_1), (a_1, a_2), \ldots, (a_{m-1}, a_m), (m > 0)$$

which is also denoted by

$$a_0 \to a_1 \to a_2 \to \cdots \to a_{m-1} \to a_m.$$

Most of the related concepts already discussed for general graphs carry over without ambiguity to the directed case; in particular we now refer to *directed chains (paths)*, *directed trails*, and *directed cycles (circuits)*. Notice that a directed cycle can have any of the lengths $1, 2, \ldots, n$.

Two vertices a and b are called *strongly connected* provided there are directed walks from a to b and from b to a. A vertex is regarded as trivially strongly connected to itself. Strong connectivity between vertices is reflexive, symmetric, and transitive. Hence strong connectivity defines an equivalence relation on the vertices of D and yields a partition

$$V_1 \cup V_2 \cup \cdots \cup V_t$$

of the vertices of D. The subdigraphs $D(V_1), D(V_2), \ldots, D(V_t)$ formed by taking the vertices in an equivalence class and the arcs incident to them are called the *strong components* of D. The general digraph D is *strongly connected (strong)* if it has exactly one strong component. Thus D is strongly connected if and only if each pair of vertices is strongly connected.

A *tournament* of order n is a digraph which can be obtained from the complete graph K_n by assigning a direction to each of its edges. Let A be the adjacency matrix of a tournament. Then A is a $(0,1)$-matrix satisfying the equation

$$A + A^T = J - I$$

and is called a *tournament matrix*.

Exercises

1. Let A be the adjacency matrix of a general digraph D. Show that there is a directed walk of length m from vertex a_i to vertex a_j if and only if the element in position (i, j) of A^m is positive.
2. Let D be a general digraph each of whose vertices has a positive indegree. Prove that D contains a directed cycle.
3. Prove that a digraph is strongly connected if and only if there is a closed directed walk which contains each vertex (at least once).

4. Prove that a tournament of order n contains a path of length $n - 1$. Conclude that the term rank of a tournament matrix of order n equals $n - 1$ or n.
5. Let A be the adjacency matrix of a digraph of order n. Prove that the term rank of A equals the maximal size of a set of vertices which can be partitioned into parts each of which is the set of vertices of a directed chain (that is, a path or directed cycle).

References

C. Berge[1973], *Graphs and Hypergraphs*, North-Holland, Amsterdam.

F. Harary[1967], Graphs and matrices, *SIAM Review*, 9, pp. 83–90.

F. Harary, R. Norman and D. Cartwright[1965], *Structural Models*, Wiley, New York.

R.J. Wilson[1972], *Introduction to graph theory*, Academic Press, New York.

3.2   Irreducible Matrices

Let $A = [a_{ij}]$, $(i, j, = 1, 2, \ldots, n)$ be a matrix of order n consisting of real or complex numbers. To A there corresponds a digraph $D = D(A)$ of order n as follows. The vertex set is the n-set $V = \{a_1, a_2, \ldots, a_n\}$. There is an arc $\alpha = (a_i, a_j)$ from a_i to a_j if and only if $a_{ij} \neq 0, (i, j = 1, 2, \ldots, n)$. We may think of a_{ij} as being a nonzero weight attached to the arc α. In the event that A is a matrix of nonnegative integers, the weight a_{ij} of α can be regarded as the multiplicity $m(\alpha)$ of α. Then D is a general digraph and A is its adjacency matrix. However, unless specified to the contrary, D is the unweighted digraph as defined above.

The matrix A of order n is called *reducible* if by simultaneous permutations of its lines we can obtain a matrix of the form

$$\begin{bmatrix} A_1 & O \\ A_{21} & A_2 \end{bmatrix}$$

where A_1 and A_2 are square matrices of order at least one. If A is not reducible, then A is called *irreducible*. Notice that a matrix of order 1 is irreducible.

Irreducibility has a direct interpretation in terms of the digraph D of A.

Theorem 3.2.1.   *Let A be a matrix of order n. Then A is irreducible if and only if its digraph D is strongly connected.*

Proof. First assume that A is reducible. Then the vertex set V of D can be partitioned into two nonempty sets V_1 and V_2 in such a way that there is no arc from a vertex in V_1 to a vertex in V_2. If a is a vertex in V_1 and b is a vertex in V_2 there is no directed walk from a to b. Hence D is not strongly connected.

Now assume that D is not strongly connected. Then there are distinct vertices a and b of D for which there is no directed walk from a to b. Let W_1 consist of b and all vertices of D from which there is a directed walk to b, and let W_2 consist of a and all vertices to which there is a directed walk from a. The sets W_1 and W_2 are nonempty and disjoint. Let W_3 be the set consisting of those vertices which belong to neither W_1 nor W_2. We now simultaneously permute the lines of A so that the lines corresponding to the vertices in W_2 come first followed by those corresponding to the vertices in W_3:

$$
\begin{array}{c}
W_2 \\
W_3 \\
W_1
\end{array}
\begin{array}{ccc}
W_2 & W_3 & W_1 \\
\left[\begin{array}{ccc}
X_{11} & X_{12} & X_{13} \\
X_{21} & X_{22} & X_{23} \\
X_{31} & X_{32} & X_{33}
\end{array}\right].
\end{array}
$$

Since there is no directed walk from a to b there is no arc from a vertex in W_2 to a vertex in W_1. Also there is no arc from a vertex c in W_3 to a vertex in W_1, because such an arc implies that c belongs to W_1. Hence $X_{13} = O$ and $X_{23} = O$, and A is reducible.   □

If A is irreducible, then for each pair of distinct vertices a_i and a_j there is a walk from a_i to a_j with length at most equal to $n - 1$. Hence if the elements of A are nonnegative numbers, then it follows from Theorem 3.2.1 that A is irreducible if and only if the elements of the matrix $(I + A)^{n-1}$ are all positive.

We return to the general case of a matrix A of order n. Let D be the digraph of A and let $D(V_1), D(V_2), \ldots, D(V_t)$ be the strong components of D. Let D^* be the digraph of order t whose vertices are the sets $V_1, V_2, \ldots, V_t$ in which there is an arc from V_i to V_j if and only if $i \neq j$ and there is an arc in D from some vertex in V_i to some vertex in V_j. The digraph D^* is the *condensation digraph* of D. The digraph D^* has no loops.

Lemma 3.2.2.   *The condensation digraph D^* of the digraph D has no closed directed walks.*

Proof. Suppose that D^* has a closed directed walk. Since D^* has no loops, its length is at least two. If V_k and V_l $(k \neq l)$ are two vertices of D^* of this walk, and $a \in V_k$ and $b \in V_l$, then a and b are strongly connected vertices of D in different strong components. This contradiction implies that D^* has no closed directed walks.   □

Lemma 3.2.3.   *Let D^* be a digraph of order t which has no closed directed walks. Then the vertices of D^* can be ordered $b_1, b_2, \ldots, b_t$ so that each arc of D^* is of the form (b_i, b_j) for some i and j with $1 \leq i < j \leq t$.*

Proof. The proof proceeds by induction on t. If $t = 1$ the digraph has no arcs. Assume that $t > 1$. The assumption that there are no closed directed walks implies that there is a vertex b_1 with indegree equal to zero. We now delete the vertex b_1 and all incident arcs and apply the induction hypothesis to the resulting digraph of order $t - 1$.   □

We now show that a square matrix can be brought to a very special form by simultaneous permutations of its lines.

Theorem 3.2.4.   *Let A be a matrix of order n. Then there exists a permutation matrix P of order n and an integer $t \geq 1$ such that*

$$PAP^T = \begin{bmatrix} A_1 & A_{12} & \cdots & A_{1t} \\ O & A_2 & \cdots & A_{2t} \\ \vdots & \vdots & \ddots & \vdots \\ O & O & \cdots & A_t \end{bmatrix}, \tag{3.1}$$

where $A_1, A_2, \ldots, A_t$ are square irreducible matrices. The matrices $A_1, A_2, \ldots, A_t$ that occur as diagonal blocks in (3.1) are uniquely determined to within simultaneous permutation of their lines, but their ordering in (3.1) is not necessarily unique.

Proof. Let $D(V_1), D(V_2), \ldots, D(V_t)$ be the strong components of the digraph D of A. By Lemma 3.2.2 the condensation graph D^* has no closed directed walks. We apply Lemma 3.2.3 to D^* and obtain an ordering $W_1, W_2, \ldots, W_t$ of $V_1, V_2, \ldots, V_t$ with the property that each arc of D^* is of the form (W_i, W_j) for some i and j with $1 \leq i < j \leq t$. We now simultaneously permute the lines of A so that the lines corresponding to the vertices in W_1 come first, followed in order by those corresponding to the vertices in $W_2, \ldots, W_t$, and then partition the resulting matrix to obtain

$$\begin{bmatrix} A_1 & A_{12} & \cdots & A_{1t} \\ A_{21} & A_2 & \cdots & A_{2t} \\ \vdots & \vdots & \ddots & \vdots \\ A_{t1} & A_{t2} & \cdots & A_t \end{bmatrix}. \tag{3.2}$$

In (3.2) $D(W_i)$ is the digraph corresponding to the matrix A_i, ($i = 1, 2, \ldots, t$). Since there is no arc of the form (W_i, W_j) with $i > j$, each of the matrices A_{ij} with $i > j$ is a zero matrix. Hence (3.2) has the form given in (3.1). Because $D(W_i)$ is strongly connected, each A_i is an irreducible matrix.

We now establish the uniqueness assertion in the theorem. Let Q be a permutation matrix of order n such that

$$QAQ^T = \begin{bmatrix} B_1 & B_{12} & \cdots & B_{1s} \\ O & B_2 & \cdots & B_{2s} \\ \vdots & \vdots & \ddots & \vdots \\ O & O & \cdots & B_s \end{bmatrix}$$

where each B_i is an irreducible square matrix. The subdigraphs $D'_1, D'_2, \ldots, D'_s$ of D corresponding to the diagonal blocks $B_1, B_2, \ldots, B_s$, respectively, are strongly connected. Moreover, since for $i > j$ there is no directed walk in D from a vertex of D'_i to a vertex of D'_j, the digraphs $D'_1, D'_2, \ldots, D'_s$ are the strong components of D. Since the strong components of a digraph are uniquely determined, The digraphs $D'_1, D'_2, \ldots, D'_s$ are $D(V_1), D(V_2), \ldots, D(V_t)$ in some order. Thus $s = t$ and to within simultaneous permutations of lines, the matrices $B_1, B_2, \ldots, B_t$ are the matrices $A_1, A_2, \ldots, A_t$ in some order. □

The form in (3.1) appears in the works of Frobenius[1912] and is called the *Frobenius normal form* of the square matrix A. The irreducible matrices $A_1, A_2, \ldots, A_t$ that occur along the diagonal are the *irreducible components* of A. By Theorem 3.2.4 the irreducible components are uniquely determined only to within simultaneous permutations of their lines. This slight ambiguity causes no difficulty. Notice that the matrix A is irreducible if and only if it has exactly one irreducible component. Whether the ordering of the irreducible components along the diagonal in the Frobenius normal form is uniquely determined depends on the matrices A_{ij} $(1 \le i < j \le t)$. For example, let

$$\begin{bmatrix} 0 & 1 & & X \\ 1 & 0 & & \\ & & 1 & 1 \\ O & & 1 & 1 \end{bmatrix}.$$

Then A is in Frobenius normal form and the irreducible components of A are

$$A_1 = \begin{bmatrix} 0 & 1 \\ 1 & 0 \end{bmatrix} \quad \text{and} \quad A_2 = \begin{bmatrix} 1 & 1 \\ 1 & 1 \end{bmatrix}.$$

If X is a zero matrix, then we may change the order of the diagonal blocks A_1 and A_2 and obtain a different Frobenius normal form for A. If, however, X is not a zero matrix, then the Frobenius normal form of A is unique. In general, the irreducible components of the square matrix A that correspond to vertices of the condensation graph D^* of $D(A)$ which are incident with no arc can always be put in the first positions along the diagonal of the Frobenius normal form (3.1).

Let A be an irreducible matrix of order n. If B is obtained from A by simultaneous line permutations, then B is also irreducible. However, a matrix B obtained from A by arbitrary line permutations may be reducible. For example, interchanging rows 1 and 2 of the irreducible matrix

$$\begin{bmatrix} 0 & 1 & 1 \\ 1 & 0 & 0 \\ 1 & 0 & 0 \end{bmatrix},$$

we obtain the reducible matrix

$$\begin{bmatrix} 1 & 0 & 0 \\ 0 & 1 & 1 \\ 1 & 0 & 0 \end{bmatrix}.$$

Thus it may be possible to permute the lines of a reducible matrix and obtain an irreducible matrix. The following theorem of Brualdi[1979] characterizes those square matrices which can be obtained from irreducible matrices by arbitrary line permutations. Since for permutation matrices P and Q, $PAQ = (PAP^T)PQ$, it suffices to consider only column permutations.

Theorem 3.2.5. *Let A be a matrix of order n. There exists a permutation matrix Q of order n such that AQ is an irreducible matrix if and only if A has at least one nonzero element in each line.*

Proof. If A has a zero line, then for each permutation matrix Q of order n, AQ has a zero line and hence is reducible. Conversely, assume that A has no zero line. Without loss of generality we assume that A is in Frobenius normal form

$$\begin{bmatrix} A_1 & A_{12} & \cdots & A_{1t} \\ O & A_2 & \cdots & A_{2t} \\ \vdots & \vdots & \ddots & \vdots \\ O & O & \cdots & A_t \end{bmatrix}.$$

If $t = 1$, then A is irreducible and we may take $Q = I$. Now assume that $t > 1$. Let $D(V_1), D(V_2), \ldots, D(V_t)$ be the strong components of the digraph D of A corresponding, respectively, to the irreducible matrices $A_1, A_2, \ldots, A_t$. Any arc that leaves the vertex set V_i enters the vertex set $V_{i+1} \cup \cdots \cup V_t, (1 \le i \le t - 1)$. For each $i = 1, 2, \ldots, t$ we choose a vertex a_i in V_i. Let B be the matrix obtained from A by cyclically permuting the columns corresponding to the vertices $a_1, a_2, \ldots, a_t$. The digraph D' of B is obtained from the digraph D of A by changing the arcs that entered

a_i into arcs that enter a_{i+1}, $(i = 1, \ldots, t - 1)$, and changing the arcs that entered a_t into arcs that enter a_1. We show that B is irreducible by proving that the digraph D' is strongly connected. For convenience of notation we define a_{t+1} to be a_1.

If each of the strong components of D either has order greater than one or has order one and contains a loop, the proof is simple to express. In this case each strong component $D(V_i)$ has a sequence $\gamma_{i1}, \gamma_{i2}, \ldots, \gamma_{ik_i}$ of $k_i \geq 1$ closed walks of nonzero length which begin and end at a_i, $(i = 1, 2, \ldots, t)$. Each vertex in V_i belongs to at least one of these walks and they may be chosen so that a_i occurs only as the first and the last vertex. By repeating walks if necessary we may assume that the k_i have a common value k. In D' the last arcs of these walks enter a_{i+1}. Let γ'_{ij} be the directed walk in D' obtained from γ_{ij} by replacing the last vertex a_i of γ_{ij} with a_{i+1} and the last arc with an arc entering a_{i+1} $(1 \leq i \leq t)$. Then

$$\gamma'_{11}, \gamma'_{12} \ldots, \gamma'_{1k}, \gamma'_{21}, \gamma'_{22}, \ldots, \gamma'_{2k}, \ldots, \gamma'_{t1}, \gamma'_{t2}, \ldots, \gamma'_{tk}$$

is a closed directed walk in D' which contains each vertex of D' at least once. It follows that D' is strongly connected in this case.

In the general case some of the strong components of D may consist of a single isolated vertex without a loop. However since A has no zero lines, neither the first component $D(V_1)$ nor the last component $D(V_t)$ can be of this form. We prove that D' is strongly connected by showing that for each vertex a there is a directed walk from a_1 to a and a directed walk from a to a_1. Suppose that a is a vertex of V_i. We may assume that $i > 1$. In order to obtain a directed walk from a_1 to a it suffices to obtain a directed walk from a_1 to a_i. Since the column of A corresponding to vertex a_{i-1} contains a 1, there is an arc in D' from some vertex b in $V_1 \cup \cdots \cup V_{i-1}$ to a_i. Arguing inductively, there is a directed walk in D' from a_1 in V_1 to b. Hence there is a directed walk in D' from a_1 to a.

An argument similar to the above, but using the assumption that each row of A has a 1, allows us to conclude that there is a directed walk in D' from a to a vertex c in V_t. It then follows that there is a directed walk in D' from a to a_1. Hence D' is strongly connected. □

For some historical remarks on the origins of the property of irreducibility of matrices, we refer the reader to Schneider[1977].

Exercises

1. Determine the special nature of the Frobenius normal form of a tournament matrix.
2. What is the Frobenius normal form of a permutation matrix of order n?
3. Determine the smallest number of nonzero elements of an irreducible matrix of order n.

4. Show by example that the product of two irreducible matrices may be reducible (even if the matrices have nonnegative elements).
5. Let A be an irreducible matrix of order n with nonnegative elements. Assume that each element on the main diagonal of A is positive. Let x be a column vector with nonnegative elements. Prove that if x contains at least one 0, then Ax has fewer 0's than x.
6. Let D be a strongly connected digraph of order n and assume that each directed cycle of D has length 2. Prove that D can be obtained from a tree of order n by replacing each edge $\{a, b\}$ with the two oppositely directed arcs (a, b) and (b, a).

References

R.A. Brualdi[1979], Matrices permutation equivalent to irreducible matrices and applications, *Linear and Multilin. Alg.*, 7, pp. 1–12.

G.F. Frobenius[1912], Über Matrizen aus nicht negativen Elementen, *Sitzungsber. Preuss. Akad. Wiss., Berl.*, pp. 476–457.

F. Harary[1959], A graph theoretic method for the complete reduction of a matrix with a view toward finding its eigenvalues, *J. Math. and Physics*, 38, pp. 104–111.

F. Harary, R. Norman and D. Cartwright[1965], *Structural Models*, Wiley, New York.

D. Rosenblatt[1957], On the graphs and asymptotic forms of finite boolean relation matrices and stochastic matrices, *Naval Research Logistics Quarterly*, 4, pp. 151–167.

H. Schneider[1977], The concept of irreducibility and full indecomposability of a matrix in the works of Frobenius, König and Markov, *Linear Alg. Applics.*, 18, pp. 139–162.

3.3 Nearly Reducible Matrices

Let D be a strong digraph of order n with vertex set $V = \{a_1, a_2, \ldots, a_n\}$, and let $A = [a_{ij}], (i, j = 1, 2, \ldots, n)$ be its (0,1)-adjacency matrix of order n. By Theorem 3.2.1 A is an irreducible matrix. The digraph D is called *minimally strong* provided each digraph obtained from D by the removal of an arc is not strongly connnected. Evidently, a minimally strong digraph has no loops. Each arc of the digraph D corresponds to a 1 in the adjacency matrix A. Thus the removal of an arc in D corresponds in the adjacency matrix A to the replacement of a 1 with a 0. The irreducible matrix A is called *nearly reducible* (Hedrick and Sinkhorn[1970]) provided each matrix obtained from A by the replacement of a 1 with a 0 is a reducible matrix. Thus the digraph D is minimally strong if and only if its adjacency matrix A is nearly reducible. A nearly reducible matrix has zero trace. The matrix

$$\begin{bmatrix} 0 & 1 & 0 & 0 \\ 0 & 0 & 1 & 0 \\ 1 & 0 & 0 & 1 \\ 0 & 1 & 0 & 0 \end{bmatrix}$$

is an example of a nearly reducible matrix.

More generally we say that an arbitrary matrix A of order n is *nearly reducible* if its digraph D is minimally strong. However, in discussing the combinatorial structure of nearly reducible matrices, it suffices to consider (0,1)-matrices.

We now investigate the structure of minimally strong digraphs as determined by Luce[1952] (see also Berge[1973]). Let D be a strong digraph of order n with vertex set V. Each vertex has indegree and outdegree at least equal to 1. A vertex whose indegree and outdegree both equal 1 is called *simple*. If all the vertices of D are simple, then D consists of the vertices and arcs of a directed cycle of length n, and D is minimally strong. A directed walk

$$a_0 \to a_1 \to \cdots \to a_{m-1} \to a_m, (m \geq 1)$$

is called a *branch* provided the following three conditions hold:

(i) a_0 and a_m are not simple vertices;
(ii) the set $W = \{a_1, \ldots, a_{m-1}\}$ contains only simple vertices;
(iii) the subdigraph $D(V - W)$ is strongly connected.

Notice that a branch may be closed and the set W may be empty. If the removal of an arc (a, b) from D results in a strongly connected digraph, then $a \to b$ is a branch. Let U be a nonempty subset of m vertices of V. The *contraction of D by U* is the general digraph $D(\otimes U)$ of order $n - m + 1$ defined as follows: The vertex set of $D(\otimes U)$ is $V - U$ with an additional vertex labeled (U). The arcs of D which have both of their endpoints in $V - U$ are arcs of $D(\otimes U)$; in addition, for each vertex a in $V - U$ there is an arc in $D(\otimes U)$ from a to (U) [respectively, from (U) to a] of multiplicity k if there are k arcs in D from a to vertices in U [respectively, from vertices in U to a].

Theorem 3.3.1. *Let D be a minimally strong digraph with vertex set V. Let U be a nonempty subset of V for which the subdigraph $D(U)$ is strongly connected. Then both $D(U)$ and $D(\otimes U)$ are minimally strong digraphs.*

Proof. If the removal of an arc of $D(U)$ leaves a strong digraph, then the removal of that arc from D leaves a strong digraph. Hence $D(U)$ is a minimally strong digraph.

It follows in an elementary way that $D(\otimes U)$ is strongly connected. Now let α be an arc of $D(\otimes U)$. First assume that α is also an arc of D. If the removal of α from $D(\otimes U)$ leaves a strong digraph, then the removal of α from D also leaves a strong digraph. Now assume that α is an arc joining the vertex (U) and some vertex a in $V - U$. If the multiplicity of α is greater than one, then since $D(U)$ is strongly connected all but one of the arcs of D that contribute to the multiplicity of α can be removed from D to leave a strong digraph. It follows that α has multiplicity one and $D(\otimes U)$ is a digraph. Let α' be the arc of D joining a and some vertex in U which corresponds to the arc α of $D(\otimes U)$. Since $D(U)$ is strongly connected, the removal of α' from D leaves a strong digraph if the removal of α from $D(\otimes U)$ leaves a strong digraph. Since D is minimally strong, we deduce that no arc can be removed from $D(\otimes U)$ to leave a strong digraph. Hence $D(\otimes)$ is a minimally strong digraph. □

A special case of Theorem 3.3.1 asserts that in a minimally strong digraph D the only arcs joining the vertices of a directed cycle in D are the arcs of the directed cycle. We now show that minimally strong digraphs contain simple vertices.

Lemma 3.3.2. *Let D be a minimally strong digraph of order $n \geq 2$. Then D has at least two simple vertices.*

Proof. Since D is strongly connected, there must be a directed cycle in D. If D is a directed cycle, then all its vertices are simple. This happens, in particular, when $n = 2$. We now assume that $n > 2$ and proceed by induction on n. First assume that all directed cycles in D have length 2. Then D can be obtained from a tree by replacing each of its edges $\{a, b\}$ by the arcs (a, b) and (b, a) in opposite directions. A tree with $n > 2$ vertices has at least two pendant vertices and each of these pendant vertices is a simple vertex of D.

Now assume that D has a directed cycle μ of length $m \geq 3$. Since D is not itself a directed cycle, we have $m \leq n - 1$. Let U be the set of vertices on the arcs of μ. Then the subdigraph $D(U)$ contains no arcs other than those of μ. The contraction $D(\otimes U)$ has order $n - m + 1 \geq 2$ and, by the induction hypothesis, has (at least) two simple vertices. A simple vertex of $D(\otimes U)$ different from the vertex (U) is a simple vertex of D. Suppose that one of the two simple vertices of $D(\otimes U)$ is (U). Then in D there is exactly one arc (a, c) with a in U and c in $V - U$, and exactly one arc (d, b) with d in $V - U$ and b in U. Since $m \geq 3$ there is a vertex e in U different from a and b. But then e is a simple vertex of D. It follows that D contains at least two simple vertices. □

In a minimally strong digraph a branch cannot have length one and hence a branch contains at least one simple vertex. A digraph D which is a directed cycle is minimally strong, but since all its vertices are simple, D contains no branches.

Lemma 3.3.3. *Let D be a minimally strong digraph of order $n \geq 3$ which is not a directed cycle. Then D has a branch of length $k \geq 2$.*

Proof. Since D is not a directed cycle, it has at least one vertex which is not simple. We define a general digraph D^* whose vertex set V^* is the set of nonsimple vertices of D. Let a and b be vertices of D^*. Each directed walk α of D from a to b all of whose vertices other than a and b are simple determines an arc $\alpha^* = (a, b)$ on D^*. The general digraph D^* is strongly connected and has no simple vertices. If D^* has a loop or a multiple arc α^*, then α is a branch of D. Otherwise D^* is a strong digraph of order at least two with no simple vertices. By Theorem 3.3.2 D^* is not minimally strong and hence there is an arc α^* whose removal from D^* leaves a strong digraph. The directed walk α is a branch of D.

$\square$

It follows from Lemma 3.3.3 that any minimally strong digraph can be constructed by beginning with a directed cycle and sequentially adding branches. However, while every digraph constructed in this way is strongly connected, it need not be minimally strong.

We now apply Lemma 3.3.3 to determine an inductive structure for nearly reducible matrices (Hartfiel[1970]).

Theorem 3.3.4. *Let A be a nearly reducible $(0, 1)$-matrix of order $n \geq 2$. Then there exists a permutation matrix P of order n and an integer m with $1 \leq m \leq n - 1$ such that*

$$
PAP^T = \begin{bmatrix}
0 & 0 & 0 & \cdots & 0 & 0 & \\
1 & 0 & 0 & \cdots & 0 & 0 & \\
0 & 1 & 0 & \cdots & 0 & 0 & \\
0 & 0 & 1 & \cdots & 0 & 0 & F_1 \\
\vdots & \vdots & \vdots & \ddots & \vdots & \vdots & \\
0 & 0 & 0 & \cdots & 1 & 0 & \\
& & & F_2 & & & A_1
\end{bmatrix} \tag{3.3}
$$

where A_1 is a nearly reducible matrix of order m. The matrix F_1 contains a single 1 and this 1 belongs to the first row and column j of F_1 for some j with $1 \leq j \leq m$. The matrix F_2 also contains a single 1 and this 1 belongs to the last column and row i of F_2 for some i with $1 \leq i \leq m$. The element in the position (i, j) of A_1 is 0.

Proof. If the digraph $D(A)$ is a directed cycle, then there is a permutation matrix P such that (3.3) holds where A_1 is a zero matrix of order 1. Assume that $D(A)$ is not a directed cycle. Then $n \geq 3$ and by Lemma 3.3.3 $D(A)$ has a branch. A direct translation of the defining properties of a branch implies the existence of a permutation matrix P such that (3.3) and the ensuing properties hold.

$\square$

A matrix A satisfying the conclusions of Theorem 3.3.4 is not necessarily nearly reducible. An example with $n = 6$ and $m = 5$ is given by

$$
A = \begin{bmatrix}
0 & 0 & 0 & 1 & 0 & 0 \\
1 & 0 & 1 & 0 & 0 & 0 \\
0 & 0 & 0 & 0 & 0 & 1 \\
0 & 0 & 1 & 0 & 0 & 0 \\
0 & 0 & 0 & 0 & 1 & 0 \\
0 & 1 & 0 & 0 & 1 & 0
\end{bmatrix}.
$$

The matrix obtained from A by replacing the 1 in row 2 and column 3 with a 0 is irreducible and hence A is not nearly reducible. However, given a nearly reducible (0,1)-matrix A_1 of order $m \geq 1$ it is possible to choose the matrices F_1 and F_2 so that the matrix in (3.3) is nearly reducible. Indeed by choosing F_1 to have a 1 in position $(1,1)$ and by choosing F_2 to have a 1 in position $(1, n - m)$, we obtain a nearly reducible matrix of order n for each integer $n > m$.

The inductive structure of minimally strong digraphs provided by Lemma 3.3.3 can be used to bound the number of arcs in a minimally strong digraph and hence the number of 1's in a nearly reducible matrix. Let T be a tree of order n. We denote by $\overleftrightarrow{T}$ the digraph of order n obtained from T by replacing each edge $\{a, b\}$ with the two oppositely directed arcs (a, b) and (b, a). The digraph $\overleftrightarrow{T}$ is called a *directed tree*. The directed tree $\overleftrightarrow{T}$ is minimally strong and has $2(n - 1)$ arcs. The following theorem is due to Gupta[1967] (see also Brualdi and Hedrick[1979]).

Theorem 3.3.5.   *Let D be a minimally strong digraph of order $n \geq 2$. Then the number of arcs of D is between n and $2(n - 1)$. The number of arcs of D is n if and only if D is a directed cycle. The number of arcs of D is $2(n - 1)$ if and only if D is a directed tree.*

Proof. The outdegree of each vertex of a strongly connected graph is at least one and hence D has at least n arcs. There are exactly n arcs in D if and only if D is a directed cycle.

The upper bound on the number of arcs of D and the characterization of equality is proved by induction on the order n. If $n = 2$, then D is a directed cycle with two arcs. Assume that $n \geq 3$. If D is a directed cycle then D has $n < 2(n - 1)$ arcs. Now suppose that D is not a directed cycle. By Lemma 3.3.3 D has a branch

$$
\alpha : a_0 \rightarrow a_1 \rightarrow \cdots \rightarrow a_{m-1} \rightarrow a_m
$$

of length $m \geq 2$. The subdigraph D_1 of D obtained by removing the vertices $a_1, \ldots, a_{m-1}$ is a minimally strong digraph of order $n - m + 1$. By the induction hypothesis D_1 has at most $2(n - m)$ arcs, and hence the number of arcs of D is at most

$$2(n - m) + m = 2n - m \leq 2n - 2.$$

Assume that D has $2n - 2$ arcs. Then $m = 2$, α is the branch $a_0 \rightarrow a_1 \rightarrow a_2$ and D_1 is a minimally strong digraph of order $n - 1$ with $2(n - 2)$ arcs. By the induction hypothesis there is a tree T_1 of order $n - 1$ such that $D_1 = \overleftrightarrow{T_1}$. Suppose that the vertex a_0 is different from the vertex a_2. Then in D_1 there is a directed chain $a_2 \rightarrow x_1 \rightarrow x_2 \rightarrow \cdots \rightarrow x_s \rightarrow a_0$ from a_2 to a_0. The arc (x_1, a_2) is an arc of D_1. Moreover,

$$x_1 \rightarrow x_2 \rightarrow \cdots \rightarrow a_0 \rightarrow a_1 \rightarrow a_2$$

is a directed chain in D from x_1 to a_2 which does not use the arc (x_1, a_2). It follows that the removal of the arc (x_1, a_2) of D leaves a strong digraph which contradicts the assumption that D is a minimally strong digraph. Hence $a_0 = a_2$. Let T be the tree obtained from T_1 by including the vertex a_1 and the edge $\{a_0, a_1\}$. Then $D = \overleftrightarrow{T}$. □

A direct translation of the preceding theorem yields the following.

Theorem 3.3.6.   *Let A be a nearly reducible $(0,1)$-matrix of order $n \geq 2$. Then the number of 1's of A is between n and $2(n-1)$. The number of 1's of A equals n if and only if there is a permutation matrix P of order n such that*

$$PAP^T = \begin{bmatrix} 0 & 0 & \cdots & 0 & 0 & 1 \\ 1 & 0 & \cdots & 0 & 0 & 0 \\ 0 & 1 & \cdots & 0 & 0 & 0 \\ \vdots & \vdots & \ddots & \vdots & \vdots & \vdots \\ 0 & 0 & \cdots & 1 & 0 & 0 \\ 0 & 0 & \cdots & 0 & 1 & 0 \end{bmatrix}.$$

The number of 1's of A equals $2(n-1)$ if and only if there is a tree T of order n such that A is the adjacency matrix of $\overleftrightarrow{T}$. □

It follows from Theorem 3.3.1 that an irreducible principal submatrix of order m of a nearly reducible $(0,1)$-matrix is itself nearly reducible and hence by Theorem 3.3.5 has at most $2(m-1)$ 1's. But a principal submatrix of a nearly reducible matrix need not be irreducible. A simple example is furnished by the leading principal submatrix of order 2 of the nearly reducible matrix

$$
\begin{bmatrix}
0 & 0 & 1 \\
1 & 0 & 0 \\
0 & 1 & 0
\end{bmatrix}.
$$

Nonetheless a principal submatrix of order m of a nearly reducible $(0,1)$-matrix has at most $2(m-1)$ 1's. This and other properties of nearly reducible matrices can be found in Brualdi and Hedrick[1979].

Lemma 3.3.3 can also be used to determine an inductive structure for strongly connected digraphs (or strongly connected general digraphs), and hence for irreducible matrices.

Theorem 3.3.7.   *Let D be a strong digraph of order $n \geq 2$ with vertex set V. Then there exists a partition of V into $m \geq 2$ nonempty sets $W_1, W_2, \ldots, W_m$ such that the subdigraphs $D(W_1), D(W_2), \ldots, D(W_m)$ are strongly connected. Let $W_{m+1} = W_1$. Each arc of D that does not belong to one of these sub digraphs issues from W_i and enters W_{i+1} for some $i = 1, 2, \ldots, m$.*

Proof. We remove arcs from D in order to obtain a minimally strong digraph D'. It follows from Lemma 3.3.3 that D' has the cyclical structure in the theorem. Indeed the partition can be chosen so that $|W_i| = 1$ ($1 \leq i \leq m-1$), $D(W_m)$ is minimally strong ($|W_m| = 1$ is possible) and there is exactly one arc issuing from W_i and entering W_{i+1} ($1 \leq i \leq m$).

Consider a digraph with the cyclical structure of the theorem. If we add a new arc, then either this cyclical structure is retained or else the arc issues from some W_i and enters some W_j where $j \neq i, i+1$. In the latter case we obtain the cyclical structure of the theorem with vertex partition

$$
W_{i+1}, \ldots, W_{j-1}, W_j \cup W_{j+1} \cup \cdots \cup W_i.
$$

Since the minimally strong digraph D' has the cyclical structure in the theorem and since D is obtained from D' by adding arcs, it now follows that D has the desired cyclical structure.   □

A direct translation of the previous theorem yields the following inductive structure for irreducible matrices.

Theorem 3.3.8.   *Let A be an irreducible matrix of order $n \geq 2$. Then there exists a permutation matrix P of order n and an integer $m \geq 2$ such that*

$$
PAP^T =
\begin{bmatrix}
A_1 & O & \cdots & O & E_1 \\
E_2 & A_2 & \cdots & O & O \\
\vdots & \vdots & \ddots & \vdots & \vdots \\
O & O & \cdots & A_{m-1} & O \\
O & O & \cdots & E_m & A_m
\end{bmatrix}
$$

where $A_1, A_2, \ldots, A_m$ are irreducible matrices and $E_1, E_2, \ldots, E_m$ are matrices having at least one nonzero entry.

Exercises

1. Show that a minimally strong regular digraph is a directed cycle.
2. Prove that the permanent of a nearly reducible (0,1)-matrix of order n equals 0 or 1. For each integer $n \geq 2$ construct an example of a nearly reducible (0,1)-matrix with permanent equal to 0 and one with permanent equal to 1 (Hedrick and Sinkhorn[1970]).
3. Let $n \geq 2$ be an integer. Show that there exists a nearly reducible (0,1)-matrix of order n with exactly k 1's for each integer k with $n \leq k \leq 2(n-1)$ (Brualdi and Hedrick[1979]).
4. Let $n \geq 3$ be an integer and let D be a minimally strong digraph of order n with exactly $2n-3$ arcs. Prove that D has a directed cycle of length 3 and does not have a directed cycle of length greater than 3 (Brualdi and Hedrick[1979]).
5. Let A be a nearly reducible (0,1)-matrix of order n. Deduce from Theorem 3.3.1 that if a principal submatrix B of A is irreducible, then in fact B is nearly reducible. Give an example of a nearly reducible matrix which has a reducible principal submatrix.
6. Let A be a nearly reducible (0,1)-matrix and let B be a principal submatrix of A of order k. Prove that the number of 1's in B is at most equal to $2(k-1)$ (Brualdi and Hedrick[1979]).

References

C. Berge[1973], *Graphs and Hypergraphs*, North-Holland, Amsterdam.

R.A. Brualdi and M.B. Hedrick[1979], A unified treatment of nearly reducible and nearly decomposable matrices, *Linear Alg. Applics.*, 24, pp. 51–73.

G. Chaty and M. Chein[1976], A note on top down and bottom up analysis of strongly connected digraphs, *Discrete Math.*, 16, pp. 309–311.

R.P. Gupta[1967], On basis diagraphs, *J. Combin. Theory*, 3, pp. 16–24.

D.J. Hartfiel[1970], A simplified form for nearly reducible and nearly decomposable matrices, *Proc. Amer. Math. Soc.*, 24, pp. 388–393.

M. Hedrick and R. Sinkhorn[1970], A special class of irreducible matrices—The nearly reducible matrices, *J. Algebra*, 16, pp. 143–150.

D.E. Knuth[1974], Wheels within wheels, *J. Combin. Theory, Ser. B*, 16, pp. 42–46.

R.D. Luce[1952], Two decomposition theorems for a class of finite oriented graphs, *Amer. J. Math.*, 74, pp. 701–722.

3.4 Index of Imprimitivity and Matrix Powers

Let D denote a strongly connected digraph of order n whose set of vertices is $V = \{a_1, a_2, \ldots, a_n\}$. Let $k = k(D)$ be the greatest common divisor of the lengths of the closed directed walks of D. (If $n = 1$ and D does not contain a loop, k is undefined.) The integer k is called the *index of imprimitivity* of D. The digraph D is *primitive* if $k = 1$ and *imprimitive* if $k > 1$. The length of a closed directed walk is the sum of the lengths of one or more directed cycles, and hence the index of imprimitivity k is also the greatest common divisor of the lengths of the directed cycles of D. The integer k does not exceed the length of any directed cycle of D. There are a number of elementary facts concerning the index of imprimitivity which we collect in the following lemma.

Lemma 3.4.1.   *Let D be a strongly connected digraph of order n with index of imprimitivity equal to k.*

(i) *For each vertex a of D, k equals the greatest common divisor of the lengths of the closed directed walks containing a.*

(ii) *For each pair of vertices a and b, the lengths of the directed walks from a to b are congruent modulo k.*

(iii) *The set V of vertices of D can be partitioned into k nonempty sets $V_1, V_2, \ldots, V_k$ where, with $V_{k+1} = V_1$, each arc of D issues from V_i and enters V_{i+1} for some i with $1 \le i \le k$.*

(iv) *For $x_i \in V_i$ and $x_j \in V_j$ the length of a directed walk from x_i to x_j is congruent to $j - i$ modulo k, $(1 \le i, j \le k)$.*

Proof. Let a and b be two vertices of D and let k_a and k_b denote the greatest common divisors of the lengths of the closed directed walks containing a and b, respectively. Let α be a closed directed walk containing a and suppose that α has length r. Since D is strongly connected there is a directed walk β from a to b of some length s and a directed walk γ from b to a of some length t. We may combine α, β and γ and obtain closed directed walks containing b with lengths $s + t$ and $r + s + t$, respectively. Thus k_b is a divisor of r, and since α was an arbitrary closed directed walk containing a, k_b is a divisor of k_a. In a similar way one proves that k_a is a divisor of k_b. Therefore $k_a = k_b$. But a and b are arbitrary vertices of D and (i) follows. (We note that (i) does not hold in general if we consider only the directed cycles containing the vertex a.)

Now let β' be another directed walk from a to b, and let s' be the length of β'. We may combine β and γ and also β' and γ to obtain closed directed walks containing a with lengths $s + t$ and $s' + t$, respectively. Since k is a divisor of $s + t$ and $s' + t$, k is a divisor of $s - s'$. Hence (ii) holds.

Let V_i denote the set of vertices x_i for which there is a directed walk from vertex a to x_i with length congruent to i modulo k, $(i = 1, 2, \ldots, k)$. By (ii) the sets $V_1, V_2, \ldots, V_k$ are mutually disjoint. Notice that the vertex a belongs to V_k. Since D is strongly connected each vertex belongs to one of the sets $V_1, V_2, \ldots, V_k$, and none of these sets can be empty. Let (x_i, x_j) be any arc of D where $x_i \in V_i$ and $x_j \in V_j$. There is a directed walk from a to x_i whose length is congruent to i modulo k and thus a directed walk from a to x_j whose length is congruent to $i + 1$ modulo k. Hence $i + 1$ is congruent to j modulo k, and (iii) follows. The proof of (iv) is quite similar to that of (iii).   □

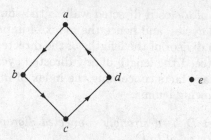

Figure 3.1

The sets $V_1, V_2, \ldots, V_k$ in (iii) of Lemma 3.4.1 are called the *sets of imprimitivity* of D. Although their construction depended on a choice of vertex a, they are uniquely determined. Indeed much more is true. Call a digraph D *cyclically r-partite with ordered partition $U_1, U_2, \ldots, U_r$* provided $U_1, U_2, \ldots, U_r$ is a partition of the vertex set V of D into r nonempty sets where, with $U_{r+1} = U_1$, each arc of D issues from U_i and enters U_{i+1} for some $i = 1, 2, \ldots, r$. If D is cyclically r-partite then r is a divisor of the length of each directed cycle of D, and in addition, D is cyclically s-partite for each positive integer s which is a divisor of r. Thus *if D is strongly connected, then D is cyclically r-partite if and only if r is a divisor of the index of imprimitivity k of D.* Let D be a strongly connected digraph with sets of imprimitivity $V_1, V_2, \ldots, V_k$. Suppose that D is cyclically r-partite with ordered partition $U_1, U_2, \ldots, U_r$. Then except for a possible cyclic rearrangement,

$$U_1 = V_1 \cup V_{r+1} \cup \cdots,$$
$$U_2 = V_2 \cup V_{r+2} \cup \cdots,$$
$$\cdots$$
$$U_r = V_r \cup V_{2r} \cup \cdots.$$

If D is not strongly connected, it may be cyclically r-partite with respect to two ordered partitions which are not cyclic rearrangements. For example, the digraph in Figure 3.1 is cyclically 4-partite with ordered partition $\{a\}, \{b\}, \{c\}, \{d, e\}$. It is also cyclically 4-partite with respect to the ordered partition $\{a\}, \{b\}, \{c, e\}, \{d\}$.

Now let A be an irreducible matrix of order n. By Theorem 3.2.1 the digraph $D(A)$ is strongly connected. We define the *index of imprimitivity* of A to be the index of imprimitivity k of $D(A)$. In addition we call A *primitive* (respectively, *imprimitive*) if $D(A)$ is primitive (respectively, imprimitive). We also say that A is *r-cyclic* if $D(A)$ is cyclically r-partite. Imprimitive matrices, more generally, r-cyclic matrices with $r > 1$ have a restrictive structure which is a consequence of the structure of cyclically r-partite digraphs discussed above.

Suppose the digraph $D(A)$ is cyclically r-partite with ordered partition $U_1, U_2, \ldots, U_r$. Let U_i contain n_i vertices $(i = 1, 2, \ldots, r)$. Then $n = n_1 + n_2 + \cdots + n_r$, and by simultaneously permuting the lines of A so that the rows corresponding to the vertices in U_1 come first, followed in order by those corresponding to the vertices in $U_2, \ldots, U_r$, we may determine a permutation matrix P of order n so that

$$PAP^T = \begin{bmatrix} O & A_{12} & O & \cdots & O \\ O & O & A_{23} & \cdots & O \\ \vdots & \vdots & \vdots & \ddots & \vdots \\ O & O & O & \cdots & A_{r-1,r} \\ A_{r1} & O & O & \cdots & O \end{bmatrix}. \tag{3.4}$$

In (3.4) the zero matrices on the diagonal are square matrices of orders $n_1, n_2, \ldots, n_r$, respectively. The matrices $A_{i,i+1}$ display the adjacencies between vertices in U_i and U_{i+1}. If $r = 1$ then (3.4) reduces to $PAP^T = A$ with P equal to the identity matrix of order n. If r is the index of imprimitivity k of A, then (3.4) holds with $r = k$. The matrices $A_{12}, A_{23}, \ldots, A_{r-1,r}, A_{r1}$ in (3.4) are called the r-cyclic components of the matrix A. The r-cyclic components may be cyclically permuted in (3.4). In addition, since the elements in the sets U_i can be given in any specified order,

$$P_1 A_{12} P_2^T, P_2 A_{23} P_3^T, \ldots, P_{r-1} A_{r-1,r} P_r^T, P_r A_{r1} P_1^T$$

can be taken as the r-cyclic components of A for any choice of permutation matrices $P_1, P_2, \ldots, P_r$ of orders $n_1, n_2, \ldots, n_r$, respectively. If A is irreducible, the r-cyclic components are uniquely determined apart from these transformations.

Suppose that A is r-cyclic and a permutation matrix P has been determined so that (3.4) holds. Then

$$PA^r P^T = \begin{bmatrix} B_1 & O & \cdots & O \\ O & B_2 & \cdots & O \\ \vdots & \vdots & \ddots & \vdots \\ O & O & \cdots & B_r \end{bmatrix} \tag{3.5}$$

where

$$B_1 = A_{12} A_{23} \cdots A_{r-1,r} A_{r1}$$
$$B_2 = A_{23} A_{34} \cdots A_{r1} A_{12}$$
$$\cdots$$
$$B_r = A_{r1} A_{12} \cdots A_{r-2,r-1} A_{r-1,r}.$$

In particular, if $r > 1$ then A^r is reducible and has at least r irreducible components. If A is a matrix whose entries are nonnegative real numbers, further information can be obtained.

We now assume that $A = [a_{ij}], (i, j = 1, 2, \ldots, n)$ is a nonnegative matrix of order n. Let t be a positive integer and let the element in the (i, j) position of A^t be denoted by $a_{ij}^{(t)}, (i, j = 1, 2, \ldots, n)$. Then $a_{ij}^{(t)}$ is positive if and only if there is a directed walk of length t from vertex a_i to vertex a_j in the digraph $D(A)$. In particular, the locations of the zeros and nonzeros in A^t are wholly determined by the digraph $D(A)$.

The following lemma is usually attributed to Schur (see Kemeny and Snell[1960]).

Lemma 3.4.2.   *Let S be a nonempty set of positive integers which is closed under addition. Let d be the greatest common divisor of the integers in S. Then there exists a positive integer N such that td is in S for every integer $t \geq N$.*

Proof. We may divide each integer in S by d and this allows us to assume that $d = 1$. There exist integers $r_1, r_2, \ldots, r_m$ in S which are relatively prime. Each integer k can be expressed as a linear combination of $r_1, r_2, \ldots, r_m$ with integral, but not necessarily nonnegative, coefficients. Let $q = r_1 + r_2 + \cdots + r_m$. Then we may determine integers c_{ij} such that

$$i = c_{i1}r_1 + c_{i2}r_2 + \cdots + c_{im}r_m, (i = 0, 1, \ldots, q - 1).$$

Let M be the maximum of the integers $|c_{ij}|$, let $N = Mq$ and let t be any integer with $t \geq N$. There exist integers p and l such that $t = pq + l$ where $p \geq M$ and $0 \leq l \leq q - 1$. Then

$$t = pq + l = p(r_1 + r_2 + \cdots + r_m) + (c_{l1}r_1 + c_{l2}r_2 + \cdots + c_{lm}r_m)$$
$$= (p + c_{l1})r_1 + (p + c_{l2})r_2 + \cdots + (p + c_{lm})r_m.$$

Since $p \geq M$, each of the integers $p + c_{lj}$ is nonnegative. Because S is closed under addition, we conclude that t is in S whenever $t \geq N$. $\square$

Let S and d satisfy the hypotheses of Lemma 3.4.2. Then there exists a smallest positive integer $\phi(S)$ such that nd is in S for every integer $n \geq \phi(S)$. The integer $\phi(S)$ is called the *Frobenius–Schur index* of S. (Frobenius was one of the first to consider the evaluation of this number.) If S consists of all nonnegative linear combinations of the positive integers $r_1, r_2, \ldots, r_m$, then we write the Frobenius–Schur index of S as $\phi(r_1, r_2, \ldots, r_m)$.

Lemma 3.4.3.   *Let D be a strongly connected digraph of order n with vertex set V. Let k be the index of imprimitivity of D and let $V_1, V_2, \ldots, V_k$ be the sets of imprimitivity of D. Then there exists a positive integer N for which the following holds: If x_i and x_j are vertices belonging respectively* *to V_i and V_j, then there are directed walks from x_i to x_j of every length $j - i + tk$ with $t \geq N, (1 \leq i, j \leq k)$.*

Proof. Let a and b be vertices in V and suppose that $a \in V_i$ and $b \in V_j$. By (iv) of Lemma 3.4.1 each directed walk from a to b has length $j - i + tk$ for some nonnegative integer k. Let t_{ab} be an integer such that $j - i + t_{ab}k$ is the length of some directed walk from a to b. The lengths of the closed directed walks containing b form a nonempty set S_b of positive integers which is closed under addition. By (i) of Lemma 3.4.1 k is the greatest common divisor of the integers in S_b. We apply Lemma 3.4.2 to S_b and obtain a positive integer N_b such that $tk \in S_b$ for every integer $t \geq N_b$. There exists a directed walk from a to b with length $j - i + tk$ for every integer $t \geq t_{ab} + N_b$. We now let N be the maximum integer in the set $\{t_{ab} + N_b | a, b \in V\}$.   □

We return to the irreducible, nonnegative matrix A of order n with index of imprimitivity equal to k. There exists a permutation matrix P of order n such that (3.4) and (3.5) hold with $r = k$. The matrices $A_{12}, A_{23}, \ldots,$ $A_{k-1,k}, A_{k1}$ are the k-cyclic components of A, and these arise from the sets of imprimitivity of the digraph $D(A)$. It follows from (3.5) that for each positive integer t we have

$$PA^{tk}P^T = \begin{bmatrix} B_1^t & O & \cdots & O \\ O & B_2^t & \cdots & O \\ \vdots & \vdots & \ddots & \vdots \\ O & O & \cdots & B_k^t \end{bmatrix}, \tag{3.6}$$

where

$$\begin{aligned} B_1 &= A_{12}A_{23}\cdots A_{k-1,k}A_{k1} \\ B_2 &= A_{23}A_{34}\cdots A_{k1}A_{12} \\ &\quad \cdots \\ B_k &= A_{k1}A_{12}\cdots A_{k-2,k-1}A_{k-1,k}. \end{aligned}$$

We apply Lemma 3.4.3 and conclude that there exists a positive integer N such that $B_1^t, B_2^t, \ldots, B_k^t$ are positive matrices for all $t \geq N$. In the special case that $k = 1$, A is primitive and A^t is a positive matrix for each integer $t \geq N$. If $k > 1$ and A is imprimitive we may apply (iv) of Lemma 3.4.1 and conclude that no positive integral power of A is a positive matrix. Hence we obtain the following characterization of primitive matrices.

Theorem 3.4.4. *Let A be a nonnegative matrix of order n. Then A is primitive if and only if some positive integral power of A is a positive matrix. If A is primitive then there exists a positive integer N such that A^t is a positive matrix for each integer $t \geq N$.*

Proof. The fact that A is primitive if and only if some positive integral power of A is a positive matrix has been proved in the above paragraph under the additional assumption that A is irreducible. Since a positive integral power of a reducible matrix can never be positive, the theorem follows.   □

Let A be a primitive nonnegative matrix. By Theorem 3.4.4 there exists a smallest positive integer $\exp(A)$ such that A^t is a positive matrix for all integers $t \geq \exp(A)$. The integer $\exp(A)$ is called the *exponent of the primitive matrix A*. The exponent is the subject of the next section. We continue now with the general development of irreducible matrices.

A matrix which is a positive integral power of a reducible matrix is reducible. However, positive integral powers of irreducible matrices may be either reducible or irreducible. In the case of a nonnegative matrix A those positive integral powers of A which are irreducible can be characterized (Dulmage and Mendelsohn[1967] and Brualdi and Lewin[1982]).

Theorem 3.4.5.   *Let A be an irreducible nonnegative matrix of order n with index of imprimitivity equal to k. Let m be a positive integer. Then A^m is irreducible if and only if k and m are relatively prime. In general there is a permutation matrix P of order n (independent of m) such that*

$$PA^mP^T = \begin{bmatrix} C_1 & O & \cdots & O \\ O & C_2 & \cdots & O \\ \vdots & \vdots & \ddots & \vdots \\ O & O & \cdots & C_r \end{bmatrix} \tag{3.7}$$

where r is the greatest common divisor of k and m. The matrices $C_1, C_2, \ldots, C_r$ in (3.7) are irreducible matrices and each has index of imprimitivity equal to k/r.

Proof. The digraph $D = D(A)$ is strongly connected with index of imprimitivity equal to k. Let $V_1, V_2 \ldots, V_k$ be the sets of imprimitivity of D. Then D is cyclically k-partite with ordered partition $V_1, V_2, \ldots, V_k$. Since r is a divisor of k, D is also cyclically r-partite with ordered partition $U_1, U_2, \ldots, U_r$ where

$$U_1 = V_1 \cup V_{r+1} \cup \cdots,$$
$$U_2 = V_2 \cup V_{r+2} \cup \cdots,$$
$$\cdots$$
$$U_r = V_r \cup V_{2r} \cup \cdots.$$

We may choose a permutation matrix P of order n such that (3.4) holds.

For this permutation matrix P, PA^rP^T has the form given in (3.5). Since r is also a divisor of m, we may write

$$PA^mP^T = (PA^rP^T)^{m/r} = \begin{bmatrix} C_1 & O & \cdots & O \\ O & C_2 & \cdots & O \\ \vdots & \vdots & \ddots & \vdots \\ O & O & \cdots & C_r \end{bmatrix}$$

where $C_1 = B_1^{m/r}, C_2 = B_2^{m/r}, \ldots, C_r = B_r^{m/r}$. If $r > 1$ then A^m is reducible. We first show that the matrices $C_1, C_2, \ldots, C_r$ are irreducible.

Let a and b be vertices in U_i where $1 \leq i \leq r$. There exist integers u and v such that $a \in V_{ur+i}$ and $b \in V_{vr+i}$. By Lemma 3.4.3 there exists a positive integer N such that there are directed walks in D from a to b of every length $(v - u)r + tk$ with $t \geq N$. Since r is the greatest common divisor of k and m, it follows from Lemma 3.4.2 that there is an integer t' such that

$$(v - u)r + t'k = ek + fm$$

for some nonnegative integers e and f. For each nonnegative integer s we have

$$(v - u)r + (t' - e + sm)k = (f + sk)m.$$

We now choose s large enough so that $t' - e + sm \geq N$. Then there is a directed walk in D from a to b with length $(f + sk)m$ and thus a directed walk in $D(A^m)$ with length $f + sk$. Since a and b are arbitrary vertices in U_i, we conclude that $D(C_i)$ is strongly connected and hence that C_i is irreducible, $(1 \leq i \leq r)$.

Let l be the length of a closed directed walk of $D(A^m)$. Then there is a closed directed walk in D with length lm. Because the index of imprimitivity of D is k, k is a divisor of lm. Since the greatest common divisor of k and m is r, k/r is a divisor of l. Therefore the index of imprimitivity of each digraph $D(C_i)$ is a multiple of k/r. We now show that the index of imprimitivity equals k/r.

We now take a and b to be the same vertex of U_i. Then $v = u$ and there are closed directed walks containing a in D of every length tk with $t \geq N$, and hence of every length

$$t\frac{m}{r}k = (tm)\frac{k}{r}$$

with $t \geq N$. It follows that in $D(C_i)$ there are closed directed walks containing a of every length $t(k/r)$ with $t \geq N$. We now take $t = N$ and $t = N + 1$ and obtain closed directed walks in $D(C_i)$ whose lengths have a greatest common divisor equal to k/r. We now conclude that the index of imprimitivity of $D(C_i)$, and thus of C_i equals k/r for each integer $i = 1, 2, \ldots, r$.   □

A matrix A of order n is *completely reducible* provided there exists a permutation matrix P of order n such that

$$PAP^T = \begin{bmatrix} A_1 & O & \cdots & O \\ O & A_2 & \cdots & O \\ \vdots & \vdots & \ddots & \vdots \\ O & O & \cdots & A_t \end{bmatrix}$$

where $t \geq 2$ and $A_1, A_2, \ldots, A_t$ are square irreducible matrices. Thus A is completely reducible if and only if A is reducible and the matrices $A_{ij}, (1 \leq i < j \leq t)$ that occur in the Frobenius normal form (2.1) are all zero matrices. The following corollary is an immediate consequence of Theorem 3.4.5.

Corollary 3.4.6. *Let A be an irreducible nonnegative matrix. Let m be a positive integer. If A^m is reducible, then A^m is completely reducible.*

Let $A = [a_{ij}], (1 \leq i, j \leq n)$ be a matrix of order n which is r-cyclic. We conclude this section by showing how the r-cyclicity of A implies a special structure for the characteristic polynomial of A. First we recall the definition of the determinant. The determinant of A is given by

$$\det(A) = \sum_{\pi} (\text{sign } \pi) a_{1\pi(1)} a_{2\pi(2)} \cdots a_{n\pi(n)}$$

where the summation extends over all permutations π of $\{1, 2, \ldots, n\}$ and $(\text{sign } \pi) = \pm 1$ is the *sign* of the permutation π. We let the set V of vertices of $D(A)$ be $\{1, 2, \ldots, n\}$ where there is an arc (i, j) from vertex i to vertex j if and only if $a_{ij} \neq 0$. Suppose that $a_{1i_1} a_{2i_2} \cdots a_{ni_n} \neq 0$. Then $U = \{(1, i_1), (2, i_2), \ldots, (n, i_n)\}$ is a set of n arcs of $D(A)$. For each vertex there is exactly one arc in U leaving the vertex and exactly one arc entering it. Thus the set U of arcs can be partitioned into nonempty sets each of which is the set of arcs of a directed cycle of $D(A)$.

Now let

$$\varphi(\lambda) = \det(\lambda I - A) = c_0 \lambda^n + c_1 \lambda^{n-1} + c_2 \lambda^{n-2} + \cdots + c_{n-1}\lambda + c_n, (c_0 = 1)$$

be the *characteristic polynomial* of A. The n roots of $\varphi(\lambda)$ are the *characteristic roots* (*eigenvalues*) of A. The coefficient c_i of λ^{n-i} equals $(-1)^i$ times the sum of the determinants of the principal submatrices of A of order i. It follows from the above discussion that $c_i \neq 0$ only if there exists in $D(A)$ a collection of directed cycles the sum of whose lengths is i which have no vertex in common. If the digraph $D(A)$ is cyclically r-partite, each directed cycle has a length which is divisible by r. Hence if $D(A)$ is cyclically r-partite, $c_i \neq 0$ only if r is a divisor of i. Hence the only powers of λ which can appear with a nonzero coefficient in $\varphi(\lambda)$ are $\lambda^n, \lambda^{n-r}, \lambda^{n-2r}, \ldots$. The following theorem is due to Dulmage and Mendelsohn[1963,1967].

Theorem 3.4.7.   *Let A be an r-cyclic matrix of order n. Let P be a permutation matrix of order n such that (3.4) and (3.5) hold. Then there exists a monic polynomial $f(\lambda)$ and nonnegative integers $p_1, p_2, \ldots, p_r$ such that the following hold:*

(i) $f(0) \neq 0$;

(ii) The characteristic polynomial of B_i is $f(\lambda)\lambda^{p_i}, (i = 1, 2, \ldots, r)$. For each root μ of $f(\lambda)$ the elementary divisors corresponding to μ are the same for each of $B_1, B_2, \ldots, B_r$;

(iii) The characteristic polynomial of A is $f(\lambda^r)\lambda^{p_1 + p_2 + \cdots + p_r}$;

(iv) The characteristic polynomial of A^r is $f(\lambda)^r \lambda^{p_1 + p_2 + \cdots + p_r}$.

Proof. Since A is r-cyclic its characteristic polynomial $\varphi(\lambda)$ can be written as $\varphi(\lambda) = f(\lambda^r)\lambda^p$ where $f(\lambda)$ is a monic polynomial with $f(0) \neq 0$ and p is a nonnegative integer. Since the eigenvalues of A^r are the rth powers of the eigenvalues of A, the characteristic polynomial of A^r is $(f(\lambda))^r \lambda^p$. Let $\varphi_i(\lambda)$ be the characteristic polynomial of B_i $(i = 1, 2, \ldots, r)$. We have

$$\varphi_1(\lambda)\varphi_2(\lambda) \cdots \varphi_r(\lambda) = (f(\lambda))^r \lambda^p, \qquad (3.8)$$

and the nonzero eigenvalues of $B_1, B_2, \ldots, B_r$ are all roots of $f(\lambda)$. Next we observe that

$$B_1 = A_{12}(A_{23} \cdots A_{r1}), \qquad \text{and} \qquad B_2 = (A_{23} \cdots A_{r1})A_{12}.$$

Standard results in matrix theory now allow us to conclude that the nonzero eigenvalues of B_1 are the same as those of B_2 and the elementary divisors of B_1 corresponding to its nonzero eigenvalues are the same as those corresponding to the nonzero eigenvalues of B_2. The same conclusions hold for B_2 and B_3, B_3 and B_4, $\ldots$, B_r and B_1. Hence $B_1, B_2, \ldots, B_r$ all have the same nonzero eigenvalues and the same elementary divisors corresponding to each of their nonzero eigenvalues. We are now able to assert that there exists a monic polynomial $g(\lambda)$ with $g(0) \neq 0$ and nonnegative integers $p_1, p_2, \ldots, p_r$ such that $\varphi_i(\lambda) = g(\lambda)\lambda^{p_i}$ $(i = 1, 2, \ldots, r)$. Substituting these last equations into (3.8) we obtain

$$(g(\lambda))^r \lambda^{p_1 + p_2 + \cdots + p_r} = (f(\lambda))^r \lambda^p. \qquad (3.9)$$

Since $g(0) \neq 0$ and $f(0) \neq 0$ we conclude from (3.9) that $f(\lambda) = g(\lambda)$ and $p = p_1 + p_2 + \cdots + p_p$. Now each of (i)-(iv) holds.   □

Most of the results in this section can be found in Dulmage and Mendelsohn[1967]. Other early papers which treat some of the topics, in some cases from a different viewpoint or with a different starting point, include Gantmacher[1959], Pták[1958], Romanovsky[1936], Pták and Sedláček[1958] and Varga[1962]. Spectral properties of irreducible nonnegative matrices relating to the index of imprimitivity will be studied in the book *Combinatorial Matrix Classes*.

Exercises

1. Show that the index of imprimitivity of a strongly connected digraph is not always equal to the greatest common divisor of the lengths of the directed cycles containing a specified vertex.
2. Prove or disprove that the product of two primitive matrices is primitive.
3. Prove that a primitive $(0,1)$-matrix of order $n \geq 2$ contains at least $n + 1$ 1's and construct an example with exactly $n + 1$ 1's.
4. Prove that the index of imprimitivity of an irreducible imprimitive symmetric matrix of order $n \geq 2$ equals 2.
5. Let A be a nonnegative matrix of order n and assume that A has no zero lines. Suppose that A is cyclically r-partite and has the form given in (3.4). Prove that A is irreducible if and only if $A_{12} \cdots A_{r-1,r} A_{r1}$ is irreducible (Dulmage and Mendelsohn[1967]; see also Minc[1974]).
6. (Continuation of Exercise 5) Prove that the number of irreducible components of A equals the number of irreducible components of $A_{12} \cdots A_{r-1,r} A_{r1}$ (Brualdi and Lewin[1982]).

References

R.A. Brualdi and M. Lewin[1982], On powers of nonnegative matrices, *Linear Alg. Applics.*, 43, pp. 87–97.

A.L. Dulmage and N.S. Mendelsohn[1963], The characteristic equation of an imprimitive matrix, *SIAM J. Appl. Math.*, 11, pp. 1034–1045.

   [1967], Graphs and matrices, *Graph Theory and Theoretical Physics* (F. Harary, ed.), Academic Press, New York, pp. 167–227.

F.R. Gantmacher[1959], *The Theory of Matrices*, vol. 2, Chelsea, New York.

J.G. Kemeny and J.L. Snell[1960], *Finite Markov Chains*, Van Nostrand, Princeton.

H. Minc[1974], The structure of irreducible matrices, *Linear Multilin. Alg.*, 2, pp. 85–90.

V. Pták[1958], On a combinatorial theorem and its applications to nonnegative matrices, *Czech. Math. J.*, 8, pp. 487–495.

V. Pták and J. Sedláček[1958], On the index of imprimitivity of non-negative matrices, *Czech. Math. J.*, 8, pp. 496–501.

V. Romanovsky[1936], Recherches sur les Chains de Markoff, *Acta. Math.*, 66, pp. 147–251.

R.S. Varga[1962], *Matrix Iterative Analysis*, Prentice-Hall, Englewood Cliffs, N.J.

3.5 Exponents of Primitive Matrices

The exponent $\exp(A)$ of a primitive nonnegative matrix A has been defined to be the smallest positive integer k such that A^t is a positive matrix for all integers $t \geq k$. The exponent of A depends only on the digraph $D(A)$ (and not on the magnitude of the elements of A) and equals the smallest positive integer k such that for each ordered pair a, b of vertices there is a directed walk from a to b of length k, and thus a directed walk from a to b of every length greater than or equal to k. In investigating the exponent there is therefore no loss in generality in considering only

(0,1)-matrices. As a result we assume throughout this section that A is a primitive (0,1)-matrix of order n. The vertex set of the digraph $D(A)$ is denoted by $V = \{a_1, a_2, \ldots, a_n\}$.

The exponent of the matrix A can be evaluated in terms of other more basic quantities. Let $\exp(A : i, j)$ equal the smallest integer k such that the element in position (i, j) of A^t is nonzero for all integers $t \geq k$, $(1 \leq i, j \leq n)$. Let $\exp(A : i)$ equal the smallest positive integer p such that all the elements in row i of A^p are nonzero, $(1 \leq i \leq n)$. Thus $\exp(A : i, j)$ equals the smallest positive integer k such that there is a directed walk of length t from a_i to a_j in $D(A)$ for all $t \geq k$, and $\exp(A : i)$ equals the smallest positive integer p such that there are directed walks of length p from a_i to each vertex of $D(A)$.

Lemma 3.5.1.   *The exponent of A equals the maximum of the integers*

$$\exp(A : i, j), \quad (i, j = 1, 2, \ldots, n).$$

It also equals the maximum of the integers

$$\exp(A : i), \quad (i = 1, 2, \ldots, n).$$

Proof. The first conclusion is an immediate consequence of the definitions involved. Suppose that there is a directed walk of length p in $D(A)$ from vertex a_i to vertex a_j for each j with $1 \leq j \leq n$. There is an arc $\alpha = (a_k, a_j)$ for some choice of vertex a_k. A directed walk from a_i to a_k of length p combined with the arc α determines a directed walk of length $p + 1$ from a_i to a_j. It follows that there are directed walks from a_i to each vertex a_j of every length $t \geq p$, and the second conclusion also holds. $\square$

Lemma 3.5.1 is useful for obtaining upper bounds for the exponent of the primitive matrix A. If $f_1, f_2, \ldots, f_n$ are integers such that there are directed walks of length f_i from a_i to every vertex of $D(A), (i = 1, 2, \ldots, n)$, then $\exp(A) \leq \max\{f_1, f_2, \ldots, f_n\}$.

An irreducible matrix with at least one nonzero element on its main diagonal is primitive, since its digraph has a directed cycle of length 1. The following theorem of Holladay and Varga[1958] gives a bound for the exponent of such a matrix.

Theorem 3.5.2.   *Let A be an irreducible matrix of order n having $p \geq 1$ nonzero elements on its main diagonal. Then A is a primitive matrix and $\exp(A) \leq 2n - p - 1$.*

Proof. The digraph $D(A)$ has p loops, and we let W be the set of p vertices which are incident with a loop. Let a_i and a_j be two vertices. There is a directed path from a_i to a vertex a_k in W whose length is at most $n - p$ and a directed path from a_k to a_j whose length is at most $n - 1$. Combining these two directed paths we obtain a directed path from a_i to

a_j of length at most equal to $2n - p - 1$. Taking advantage of the loop at vertex a_k we obtain a directed walk from a_i to a_j whose length is exactly $2n - p - 1$. □

If the irreducible matrix A in Theorem 3.5.2 has no zeros on its main diagonal, then the exponent of A is at most $n - 1$. This special case of Theorem 3.5.2 is equivalent to the property noted in Section 2 that for an irreducible nonnegative matrix A of order n, $(I + A)^{n-1}$ is a positive matrix.

The characterization of those matrices achieving the bound in the following theorem is due to Shao[1987].

Theorem 3.5.3.   *Let A be a symmetric irreducible $(0, 1)$-matrix of order $n \geq 2$. Then A is primitive if and only if its associated digraph $D(A)$ has a directed cycle of odd length. If the symmetric matrix A is primitive, then $\exp(A) \leq 2n - 2$ and equality occurs if and only if there exists a permutation matrix P of order n such that*

$$PAP^T = \begin{bmatrix} 0 & 1 & 0 & \cdots & 0 & 0 \\ 1 & 0 & 1 & \cdots & 0 & 0 \\ 0 & 1 & 0 & \cdots & 0 & 0 \\ \vdots & \vdots & \vdots & \ddots & \vdots & \vdots \\ 0 & 0 & 0 & \cdots & 0 & 1 \\ 0 & 0 & 0 & \cdots & 1 & 1 \end{bmatrix}.$$

Proof. The digraph $D(A)$ is a symmetric digraph and has a directed cycle of length 2. Hence A is primitive if and only if $D(A)$ has a directed cycle of odd length. Assume that A is primitive. The matrix A^2 is primitive and has no zeros on its main diagonal. By Theorem 3.5.2 $(A^2)^{n-1}$ is a positive matrix and hence $\exp(A) \leq 2n - 2$. First suppose that A is the matrix displayed in the theorem. The smallest *odd* integer k for which there is a directed walk from vertex a_1 to itself is $2n - 1$. Hence $\exp(A) \geq \exp(A : 1, 1) \geq 2n - 2$. Now suppose that $\exp(A) = 2n - 2$. Examining the proof of Theorem 3.5.2 as applied to the matrix A^2 we see that there are two vertices whose distance in $D(A^2)$ is $n - 1$. Thus $D(A^2)$ is a directed chain of length $n - 1$ with a loop incident at each vertex. If there were three vertices a, b and c each two of which were adjacent in the symmetric digraph $D(A)$, then $a \to b \to c \to a$ would be a directed cycle in $D(A^2)$. It follows that $D(A)$ is a directed chain of length $n - 1$ with at least one vertex incident with a loop. Since $\exp(A) = 2n - 2$ there is exactly one loop in $D(A)$ and it is incident with one of the end vertices of the directed chain. □

Theorem 3.5.2 gives the upper bound $2n - 2$ for the exponent of a primitive matrix whose digraph has a directed cycle of length 1. The following

theorem of Sedláček[1959] and Dulmage and Mendelsohn[1964] furnishes a
bound for the exponent in terms of the lengths of directed cycles.

Theorem 3.5.4.   *Let A be a primitive $(0,1)$-matrix of order n. Let s be
the smallest length of a directed cycle in the digraph $D(A)$. Then*

$$\exp(A) \leq n + s(n-2).$$

Proof. The matrix A^s has at least s positive elements on its main diagonal. Let W be the set of vertices of $D(A^s)$ which are incident with a loop.
Then $|W| \geq s$ and each vertex can be reached by a directed walk in $D(A^s)$
of length $n-1$ starting from any vertex in W. Let the set of vertices of
$D(A)$ be $V = \{a_1, a_2, \ldots, a_n\}$. In $D(A)$ each vertex a_j can be reached by
a directed walk of length $s(n-1)$ starting from any vertex in W. For each
vertex a_i there is a directed walk of length l_i to some vertex in W where
$l_i \leq n - s$. It follows that

$$\exp(D(A) : i) \leq l_i + s(n-1) \leq n + s(n-2), \quad (i = 1, 2, \ldots, n).$$

We now apply Lemma 3.5.1 and obtain the conclusion of the theorem.   □

Shao[1985] has characterized the $(0,1)$-matrices A whose exponent $\exp(A)$
achieves the upper bound $n + s(n-2)$ in the theorem.

Theorem 3.5.4 can be used to determine the largest exponent possible
for a primitive matrix of order n. First we determine the Frobenius–Schur
index of two relatively prime integers.

Lemma 3.5.5.   *Let p and q be relatively prime positive integers. Then
$\phi(p,q) = (p-1)(q-1) = pq - p - q + 1$.*

Proof. We first show that $\phi(p,q) \geq pq - p - q + 1$. Suppose that there are
nonnegative integers a and b such that $pq - p - q = ap + bq$. The relative
primeness of p and q implies that p is a divisor of $b+1$ and q is a divisor
of $a+1$. Hence

$$pq - p - q = ap + bq \geq (q-1)p + (p-1)q = pq - p - q + pq,$$

a contradiction.

We next show that every integer $m > pq$ can be expressed as a *positive*
integral linear combination of p and q. There exists an integer a with $1 \leq
a \leq q$ such that $m \equiv ap \pmod{q}$. Let $b = (m - ap)/q$. Then b is a positive
integer and $m = ap + bq$. It now follows that every integer $m > pq - p - q$
can be expressed as a nonnegative linear combination of p and q.   □

The following inequality was stated by Wielandt[1950]; the first published proofs appeared in Rosenblatt[1957], Holladay and Varga[1958] and
Pták[1958].

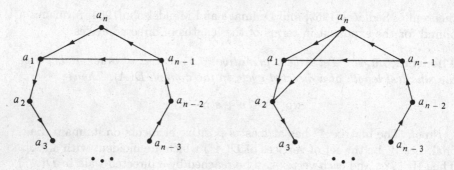

Figure 3.2

Theorem 3.5.6. *Let A be a primitive $(0, 1)$-matrix of order $n \geq 2$. Then*

$$\exp(A) \leq (n-1)^2 + 1. \tag{3.10}$$

Equality holds in (3.10) if and only if there exists a permutation matrix P of order n such that

$$PAP^T = \begin{bmatrix} 0 & 1 & 0 & \cdots & 0 \\ 0 & 0 & 1 & \cdots & 0 \\ \vdots & \vdots & \vdots & \ddots & \vdots \\ 1 & 0 & 0 & \cdots & 1 \\ 1 & 0 & 0 & \cdots & 0 \end{bmatrix}. \tag{3.11}$$

Proof. Let s denote the smallest length of a directed cycle in the digraph $D(A)$. Since A is primitive we have $s \leq n - 1$. Equation (3.10) now follows from Theorem 3.5.4. Assume that $\exp(A) = (n-1)^2 + 1$. Then $s = n - 1$ and the primitivity of A implies that $D(A)$ also has a directed cycle of length n. [If $n = 2$ we use the fact that $D(A)$ is strongly connected.] Since $D(A)$ does not have a directed cycle with length smaller than $n - 1$, it follows readily that apart from the labeling of the vertices $D(A)$ is one of the two digraphs D_1 and D_2 shown in Figure 3.2.

The digraph D_1 is the digraph of the matrix displayed in (3.11).

First assume that the digraph $D(A)$ equals D_1. Every closed directed walk from a_n to a_n has length $n + a(n-1) + bn$ for some nonnegative integers a and b. It follows from Lemma 3.5.5 with $p = n - 1$ and $q = n$ that the integer $(n-2)(n-1) - 1$ cannot be expressed as $a(n-1) + bn$ for any choice of nonnegative integers a and b. Hence there is no directed walk from a_n to a_n of length $(n-2)(n-1) + n - 1 = (n-1)^2$. Using Lemma 3.5.1 we now see that $\exp(A) \geq \exp(A : n) \geq (n-1)^2 + 1$. Thus $\exp(A) = (n-1)^2 + 1$.

Now assume that the digraph $D(A)$ equals D_2. Every directed walk from a_1 to a_n has length $(n-1) + a(n-1) + bn$ for some nonnegative integers a

and b. Lemma 3.5.5 implies that there is no directed walk from a_1 to a_n with length $(n-1)+(n-2)(n-1)-1=(n-1)^2-1$. Hence $\exp(A) \geq (n-1)^2$. We now show that $\exp(A) = (n-1)^2$. Since each vertex is on a directed cycle of length n and is also on a directed cycle of length $n-1$, we apply Lemma 3.5.5 again and conclude that each vertex belongs to a closed directed walk of length t for each integer $t \geq (n-2)(n-1)$. Let a_i and a_j be any two vertices. There is a directed walk from a_i to a_j with length $l_{ij} \leq n-1$ and hence a directed walk from a_i to a_j of length $l_{ij}+t$ for each integer $t \geq (n-2)(n-1)$. Thus

$$\exp(A:i,j) \leq l_{ij} + (n-2)(n-1) \leq (n-1) + (n-2)(n-1) = (n-1)^2.$$

We now apply Lemma 3.5.1 to obtain $\exp(A) \leq (n-1)^2$. Hence $\exp(A) = (n-1)^2$. Thus the primitive matrix of order n has exponent equal to $(n-1)^2+1$ if and only if there is a permutation matrix P of order n such that (3.11) holds.   □

In the proof of Theorem 3.5.6 we have also established the fact that a primitive (0,1)-matrix A of order n has exponent equal to $(n-1)^2$ if and only if the digraph $D(A)$ is, apart from the labeling of its vertices, the digraph D_2 of Figure 3.2. Thus A has exponent $(n-1)^2$ if and only if there is a permutation matrix P of order n such that

$$PAP^T = \begin{bmatrix} 0 & 1 & 0 & \cdots & 0 \\ 0 & 0 & 1 & \cdots & 0 \\ \vdots & \vdots & \vdots & \ddots & \vdots \\ 1 & 0 & 0 & \cdots & 1 \\ 1 & 1 & 0 & \cdots & 0 \end{bmatrix}.$$

Let n be a positive integer and let E_n denote the set of integers t for which there is a primitive matrix of order n with exponent equal to t. By Theorem 3.5.6, $E_n \subseteq \{1, 2, \ldots, w_n\}$ where $w_n = (n-1)^2 + 1$. The exponent of a positive matrix is 1, and thus $1 \in E_n$. By Theorem 3.5.6 and the discussion immediately following its proof, w_n and $w_n - 1 = (n-1)^2$ are in E_n. The following theorem of Shao[1985] shows that the exponent sets E_n form a nondecreasing chain of sets.

Theorem 3.5.7.   *For all $n \geq 1$,*

$$E_n \subseteq E_{n+1}.$$

Proof. Let A be a primitive (0,1)-matrix of order n with exponent equal to t. Let B be the matrix of order $n+1$ whose leading principal submatrix of order n equals A and whose last two rows and last two columns, respectively, are equal. The digraph $D(B)$ is obtained from the digraph $D(A)$ by adjoining a new vertex a_{n+1} to the vertex set $V = \{a_1, a_2, \ldots, a_n\}$ of

$D(A)$ and then adding arcs (a_i, a_{n+1}) and (a_{n+1}, a_j) whenever (a_i, a_n) and (a_n, a_j) are, respectively, arcs of $D(A)$. In addition, (a_{n+1}, a_{n+1}) is an arc of $D(B)$ if and only if (a_n, a_n) is an arc of $D(A)$.

An elementary argument based on these digraphs reveals that B is a primitive matrix and that B has exponent equal to t.   □

Dulmage and Mendelsohn[1964] showed that if $n \geq 4$, E_n is a proper subset of $\{1, 2, \ldots, w_n\}$. Specifically they showed that there is no primitive matrix A of odd order n such that

$$n^2 - 3n + 5 \leq \exp(A) \leq (n-1)^2 - 1$$

or

$$n^2 - 4n + 7 \leq \exp(A) \leq n^2 - 3n + 1.$$

If n is even, there is no primitive matrix A with

$$n^2 - 4n + 7 \leq \exp(A) \leq (n-1)^2 - 1.$$

Intervals in the set $\{1, 2, \ldots, w_n\}$ containing no integer which is the exponent of a primitive matrix of order n have been called *gaps* in E_n. Lewin and Vitek[1981] obtained a family of gaps in E_n and in doing so obtained a test for deciding whether or not an integer m satisfying

$$\lfloor \frac{w_n}{2} \rfloor + 2 \leq m \leq w_n$$

belongs to E_n. The following theorem plays an important role in this test.

Theorem 3.5.8.   *Let the exponent of a primitive $(0,1)$-matrix A of order n satisfy*

$$\exp(A) \geq \lfloor \frac{w_n}{2} \rfloor + 2.$$

Then the digraph $D(A)$ has directed cycles of exactly two different lengths.

Theorem 3.5.8 puts severe restrictions on a primitive matrix of order n whose exponent is at least $\lfloor w_n/2 \rfloor + 2$. Such a matrix must contain a large number of zeros. Lewin and Vitek[1981] conjectured that every positive integer m with $m \leq \lfloor w_n/2 \rfloor + 1$ is the exponent of at least one primitive matrix of order n. Shao[1985] disproved this conjecture by showing that the integer 48 which satisfies $48 < \lfloor w_{11}/2 \rfloor + 1 = 51$ is not the exponent of a primitive matrix of order 11. In addition he proved that

$$\{1, 2, \ldots, \lfloor \frac{w_n}{4} \rfloor + 1\} \subseteq E_n$$

for all n and that

$$\{1, 2, \ldots, \lfloor \frac{w_n}{2} \rfloor + 1\} \subseteq E_n$$

for all sufficiently large n. In doing so he showed that the conjecture of Lewin and Vitek would be true if one could establish the validity of a certain number theoretical question. Zhang[1987] proved the validity of the number theoretical question thereby settling the question which arose from the conjecture of Lewin and Vitek.

Theorem 3.5.9.   *Let n be an integer with $n \geq 2$. Then for each positive integer m with $m \leq \lfloor w_n/2 \rfloor + 1$ there is a primitive matrix of order n with exponent equal to m with the exception of the integer $m = 48$ when $n = 11$.*

Let E_n^s denote the set of integers t for which there exists a symmetric primitive $(0,1)$-matrix of order n with exponent equal to t. By Theorem 3.5.3, $E_n^s \subseteq \{1, 2, \ldots, 2n - 2\}$. Shao[1987] proved that $E_n^s = \{1, 2, \ldots, 2n - 2\} - S$ where S is the set of odd integers m with $n \leq m \leq 2n - 2$. Liu, McKay, Wormald and Zhang[1990] proved that the set of exponents of primitive $(0,1)$-matrices with zero trace is $\{2, 3, \ldots, 2n - 4\} - S'$ where S' is the set of odd integers m with $n - 2 \leq m \leq 2n - 5$. Liu[1990] proved that for $n \geq 4$ every integer $m > 1$ which is the exponent of a primitive $(0,1)$-matrix of order n is the exponent of a primitive $(0,1)$-matrix of order n with zero trace.

Let A be a primitive matrix of order n. A matrix obtained from A by simultaneous line permutations is also primitive. But a matrix obtained from A by arbitrary line permutations need not be primitive even if it is irreducible. For example, the matrix

$$\begin{bmatrix} 0 & 0 & 1 & 1 & 0 \\ 1 & 0 & 0 & 0 & 0 \\ 0 & 1 & 0 & 0 & 0 \\ 0 & 0 & 0 & 0 & 1 \\ 0 & 1 & 0 & 0 & 0 \end{bmatrix}$$

is a primitive matrix. Suppose we move column 1 so that it is between columns 3 and 4. We obtain the matrix

$$\begin{bmatrix} 0 & 1 & 0 & 1 & 0 \\ 0 & 0 & 1 & 0 & 0 \\ 1 & 0 & 0 & 0 & 0 \\ 0 & 0 & 0 & 0 & 1 \\ 1 & 0 & 0 & 0 & 0 \end{bmatrix}$$

which is irreducible but not primitive.

The following theorem of Shao[1985] characterizes those matrices which can be obtained from primitive matrices by arbitrary line permutations. As in the case of irreducible matrices in Theorem 3.2.5, it suffices to consider only column permutations.

Theorem 3.5.10.   *Let A be a $(0,1)$-matrix of order n. There exists a permutation matrix Q of order n such that AQ is a primitive matrix if and only if the following three conditions hold:*

(i) A has at least one 1 in each row and column;
(ii) A is not a permutation matrix;
(iii) A is not the matrix

$$\begin{bmatrix} 0 & 1 & \cdots & 1 \\ 1 & & & \\ \vdots & & O & \\ 1 & & & \end{bmatrix}$$

or any matrix obtained from it by line permutations.

We now discuss a theorem of Moon and Moser[1966]. Let $\mathcal{A}_n$ denote the set of all $(0,1)$-matrices of order n. Let $\mathcal{P}_n$ denote the subset of $\mathcal{A}_n$ consisting of the primitive matrices. The proportion of primitive matrices among all the $(0,1)$-matrices of order n is

$$\frac{|\mathcal{P}_n|}{|\mathcal{A}_n|} = \frac{|\mathcal{P}_n|}{2^{n^2}}.$$

Theorem 3.5.11.   *Almost all $(0,1)$-matrices of order n are primitive, that is*

$$\lim_{n \to \infty} \frac{|\mathcal{P}_n|}{2^{n^2}} = 1.$$

Indeed almost all $(0,1)$-matrices of order n are primitive and have exponent equal to 2.

Since a primitive matrix is irreducible, almost all $(0,1)$-matrices of order n are irreducible.

The study of the exponent of primitive, nearly reducible matrices was initiated by Brualdi and Ross[1980] and further investigated by Ross[1982], Yang and Barker[1988] and Li[1990]. Other bounds for the exponent are in Heap and Lynn [1964], Lewin[1971] and Lewin[1974]. Some generalizations of the exponent are considered in Brualdi and Li[1990], Chao[1977], Chao and Zhang[1983] and Schwarz[1973].

Exercises

1. Let A be primitive symmetric $(0,1)$-matrix of order $n \geq 2$ having $p \geq 1$ 1's on its main diagonal. Prove that

$$\exp(A) \leq \max\{n - 1, 2(n - p)\}$$

(Lewin[1971]).

2. Let A be an irreducible $(0,1)$-matrix of order $n > 2$. Consider the digraph $D(A)$ and assume that there are vertices a_i and a_j (possibly the same vertex) such that there are directed walks from a_i to a_j of each of the lengths $1, 2, \ldots, n-1$. Prove that A is primitive and that the exponent of A does not exceed $2d + 1$ where d is the diameter of $D(A)$ (Lewin[1971]).

3. Prove that there does not exist a primitive matrix of order $n \geq 5$ whose exponent equals $n^2 - 2n$ (Dulmage and Mendelsohn[1964]).

4. Let A be a $(0,1)$-matrix of order n. Prove that the conditions (i), (ii) and (iii) in Theorem 3.5.10 must be satisfied if there is to be a permutation matrix Q such that AQ is a primitive matrix.

5. Let A be the matrix (3.11) displayed in Theorem 3.5.6. Determine the numbers $\exp(A; i), (i = 1, 2, \ldots, n)$ (Brualdi and Li[1990]).

6. Let n be a positive integer and let E_n^s denote the set of integers t for which there exists a primitive symmetric matrix of order n and exponent t. Prove that $E_n^s \subseteq E_{n+1}^s, (n \geq 1)$. Use this fact and the displayed matrix in Theorem 3.5.3 to show that $k \in E_n^s$ for each even positive integer $k \leq 2n - 2$ (Shao[1987]).

7. Prove that the exponent of a primitive, nearly reducible matrix of order n is at least 4. [In fact, it is at least 6, and for each $n \geq 4$ there exists a primitive, nearly reducible matrix of order n whose exponent equals 6 (Brualdi and Ross[1980]).]

8. Let A be a tournament matrix of order n. If $n \geq 4$, prove that A is primitive if and only if A is irreducible. If $n \geq 5$ and A is primitive, prove that the exponent of A is at most equal to $n + 2$ (Moon and Pullman[1967]).

References

R.A. Brualdi and J.A. Ross[1980], On the exponent of a primitive, nearly reducible matrix, *Math. Oper. Res.*, 5, pp. 229–241.

R.A. Brualdi and B. Liu[1991], Fully indecomposable exponents of primitive matrices, *Proc. Amer. Math. Soc.*, to be published.

[1991], Hall exponents of Boolean matrices, *Czech. Math. J.*, to be published.

[1990], Generalized exponents of primitive directed graphs, *J. Graph Theory*, 14, pp. 483–499.

C.Y. Chao[1977], On a conjecture of the semigroup of fully indecomposable relations, *Czech. Math. J.*, 27, pp. 591–597.

C.Y. Chao and M.C. Zhang[1983], On the semigroup of fully indecomposable relations, *Czech. Math. J.*, 33, pp. 314–319.

A.L. Dulmage and N.S. Mendelsohn[1962], The exponent of a primitive matrix, *Canad. Math. Bull.*, 5, pp. 642–656.

[1964], Gaps in the exponent set of primitive matrices, *Illinois J. Math.*, 8, pp. 642–656.

[1964], The exponents of incidence matrices, *Duke Math. J.*, 31, pp. 575–584.

B.R. Heap and M.S. Lynn[1964], The index of primitivity of a non-negative matrix, *Numer. Math.*, 6, pp. 120–141.

J.C. Holladay and R.S. Varga[1958], On powers of non-negative matrices, *Proc. Amer. Math. Soc.*, 9, pp. 631–634.

M. Lewin[1971], On exponents of primitive matrices, *Numer. Math.*, 18, pp. 154–161.

[1974], Bounds for the exponents of doubly stochastic matrices, *Math. Zeit.*, 137, pp. 21–30.

M. Lewin and Y. Vitek[1981], A system of gaps in the exponent set of primitive matrices, *Illinois J. Math.*, 25, pp. 87–98.

B. Liu[1991], New results on the exponent set of primitive, nearly reducible matrices, *Linear Alg. Applics*, to be published.

[1990], A note on the exponents of primitive (0,1)-matrices, *Linear Alg. Applics.*, 140, pp. 45–51.

B. Liu, B. McKay, N. Wormwald and K.M. Zhang[1990], The exponent set of symmetric primitive (0,1)-matrices with zero trace, *Linear Alg. Applics*, 133, pp. 121–131.

J.W. Moon and L. Moser[1966], Almost all (0,1)-matrices are primitive, *Studia Scient. Math. Hung.*, 1, pp. 153–156.

J.W. Moon and N.J. Pullman[1967], On the powers of tournament matrices, *J. Combin. Theory*, 3, pp. 1–9.

D. Rosenblatt[1957], On the graphs and asymptotic forms of finite Boolean relation matrices and stochastic matrices, *Naval Res. Logist. Quart.*, 4, pp. 151–167.

V. Pták[1958], On a combinatorial theorem and its application to nonnegative matrices, *Czech. Math. J.*, 8, pp. 487–495.

J.A. Ross[1982], On the exponent of a primitive, nearly reducible matrix II, *SIAM J. Alg. Disc. Meth.*, 3, pp. 385–410.

S. Schwarz[1973], The semigroup of fully indecomposable relations and Hall relations, *Czech. Math. J.*, 23, pp. 151–163.

J. Sedláček[1959], O incidenčnich matiach orientirovaných grafò, *Časop. Pěst. Mat.*, 84, pp. 303–316.

J.Y. Shao[1985], On a conjecture about the exponent set of primitive matrices, *Linear Alg. Applics.*, 65, pp. 91–123.

[1985], On the exponent of a primitive digraph, *Linear Alg. Applics.*, 64, pp. 21–31.

[1985], Matrices permutation equivalent to primitive matrices, *Linear Alg. Applics.*, 65, pp. 225–247.

[1987], The exponent set of symmetric primitive matrices, *Scientia Sinica Ser. A*, vol. XXX, pp. 348–358.

H. Wielandt[1950], Unzerlegbare, nicht negative Matrizen, *Math. Zeit.*, 52, pp. 642–645.

S. Yang and G.P. Barker[1988], On the exponent of a primitive, minimally strong digraph, *Linear Alg. Applics.*, 99, pp. 177–198.

K.M. Zhang[1987], On Lewin and Vitek's conjecture about the exponent set of primitive matrices, *Linear Alg. Applics.*, 96, pp. 101–108.

3.6 Eigenvalues of Digraphs

Let D be a general digraph of order n and let its vertex set be the n-set $V = \{a_1, a_2, \ldots, a_n\}$. Let A be the adjacency matrix of D. The characteristic polynomial of A is called the *characteristic polynomial* of D and the collection of the n eigenvalues of A is called the *spectrum* of D. If D is a symmetric digraph, then the spectrum of D consists of n real numbers. In general, the spectrum of D consists of n complex numbers. Information about the eigenvalues in the spectrum of the digraph D is not in general easy to obtain. In addition, the combinatorial significance of the eigenvalues is less evident than it is for the eigenvalues of a graph. The directed cycles of D can be used to obtain regions in the complex plane which contain the spectrum of D. Such *eigenvalue inclusion regions* can be obtained more generally for complex matrices and their associated digraphs.

Let $A = [a_{ij}]$, $(i, j = 1, 2, \ldots, n)$ be a complex matrix of order n. Inclusion regions for the eigenvalues of A are closely connected with conditions which guarantee that A is nonsingular. The most classical results of this type are the Gerŝgorin theorem and the Lévy–Desplanques theorem.

Let

$$R_i = |a_{i1}| + \cdots + |a_{i,i-1}| + |a_{i,i+1}| + \cdots + |a_{in}|, \quad (i = 1, 2, \ldots, n) \quad (3.12)$$

denote the sum of the moduli of the off-diagonal elements in row i of A. The matrix A is called *diagonally dominant* provided

$$|a_{ii}| > R_i, \quad (i = 1, 2, \ldots, n).$$

Notice that a diagonally dominant matrix can have no zeros on its main diagonal. The theorem of Lévy[1881] and Desplanques[1887] (see Marcus and Minc[1964]) gives a sufficient condition for A to be nonsingular.

Theorem 3.6.1. *If the matrix A is diagonally dominant, then $\det(A) \neq 0$.*

The theorem of Gerŝgorin[1931] (see also Taussky[1949]) determines an inclusion region for the eigenvalues of A.

Theorem 3.6.2. *The eigenvalues of the matrix A of order n lie in the region of the complex plane determined by the union of the n closed discs*

$$Z_i = \{z : |z - a_{ii}| \leq R_i\}, \quad (i = 1, 2, \ldots, n).$$

It is straightforward to derive either of these theorems from the other. Let λ be an eigenvalue of A. Then $\det(\lambda I - A) = 0$ and hence by Theorem 3.6.1, $\lambda I - A$ is not diagonally dominant. Thus for at least one integer i with $1 \leq i \leq n$, $|\lambda - a_{ii}| \leq R_i$. Thus Theorem 3.6.2 follows from Theorem 3.6.1. If A is diagonally dominant, then none of the discs Z_i contains the number 0. Therefore Theorem 3.6.2 implies that a diagonally dominant matrix A has no eigenvalue equal to 0 and so $\det(A) \neq 0$.

In order to obtain generalizations of Theorems 3.6.1 and 3.6.2 which utilize the digraph of a matrix, we need one elementary result about digraphs. Let D be a digraph of order n with vertices $\{a_1, a_2, \ldots, a_n\}$. Corresponding to each vertex a_i let there be given a real number w_i, $(i = 1, 2, \ldots, n)$. Under these circumstances we call D a *vertex-weighted digraph*. An arc (a_i, a_j) issuing from vertex a_i is a *dominant arc from vertex a_i* provided $w_j \geq w_k$ for all integers k for which (a_i, a_k) is an arc of D. There may be more than one dominant arc from a given vertex. A directed cycle

$$a_{i_1} \to a_{i_2} \to \cdots \to a_{i_p} \to a_{i_{p+1}} = a_{i_1}$$

in which $(a_{i_j}, a_{i_{j+1}})$ is a dominant arc from vertex a_{i_j} for each $j = 1, 2,$ $\ldots, p$ is called a *dominant directed cycle* (with respect to the given vertex-weighting).

Lemma 3.6.3.   *Assume that each vertex of the digraph D has an arc issuing from it. Then D has a dominant directed cycle.*

Proof. Let a_{k_1} be any vertex of D. Choose a dominant arc (a_{k_1}, a_{k_2}) issuing from a_{k_1}, then a dominant arc (a_{k_2}, a_{k_3}) issuing from a_{k_2} and so on. In this way we obtain a directed walk

$$a_{k_1} \to a_{k_2} \to a_{k_3} \to \cdots.$$

Let s be the smallest integer for which there is an integer r with $1 \leq r < s$ such that $k_r = k_s$. Then

$$a_{k_r} \to a_{k_{r+1}} \to \cdots \to a_{k_s}$$

is a directed cycle each of whose arcs is dominant.   □

We return to the complex matrix $A = [a_{ij}]$ of order n and denote by $D_0(A)$ the digraph obtained from $D(A)$ by removing all loops. If a digraph is vertex-weighted with weights $w_1, w_2, \ldots, w_n$ and γ is a directed cycle, then $\prod_\gamma w_i$ denotes the product of the weights of the vertices that are on γ. We may regard the digraphs $D(A)$ and $D_0(A)$ as vertex-weighted by the numbers $|a_{11}|, |a_{22}|, \ldots, |a_{nn}|$ as well as by the numbers $R_1, R_2, \ldots, R_n$.

Theorem 3.6.4.   *Let all the elements on the main diagonal of the complex matrix $A = [a_{ij}]$ of order n be different from zero. If*

$$\prod_\gamma |a_{ii}| > \prod_\gamma R_i \qquad\qquad (3.13)$$

for all directed cycles of $D(A)$ with length at least 2, then

$$\det(A) \neq 0.$$

Proof. Since the determinant of A is the product of the determinants of its irreducible components and since no main diagonal element of A equals zero, we assume that A is an irreducible matrix of order at least two. Suppose that $\det(A) = 0$. There exists a nonzero vector $x = (x_1, x_2, \ldots, x_n)^T$ such that

$$Ax = 0. \qquad\qquad (3.14)$$

Let W consist of those vertices a_i for which $x_i \neq 0$. Let D_0' be the subdigraph of $D_0(A)$ obtained by deleting the vertices not in W and all arcs incident with at least one of the deleted vertices. The equation (3.14) implies that each vertex of D_0' has an arc issuing from it. We weight the vertices of W by assigning to the vertex a_i the weight $|x_i|$ and apply Lemma 3.6.2 to obtain a dominant directed cycle

$$\gamma' : a_{i_1} \to a_{i_2} \to \cdots \to a_{i_p} \to a_{i_{p+1}} = a_{i_1}$$

in D'_0 of length $p \geq 2$. Let j be any integer with $1 \leq j \leq p$. By (3.14) and the definition of W we obtain

$$a_{i_j i_j} x_{i_j} = -\sum_{k \neq i_j} a_{i_j k} x_k = -\sum_{\{k : a_k \in W - \{a_{i_j}\}\}} a_{i_j k} x_k.$$

Since γ' is a dominant directed cycle, we obtain

$$|a_{i_j i_j}||x_{i_j}| \leq \sum_{\{k : a_k \in W - \{a_{i_j}\}\}} |a_{i_j k}||x_k|$$

$$\leq \left(\sum_{\{k : a_k \in W - \{a_{i_j}\}\}} |a_{i_j k}| \right) |x_{i_{j+1}}|.$$

Hence

$$|a_{i_j i_j}||x_{i_j}| \leq \left(\sum_{k \neq i_j} |a_{i_j k}| \right) |x_{i_{j+1}}|, (1 \leq j \leq p). \tag{3.15}$$

By (3.12)

$$|a_{i_j i_j}||x_{i_j}| \leq R_{i_j} |x_{i_{j+1}}|, (1 \leq j \leq p). \tag{3.16}$$

We multiply the p inequalities in (3.16) and use the fact that $i_{p+1} = i_1$ and obtain

$$\prod_{\gamma'} |a_{ii}| \prod_{\gamma'} |x_i| \leq \prod_{\gamma'} R_i \prod_{\gamma'} |x_i|.$$

Since $x_i \neq 0$ if $a_i \in W$, we obtain

$$\prod_{\gamma'} |a_{ii}| \leq \prod_{\gamma'} R_i. \tag{3.17}$$

Inequality (3.17) is in contradiction to our assumption (3.13), and hence $\det(A) \neq 0$. $\qquad\square$

If the matrix $A = [a_{ij}]$ of order n does not have an irreducible component of order 1, then each vertex of $D(A)$ belongs to at least one directed cycle of length at least 2 and hence (3.13) implies that $a_{ii} \neq 0$ for $i = 1, 2, \ldots, n$. It follows that the assumption in Theorem 3.6.3 that the elements on the main diagonal of A are different from zero is not needed if A is an irreducible matrix of order at least 2. A zero matrix shows that the assumption cannot be removed in general.

We may apply Theorem 3.6.4 to obtain an eigenvalue inclusion region.

Theorem 3.6.5. *Let $A = [a_{ij}]$ be a complex matrix of order n. Then the eigenvalues of A lie in that part of the complex plane determined by the union of the regions*

$$Z_\gamma = \left\{ z : \prod_\gamma |z - a_{ii}| \leq \prod_\gamma R_i \right\}$$

over all directed cycles γ of $D(A)$ having length at least 2.

Proof. Since the vertices of a directed cycle all belong to the same strong component, it suffices to prove the theorem for an irreducible matrix A. We assume that A is irreducible matrix of order $n \geq 2$ and proceed as in the derivation of Theorem 3.6.2 from Theorem 3.6.1. Let λ be an eigenvalue of A. Then $\det(\lambda I - A) = 0$. The digraphs $D(A)$ and $D(\lambda I - A)$ have the same set of directed cycles of length at least 2. Moreover, R_i is also the sum of the moduli of the off-diagonal elements in row i of $\lambda I - A, (1 \leq i \leq n)$. We apply Theorem 3.6.4 to $\lambda I - A$, and noting the discussion following its proof, we conclude that there is a directed cycle γ of $D(A)$ with length at least 2 such that

$$\prod_\gamma |\lambda - a_{ii}| \leq \prod_\gamma R_i.$$

Hence the eigenvalue λ is in Z_γ.   □

Notice that Theorem 3.6.1 and hence Theorem 3.6.2 are direct consequences of Theorem 3.6.4. Another consequence of Theorem 3.6.4 is the following theorem of Ostrowski[1937] and Brauer[1947].

Theorem 3.6.6. *Let $A = [a_{ij}]$ be a complex matrix of order $n \geq 2$. If*

$$|a_{ii}||a_{jj}| > R_i R_j, \quad (i, j = 1, 2, \ldots, n; i \neq j), \qquad (3.18)$$

then $\det(A) \neq 0$. The eigenvalues of the matrix A lie in the region of the complex plane determined by the union of the ovals

$$Z_{ij} = \{z : |z - a_{ii}||z - a_{jj}| \leq R_i R_j\}, \quad (i, j = 1, 2, \ldots, n; i \neq j).$$

Proof. Assume that (3.18) holds. Then $a_{ii} \neq 0, (i = 1, 2, \ldots, n)$ and (3.13) holds for all directed cycles γ of $D(A)$ of length 2. We show that (3.13) holds also for all directed cycles of length at least 3. If

$$|a_{ii}| > R_i \qquad (3.19)$$

holds for all i, then (3.13) is satisfied. It follows from (3.18) that the only other possibility is for (3.19) to fail for exactly one integer i. Let γ be a directed cycle of length $p \geq 3$. We choose distinct integers i and j such that a_i and a_j are vertices of γ and $|a_{kk}| > R_k$ whenever a_k is a vertex of γ with $k \neq i, j$. It now follows from (3.18) that (3.13) holds for γ. We now apply Theorem 3.6.4 and obtain $\det(A) \neq 0$. The second conclusion of the theorem follows by applying the first conclusion to the matrix $\lambda I - A$ for each eigenvalue λ of A. $\square$

The *girth* of a digraph D is defined to be the smallest integer $k \geq 2$ such that D has a directed cycle with length k. Notice that we exclude loops in the calculation of the girth. If D has no directed cycle with length at least 2, the girth of D is undefined.

If the square matrix A has only zeros on its main diagonal, then Theorem 3.6.5 simplifies considerably. The numbers $R_1, R_2, \ldots, R_n$ defined in (3.12) are then the sum of the moduli of the elements in the rows of A.

Corollary 3.6.7.   *Let $A = [a_{ij}]$ be a complex matrix of order n with only 0's on its main diagonal. Suppose that the numbers $R_1, R_2, \ldots, R_n$ defined in (3.12) satisfy $R_1 \leq R_2 \leq \cdots \leq R_n$. Then each eigenvalue λ of A satisfies*

$$|\lambda| \leq \sqrt[g]{R_{n-g+1} \cdots R_n}$$

where g is the girth of $D(A)$.

Proof. Let λ be an eigenvalue of A. By Theorem 3.6.5 there exists a directed cycle γ of $D(A)$ of length $p \geq g$ such that

$$|\lambda|^p \leq \prod_{\gamma} R_i \leq R_{n-p+1} \cdots R_n.$$

Hence

$$|\lambda| \leq \sqrt[p]{R_{n-p+1} \cdots R_n} \leq \sqrt[g]{R_{n-g+1} \cdots R_n}.$$ $\square$

Applying Corollary 3.6.7 to the adjacency matrix of a digraph D of girth g with no loops, we conclude that the absolute value of each eigenvalue of D does not exceed the gth root of the product of the g largest outdegrees of its vertices.

If the matrix A is irreducible, the sufficient conditions obtained for the nonvanishing of the determinant and the corresponding eigenvalue inclusion regions can be improved. The improvement of Theorems 3.6.1 and 3.6.2 as given in the next theorem is due to Taussky[1949].

Theorem 3.6.8.   *Let $A = [a_{ij}]$ be an irreducible complex matrix of order n. If*

$$|a_{ii}| \geq R_i, \quad (i = 1, 2, \ldots, n)$$

with strict inequality for at least one i, then $\det(A) \neq 0$. A boundary point w of the union of the n closed discs

$$Z_i = \{z : |z - a_{ii}| \leq R_i\}, \quad (i = 1, 2, \ldots, n)$$

can be an eigenvalue of A only if w is a boundary point of each of the n discs.

The more general Theorems 3.6.4 and 3.6.5 admit a similar improvement.

Theorem 3.6.9.   *Let $A = [a_{ij}]$ be an irreducible complex matrix of order $n \geq 2$. If*

$$\prod_\gamma |a_{ii}| \geq \prod_\gamma R_i \qquad (3.20)$$

for all directed cycles γ of $D(A)$ with length at least 2, with strict inequality for at least one such directed cycle, then $\det(A) \neq 0$. A boundary point w of the union of the regions

$$Z_\gamma = \left\{ z : \prod_\gamma |z - a_{ii}| \leq \prod_\gamma R_i \right\}$$

can be an eigenvalue of A only if w is a boundary point of each Z_γ.

Proof. Assume that (3.20) holds for all γ. Since A is irreducible, $R_i \neq 0, (i = 1, 2, \ldots, n)$ and (3.20) implies that $a_{ii} \neq 0, (i = 1, 2, \ldots, n)$. We suppose that $\det(A) = 0$ and proceed as in the proof of Theorem 3.6.4. However, since we only assume the weaker inequality (3.20), when we reach (3.17) we can only conclude that equality holds, that is

$$\prod_{\gamma'} |a_{ii}| = \prod_{\gamma'} R_i. \qquad (3.21)$$

It follows that equality holds in (3.16). We conclude from the derivation of (3.16) that for each integer $j = 1, 2, \ldots, p$

$$a_{i_j k} \neq 0 \ \text{ implies } \ |x_k| = |x_{i_{j+1}}|, \quad (k = 1, 2, \ldots, n; k \neq i_j).$$

Thus for each vertex a_{i_j} of γ', the weights of the vertices to which there is an arc from a_{i_j} are constant. Suppose there is a vertex of $D(A)$ which is not a vertex of γ. Since A is irreducible, $D(A)$ is strongly connected and hence there is an arc (a_{i_j}, a_k) from some vertex a_{i_j} of γ to some vertex a_k not belonging to γ. We may argue as in Lemma 3.6.3 and obtain a dominant directed cycle γ'' which has at least one vertex different from a vertex of γ'. Replacing γ' with γ'' we conclude as above that the weights are constant over those vertices to which there is an arc from a specified vertex of γ''. Continuing like this, we conclude that each vertex a_i has the property that the weights of the vertices to which there is an arc from a_i are constant. This implies that every directed cycle of $D(A)$ of length at least 2 is a dominant directed cycle. Hence the proof of Theorem 3.6.4 applies to all directed cycles of length at least 2. This means that (3.21) holds for all directed cycles γ' of length at least 2, contradicting the assumption in the theorem.   □

Consider once again the complex matrix $A = [a_{ij}]$ of order n and let

$$S_i = |a_{1i}| + \cdots + |a_{i-1,i}| + |a_{i+1,i}| + \cdots + |a_{ni}|, \quad (i = 1, 2, \ldots n),$$

the sum of the moduli of the off-diagonal elements in column i of A. The theorems proved in this section remain true if the numbers $R_1, R_2, \ldots, R_n$ are replaced, respectively, by the numbers $S_1, S_2, \ldots, S_n$. This is because we may apply the theorems to the transposed matrix A^T. The eigenvalues of A^T-are identical to those of A. The digraph $D(A^T)$ is obtained from the digraph $D(A)$ by reversing the directions of all arcs. As a result the directed cycles of $D(A^T)$ are obtained from those of $D(A)$ by reversing the cyclical order of the vertices.

Ostrowski[1951] combined the sequences $R_1, R_2, \ldots, R_n$ and $S_1, S_2, \ldots, S_n$ to obtain extensions of Theorems 3.6.1 and 3.6.2.

Theorem 3.6.10. *Let $A = [a_{ij}]$ be a complex matrix of order n, and let p be a real number with $0 \le p \le 1$. If*

$$|a_{ii}| > R_i^p S_i^{1-p}, \quad (1 \le i \le n)$$

then $\det(A) \ne 0$. The eigenvalues of the matrix A lie in the region of the complex plane determined by the union of the n discs

$$\{z : |z - a_{ii}| \le R_i^p S_i^{1-p}\}, \quad (1 \le i \le n).$$

Theorems 3.6.4 and 3.6.5 can be similarly extended (Brualdi[1982]).

Exercises

1. Let $A = [a_{ij}]$ be a real matrix of order n such that $a_{ii} \ge 0, (i = 1, 2, \ldots, n)$ and $a_{ij} \le 0, (i, j = 1, 2, \ldots, n; i \ne j)$. Assume that each row sum of A is positive. Prove that the determinant of A is positive and that the real part of each eigenvalue of A is positive.
2. For each integer $n \ge 2$ determine a matrix A of order n whose eigenvalues are not contained in the union of the regions

$$\{z : |z - a_{ii}||z - a_{jj}||z - a_{kk}| \le R_i R_j R_k\}.$$

Here i, j and k are distinct integers between 1 and n and R_i denotes the sum of the moduli of the off-diagonal elements in row i of A.
3. Let D be a digraph of order n and let D' be the digraph obtained from D by reversing the direction of each arc. The digraphs D and D' have the same spectrum. Construct a digraph D such that D' is not isomorphic to D and thereby obtain a pair of cospectral digraphs.
4. Let A be a tournament matrix of order n, that is a $(0,1)$-matrix satisfying $A + A^T = J - I$. Prove that the real part of each eigenvalue of A lies between $-1/2$ and $(n-1)/2$ (Brauer and Gentry[1968] and de Oliveira[1974]).

References

A. Brauer[1947], Limits for the characteristic roots of a matrix II, *Duke Math. J.*, 14, pp. 21–26.

A. Brauer and I.C. Gentry[1968], On the characteristic roots of tournament matrices, *Bull. Amer. Math. Soc.*, 74, pp. 1133–1135.

R.A. Brualdi[1982], Matrices, eigenvalues, and directed graphs, *Linear and Multilin. Alg.*, 11, pp. 143–165.

M. Marcus and H. Minc[1964], *A Survey of Matrix Theory and Matrix Inequalities*, Allyn and Bacon, Boston.

G.N. de Oliveira[1974], Note on the characteristic roots of tournament matrices, *Linear Alg. Applics.*, 8, pp. 271–272.

A. Ostrowski[1937], Über die Determinanten mit überwiegender Hauptdiagonale, *Comm. Math. Helv.*, 10, pp. 69–96.

[1951], Über das Nichtverschwinden einer Klasse von Determinanten und die Lokalisierung der charakteristischen Wurzeln von Matrizen, *Compositio Math.*, 9, pp. 209–226.

O. Taussky[1949], A recurring theorem on determinants, *Amer. Math. Monthly*, 10, pp. 672–676.

3.7   Computational Considerations

Let $A = [a_{ij}], (i, j = 1, 2, \ldots, n)$ be a matrix of order n. In Theorem 3.2.4 we have established the existence of a permutation matrix P of order n such that PAP^T is in the Frobenius normal form given in (3.1). The diagonal blocks $A_1, A_2, \ldots, A_t$ in (3.1) are the irreducible components of A. By Theorem 3.2.4 they are uniquely determined apart from simultaneous permutations of their lines. In this section we discuss two algorithms. The first algorithm, due to Tarjan[1972] (see also Aho, Hopcroft and Ullman[1975]), obtains the irreducible components $A_1, A_2, \ldots, A_t$ of A including their ordering in (3.1). The second algorithm, due to Denardo[1977] and Atallah[1982], determines the index of imprimitivity k of an irreducible matrix A and the k-cyclic components $A_{12}, A_{23}, \ldots, A_{k-1,k}, A_{k1}$ as given in (3.4) with $r = k$. These algorithms are best discussed in the language of digraphs. The equivalent formulation of these considerations in terms of the digraph $D(A)$ has been discussed in sections 3.2 and 3.4.

We begin by recalling some definitions from the theory of digraphs. A *directed tree with root* r is a digraph with a distinguished vertex r having the property that for each vertex a different from r there is a unique directed chain from r to a. It follows that a directed tree with root r can be obtained from a tree T by labeling one vertex of T with r, thereby obtaining a tree rooted at r, and directing all edges of T away from r. In particular, a directed tree of order n has exactly $n-1$ arcs. A *directed forest* is a digraph consisting of one or more directed trees no two of which have a vertex in common.

Throughout we let D be a digraph of order n with vertex set V. Let F be a subset of the arcs of D. The subdigraph of D consisting of the arcs in F and all vertices of D which are incident with at least one arc in F is denoted by $\langle F \rangle$. If the vertex set of $\langle F \rangle$ is V, then $\langle F \rangle$ is a *spanning subdigraph* of D. The spanning subdigraph of D whose set of vertices is V and whose set of arcs is F is denoted by $\langle F \rangle^*$. A spanning subdigraph of D which is a directed tree or a directed forest is called, respectively, a *spanning directed tree* or a *spanning directed forest* of D. For each vertex $a \in V$ we define the *out-list* of a to be the set $L(a)$ of vertices b for which there is an arc (a, b) from a to b.

We first discuss an algorithm for obtaining a spanning directed forest of D. The algorithm is based on a technique called *depth-first search* for visiting all the vertices of D, and as a result the spanning directed forest that it determines is called a *depth-first spanning directed forest*. As the name suggests, in the search for new vertices preference is given to the *deepest* (or *forward*) direction. In the algorithm each vertex $a \in V$ is assigned a positive integer between 1 and n which is called its *depth-first number* and is denoted by $df(a)$. The depth-first numbers give the order in which the vertices of D are visited in the search.

Initially, all vertices of D are labeled *new*, F is empty and a function *COUNT* has the value 0. We choose a vertex a and apply the following procedure *Search(a)* described below.

Search(a)

1. $COUNT \leftarrow COUNT + 1$.
2. $df(a) \leftarrow Count$.
3. Change the label of a to *old*.
4. For each vertex b in $L(a)$, do
       (i) If b is *new*, then
           (a) Put the arc (a, b) in F.
           (b) Do *Search(b)*.

If upon the completion of *Search(a)*, all vertices of D are now labeled *old*, then $\langle F \rangle$ is a spanning directed tree of D rooted at a. If vertices labeled *new* remain, we choose any *new* vertex c and proceed with *Search(c)*. We continue in this way until every vertex has the label *old*. Then $\langle F \rangle^*$ is a spanning directed forest of D consisting of l directed trees $T_1, T_2, \ldots, T_l$ for some integer $l \geq 1$. We assume that these directed trees have been obtained in the order specified by their subscripts, and we speak of T_i as being *to the left* of T_j if $i < j$.

We illustrate the depth-first procedure with the digraph D in Figure 3.3.

A depth-first spanning directed forest $\langle F \rangle^*$ produced by the algorithm is illustrated in Figure 3.4.

The unbroken lines designate arcs of F; the broken lines designate the remaining arcs of D. The numbers in the parentheses designate the depth-first numbers of the vertices.

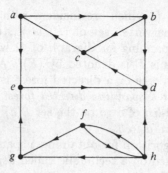

Figure 3.3

Application of the depth-first algorithm to a digraph D determines a partition of the arcs of D into four (possibly empty) sets specified below:

forest arcs: these are the arcs in F produced by the algorithm;

forward arcs: these are the arcs which go from a vertex to a proper descendant of that vertex in one of the directed trees of $\langle F \rangle^*$, but which are not forest arcs. (For the determination of the strong components of D, these arcs are of no importance and can be ignored.)

back arcs: these are the arcs which go from a vertex to an ancestor of that vertex in one of the directed trees of $\langle F \rangle^*$. (Here we can include the loops of D.)

cross arcs: these are the arcs which join two vertices neither of which is an ancestor of the other. The vertices may belong to the same directed tree or different directed trees of $\langle F \rangle^*$.

Suppose that (c, d) is an arc of D. If (c, d) is either a forest arc or a forward arc then $df(c) < df(d)$. If (c, d) is a back arc then $df(c) \geq df(d)$ (equality can hold only if $c = d$). For cross arcs we have the following.

Lemma 3.7.1.   *If (c, d) is a cross arc of D then $df(c) > df(d)$.*

Proof. Let (c, d) be an arc of D satisfying $df(c) < df(d)$. When c is changed from a *new* vertex to an *old* vertex, d is still *new*. Since d is in $L(c)$, $Search(c)$ cannot end until d is reached. It follows that (c, d) is either a forest arc or a forward arc.   □

Let the strong components of D be $D(V_1), D(V_2), \ldots, D(V_k)$. The next lemma is the first step in the identification of $V_1, V_2, \ldots, V_k$.

Lemma 3.7.2.   *The vertex set of a strong component of D is the set of vertices of some directed subtree of one of the directed trees $T_1, T_2, \ldots, T_l$ of the depth-first spanning directed forest $\langle F \rangle^*$ of D.*

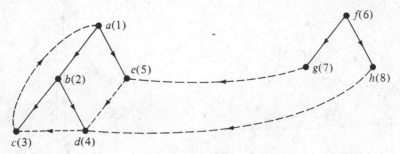

Figure 3.4

Proof. Let c and d be two vertices of D which belong to the same strong component $D(V_i)$. We first show that there is a vertex in V_i which is a common ancestor of c and d. Assume that $df(c) < df(d)$. Since c and d are in the same strong component of D, there is a directed chain γ from c to d all of whose vertices belong to V_i. Let x be the vertex of γ with the smallest depth-first number. The vertices which come after x in γ are each descendants of x in one of the directed trees $T_1, T_2, \ldots, T_l$. This is true because by Lemma 3.7.1 each of the arcs of γ beginning with the one leaving x is either a forest arc or a forward arc. In particular, d is a descendant of x. Since $df(x) \leq df(c) < df(d)$, it follows from the way that depth-first search is carried out, that c is also a descendant of x. Therefore each pair of vertices in V_i have a common ancestor which also belongs to V_i. We conclude that there is a vertex s_i in V_i such that s_i is a common ancestor of all vertices in V_i. In particular, V_i is a subset of the vertex set of one of the directed trees of $\langle F \rangle^*$.

Now let c be any vertex in V_i and let d be a vertex on the directed chain in $\langle F \rangle^*$ from s_i to c. Since c and s_i belong to the strong component $D(V_i)$ of D, there is a directed chain in D from c to s_i. Hence there are directed chains from s_i to d and from d to s_i, and we conclude that d is also in V_i. It follows that V_i is the set of vertices of a directed subtree of one of the directed trees of $\langle F \rangle^*$.   □

By Lemma 3.7.2 the vertex sets $V_1, V_2, \ldots, V_k$ of the strong components of D are the vertex sets of directed subtrees of the depth-first spanning forest $\langle F \rangle^*$. Let the roots of these directed subtrees be $s_1, s_2, \ldots, s_k$, respectively, where we have chosen the ordering of $V_1, V_2, \ldots, V_k$ in which depth-first search of their roots has terminated. Thus $Search(s_i)$ terminates before $Search(s_{i+1})$ for $i = 1, 2, \ldots, k-1$.

The strong components of D can be determined from the roots as follows.

Lemma 3.7.3.   *For each integer i with $1 \leq i \leq k$, the vertex set V_i of the strong component $D(V_i)$ of D consists of those vertices which are descendants of s_i but which are not descendants of any of $s_1, s_2, \ldots, s_{i-1}$.*

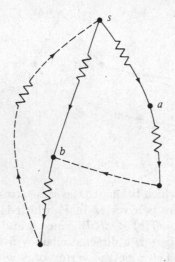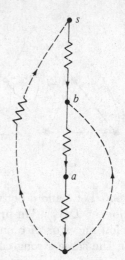

Figure 3.5

Proof. We first observe that the vertex set V_1 consists of all descendants of s_1. This is because $Search(s_1)$ terminates before $Search(s_i)$ for $i = 2, 3, \ldots, k$ and hence by definition of depth-first search, no s_i with $i > 1$ can be a descendant of s_1. Similarly, for $j > 1$, s_j cannot be a descendant of $s_1, \ldots, s_{j-1}$ and the lemma follows.   □

By Lemma 3.7.3 the vertex sets of the strong components of D can be determined once the roots of the strong components are known. To find the roots, a new function g, called $LOWLINK$, defined on the vertex set V of D is introduced. If a is a vertex in V, $g(a)$ is the smallest depth-first number of the vertices in the set consisting of a and those vertices b satisfying the property: there is a cross arc or back arc from a descendant of a (possibly a itself) to b where the root s of the strong component containing b is an ancestor of a. The two possibilities, namely those of a cross arc and a back arc, are illustrated in Figure 3.5. In that figure s and b are in the same strong component and hence there must be a directed chain from a descendant of b to s. Notice that for all vertices a we have $g(a) \leq df(a)$.

The function $LOWLINK$ provides a characterization of the strong components.

Lemma 3.7.4.   *A vertex a of the digraph D is the root of one of its strong components if and only if $g(a) = df(a)$.*

Proof. We first assume that $g(a) \neq df(a)$. Then there is a vertex b satisfying

(i) there is a back arc or cross arc from a descendant of a to b;

(ii) the root s of the strong component containing b is an ancestor of a;

(iii) $df(b) < df(a)$.

We have $df(s) \le df(b) < df(a)$ and hence $s \ne a$. Since there is a directed chain from s to a and a directed chain from a to s (through b), a, b and s are in the same strong component of D. Thus a is not the root of a strong component of D.

We now assume that $g(a) = df(a)$. Let r be the root of the strong component containing a and suppose that $r \ne a$. There is a directed chain γ from a to r. Since r is an ancestor of a, there is a first arc α of γ which goes from a descendant of a to a vertex b which is not a descendant of a. The arc α is either a cross arc or a back arc. In either case $df(b) < df(a)$. The directed chain γ implies the existence of a directed chain from b to r. Since there is a directed chain from r to a, there is also a directed chain from r to b. Hence r and b are in the same strong component. By definition of $LOWLINK$ we have $g(a) \le df(b)$, contradicting $g(a) = df(a) > df(b)$.   □

The computation of $LOWLINK$ can be incorporated into the depth-first search algorithm by replacing *Search* with *SearchComp*. This enhancement allows us to obtain the vertex sets of the strong components.

SearchComp(a)

1. *Count* ← *Count* +1.
2. $df(a)$ ← *Count*.
3. Change the label of a to *old*.
4. $g(a)$ ← $df(a)$.
5. Push a on a *Stack*.
6. For each vertex b in $L(a)$, do
    (i) if b is *new*, then
        (a) Do *SearchComp(b)*.
        (b) $g(a)$ ← $\min\{g(a), g(b)\}$.
    (ii) if b is *old* (hence $df(b) < df(a)$), then
        (a) if b is on the *Stack*, $g(a)$ ← $\min\{g(a), df(b)\}$.
7. If $g(a) = df(a)$, then *pop* x from the *Stack* until $x = a$. The vertices popped are declared the set of vertices of a strong component of D and a is declared its root.

With this enhancement we obtain the following.

Theorem 3.7.5.   *The enhanced depth-first search algorithm correctly determines the vertex sets of the strong components of the digraph D.*

Proof. We first observe that a vertex a is declared a root of a strong component if and only if $g(a) = df(a)$. Hence by Lemma 3.7.4 the roots of the strong components are computed correctly provided the function *LOWLINK* is. Moreover, when a is declared a root, the vertices put into the strong component with a are precisely the vertices above a on the *Stack*, that is the descendants of a which have not yet been put into a strong component. This is in agreement with Lemma 3.7.3. It thus remains to prove that *LOWLINK* is correctly computed. We accomplish this by using induction on the number of calls to *SearchComp* that have terminated.

We first show that the computed value for $g(a)$ is at least equal to the correct value. There are two places in *SearchComp* where the computed value of $g(a)$ could be less than $df(a)$. In $6(i)$ it can happen if b is a child of a and $g(b) < df(a)$. In this case, since $df(a) < df(b)$, there is a vertex x with $df(x) = g(b)$ such that x can be reached from a descendant y of b by either a cross arc or a back arc. The vertex x has the additional property that the root r of the strong component containing x is an ancestor of b and hence of a. Thus the correct value of $g(a)$ should be at least as low as $g(b) = df(x)$. In $6(ii)$ the computed value of $g(a)$ could be less than $df(a)$ if there is a cross arc or back arc from a to b and the strong component C containing b has not yet been found. In this case the call of *SearchComp* on the root r of C has not yet terminated, so that r is an ancestor of a. As in the previous case $g(a)$ should be at least as low as $df(b)$.

Now we show that the computed value for $g(a)$ is at most equal to the correct value. Suppose that x is a descendant of a for which there is a cross arc or a back arc from x to a vertex y where the root r of the strong component containing y is an ancestor of a. We need to show that the computed value of $g(a)$ is at least as small as $df(y)$. We distinguish two cases. In the first case $x = a$. By the inductive assumption, all strong components found thus far are correct. Since *SearchComp*(a) has not yet terminated, neither has *SearchComp*(r). Hence y is still on the *Stack*. Thus $6(ii)$ sets $g(a)$ to $df(y)$ or lower. In the second case, $x \neq a$. Then there exists a child z of a of which x is a descendant. By the inductive assumption, when *SearchComp*(z) terminates, $g(z)$ has been set to $df(y)$ or lower. In $6(ii)$, $g(a)$ is set this low unless it is already lower. Thus it follows by induction that the computed values of *LOWLINK* are correct.

□

The number of steps used in the preceding algorithm for determining the strong components of a digraph D of order n is bounded by $c \max\{n, e\}$ where c is a constant independent of the number n of vertices and e is the number of arcs of D.

We return now to a matrix A of order n. Let the vertex sets of the strong components of the digraph $D(A)$ be determined by the algorithm in the order $V_1, V_2, \ldots, V_k$. We simultaneously permute the lines of A so that the lines corresponding to the vertices in V_i come before the lines corresponding to the vertices in V_{i-1} $(2 \leq i \leq k)$. The Frobenius normal form is then

$$
\begin{bmatrix}
A_1 & A_{12} & \cdots & A_{1k} \\
O & A_2 & \cdots & A_{2k} \\
\vdots & \vdots & \ddots & \vdots \\
O & O & \cdots & A_k
\end{bmatrix}
$$

where A_i is the adjacency matrix of the strong component $D(V_{k+1-i})$, $(i = 1, 2, \ldots, k)$.

We now discuss an algorithm for determining the index of imprimitivity k of an *irreducible* matrix A and for determining the k-cyclic components $A_{12}, A_{23}, \ldots, A_{k-1,k}, A_{k1}$ of A. As in the previous algorithm we frame our discussion in the language of digraphs and determine the index of imprimitivity k and the sets of imprimitivity of a strongly connected digraph.

Let D be a strongly connected digraph of order n with vertex set V. We recall that for a vertex a of V, $L(a)$ denotes the set of vertices b for which (a, b) is an arc of D. Since D is strongly connected, a depth-first spanning directed forest of D is a directed tree. In the algorithm we assume that a spanning directed tree T with root r has been determined. We also assume that the length $d(a)$ of the unique directed chain in T from r to a has also been computed for each vertex a [we define $d(r) = 0$]. The algorithm *INDEX* computes the index of imprimitivity of D and its sets of imprimitivity.

INDEX

1. $\delta \leftarrow 0$.
2. For each vertex a in V, do
        (i) For each b in $L(a)$, do
                (a) $\delta \leftarrow \gcd\{\delta, d(a) - d(b) + 1\}$.
3. $W_1 \leftarrow \{a : d(a) \equiv 0 \pmod{\delta}\}$,
    $W_2 \leftarrow \{a : d(a) \equiv 1 \pmod{\delta}\}$,
    $\vdots$
    $W_\delta \leftarrow \{a : d(a) \equiv \delta - 1 \pmod{\delta}\}$.

The greatest common divisor gcd in the algorithm is always taken to be a nonnegative integer. We use the convention that $\gcd\{0, 0\} = 0$.

We show that upon termination of *INDEX*, the value of δ is the index of imprimitivity of D, and D is cyclic with respect to the ordered partition $W_1, W_2, \ldots, W_\delta$.

Lemma 3.7.6.   *The strong digraph D is cyclically r-partite if and only if for each arc (a, b) of D, r is a divisor of $d(a) - d(b) + 1$.*

Proof. First we assume that D is cyclically r-partite with ordered partition $U_1, U_2, \ldots, U_r$. Let (a, b) be an arc of D. In T there are directed chains α_a and α_b from r to a and b with lengths $d(a)$ and $d(b)$, respectively. Because D is strongly connected there is a directed chain β in D from b to r. Let the length of β be p. These directed chains along with the arc (a, b) determine closed directed walks of lengths $d(a) + 1 + p$ and $d(b) + p$, respectively. Since D is cyclically r-partite, we have

$$d(a) + 1 + p \equiv 0 \pmod{r},$$

and

$$d(b) + p \equiv 0 \pmod{r}.$$

Hence

$$d(a) - d(b) + 1 \equiv 0 \pmod{r}.$$

Conversely, suppose that r is a divisor of $d(a) - d(b) + 1$ for each arc (a, b) of D. Let $W_1, W_2, \ldots, W_r$ be defined as in Step 3 of *INDEX* with r replacing δ. Let (a, b) be any arc of D and suppose that a is in W_i and b is in W_j. We then have

$$d(a) - d(b) + 1 \equiv 0 \pmod{r},$$

$$d(a) \equiv i - 1 \pmod{r},$$

and

$$d(b) \equiv j - 1 \pmod{r}.$$

From these three relations it follows that $j \equiv i + 1 \pmod{r}$. Hence D is cyclically r-partite with respect to the ordered partition $W_1, W_2, \ldots, W_r$.
□

Theorem 3.7.7.   *Let D be a strongly connected digraph of order n. The number δ computed by the algorithm INDEX is the index of imprimitivity of D. Moreover, D is cyclically δ-partite with respect to the ordered partition $W_1, W_2, \ldots, W_\delta$.*

Proof. As shown in section 3.4, the index of imprimitivity of D equals the largest integer k such that D is cyclically k-partite. It thus follows from Lemma 3.7.6 that

$$k = \gcd\{d(a) - d(b) + 1 : (a, b) \text{ is an arc of } D\}.$$

Hence when algorithm *INDEX* terminates, δ has the value k. The proof that D is cyclically δ-partite with respect to the ordered partition $W_1, W_2, \ldots, W_\delta$ is the same as the one used in the proof of Lemma 3.7.6.   □

The algorithm *INDEX* can be implemented so that the number of steps taken is bounded by $c \max\{n, e\}$ where e is the number of arcs of the digraph D. For a strongly connected digraph, $e \geq n$ and hence this bound is ce.

Let A be an irreducible matrix of order n with index of imprimitivity equal to k. We apply the algorithm *INDEX* to the strong digraph $D(A)$. The computed value of δ is k. Let $W_1, W_2, \ldots, W_k$ be the partition of the vertex set of $D(A)$ produced by *INDEX*. If we simultaneously permute the lines of A so that the lines corresponding to the vertices in W_i come before those corresponding to $W_{i+1}, (i = 1, 2, \ldots, k-1)$, we obtain

$$
\begin{array}{c}
\\
W_1 \\
W_2 \\
\vdots \\
W_{k-1} \\
W_k
\end{array}
\begin{array}{c}
\begin{array}{ccccc}
W_1 & W_2 & W_3 & \cdots & W_k
\end{array} \\
\left[
\begin{array}{ccccc}
O & A_{12} & O & \cdots & O \\
O & O & A_{23} & \cdots & O \\
\vdots & \vdots & \vdots & \ddots & \vdots \\
O & O & O & \cdots & A_{k-1,k} \\
A_{k1} & O & O & \cdots & O
\end{array}
\right]
\end{array} .
$$

The matrices $A_{12}, A_{23}, \ldots, A_{k-1,k}, A_{k1}$ are the k-cyclic components of A.

Exercise

1. Use the algorithms in this section to show that the matrix below is irreducible and to determine its index k of imprimitivity and its k-cyclic components:

$$
\left[
\begin{array}{cccccccccc}
0 & 0 & 0 & 1 & 0 & 0 & 0 & 1 & 0 & 0 \\
0 & 0 & 0 & 0 & 0 & 0 & 0 & 1 & 0 & 0 \\
1 & 1 & 0 & 0 & 1 & 0 & 0 & 0 & 0 & 0 \\
0 & 0 & 0 & 0 & 0 & 1 & 0 & 0 & 0 & 1 \\
0 & 0 & 0 & 1 & 0 & 0 & 0 & 0 & 0 & 0 \\
0 & 0 & 0 & 0 & 0 & 0 & 0 & 0 & 1 & 0 \\
0 & 0 & 1 & 0 & 0 & 0 & 0 & 0 & 0 & 0 \\
0 & 0 & 0 & 0 & 0 & 0 & 1 & 0 & 0 & 1 \\
0 & 1 & 0 & 0 & 1 & 0 & 0 & 0 & 0 & 0 \\
0 & 0 & 1 & 0 & 0 & 0 & 0 & 0 & 1 & 0
\end{array}
\right] .
$$

References

A.V. Aho, J.E. Hopcroft and J.D. Ullman[1975], *The Design and Analysis of Computer Algorithms*, Addison-Wesley, Reading, Mass.

M.J. Atallah[1982], Finding the cycle index of an irreducible, nonnegative matrix, *SIAM J. Computing*, 11, pp. 567–570.

E.V. Denardo[1977], Periods of connected networks, *Math. Oper. Res.*, 2, pp. 20–24.

R.E. Tarjan[1972], Depth first search and linear graph algorithms, *SIAM J. Computing*, 1, pp. 146–160.

4

Matrices and Bipartite Graphs

4.1 Basic Facts

Bipartite graphs are defined in section 2.6. A multigraph G is *bipartite* provided that its vertices may be partitioned into two subsets X and Y such that every edge of G is of the form $\{a, b\}$ where a is in X and b is in Y. The pair $\{X, Y\}$ is called a *bipartition* of G. If G is connected its bipartition is unique.

The bipartite multigraph G is characterized by an m by n nonnegative integral matrix

$$B = [b_{ij}], (i = 1, 2, \ldots, m; j = 1, 2, \ldots, n),$$

where m is the number of vertices in X and n is the number in Y. Let $X = \{x_1, x_2, \ldots, x_m\}$ and $Y = \{y_1, y_2, \ldots, y_n\}$. The element b_{ij} equals the multiplicity $m\{x_i, y_j\}$ of the edges of the form $\{x_i, y_j\}$. The adjacency matrix of G is the $m + n$ by $m + n$ matrix

$$\begin{bmatrix} O & B \\ B^T & O \end{bmatrix}. \tag{4.1}$$

We call B the *reduced adjacency matrix* of the bipartite multigraph G. Every m by n nonnegative integral matrix is the reduced adjacency matrix of some bipartite multigraph.

We begin with two elementary but fundamental characterizations of bipartite multigraphs.

Theorem 4.1.1. *A multigraph G is bipartite if and only if every cycle of G has even length.*

Proof. The definition of a bipartite graph implies at once that every cycle has even length.

It suffices to prove the converse proposition for a connected component G' of G. We select an arbitrary vertex a in G'. Let X be the set of vertices of G' whose distance from a is even, and let Y be the set of vertices of G' whose distance from a is odd. Let p and q be two vertices in X. We show that G' does not contain an edge of the form $\{p, q\}$. Let $a \to \cdots \to p$ and $a \to \cdots \to q$ denote walks of minimal length from a to p and a to q, respectively. Let b be the last common vertex in these two walks. Then the walks $b \to \cdots \to p$ and $b \to \cdots \to q$ are both of even length or both of odd length. But an edge of the form $\{p, q\}$ implies the existence of a cycle of odd length

$$b \to \cdots \to p \to q \to \cdots \to b,$$

contrary to hypothesis. In the same way one shows that two distinct vertices in Y are not connected by an edge. Hence G' is bipartite.   □

Theorem 4.1.2.   *Let G be a multigraph and let A be the incidence matrix of G. Then G is bipartite if and only if A is totally unimodular.*

Proof. Let G be bipartite. Then G has a bipartition $\{X, Y\}$, and this implies that A^T satisfies the requirements of Theorem 2.3.3. Hence A is totally unimodular.

Now let A be totally unimodular and suppose that G is not bipartite. By Theorem 4.1.1 G has a cycle of odd length r. But this implies that A contains a submatrix A' of order r such that $\det(A') = \pm 2$. This contradicts the hypothesis that A is totally unimodular.   □

Let G be a bipartite graph with bipartition $\{X, Y\}$ where X is an m-set and Y is an n-set. If for each x in X and each y in Y, G contains exactly one edge of the form $\{x, y\}$, then G is called a *complete bipartite graph* and is denoted by $K_{m,n}$.

Let K_n denote the complete graph of order n. Let $G_1, G_2, \ldots, G_r$ denote complete bipartite subgraphs of K_n. Suppose that the graphs $G_1, G_2, \ldots, G_r$ are edge disjoint and between them contain all of the edges of K_n. Then we say that $G_1, G_2, \ldots, G_r$ form a *decomposition* of K_n. It is easy to construct decompositions of K_n for which $r = n - 1$ and $G_1, G_2, \ldots, G_{n-1}$ is $K_{1,n-1}, K_{1,n-2}, \ldots, K_{1,1}$. The following theorem of Graham and Pollak[1971] tells us that it is not possible to form a decomposition of K_n into complete bipartite subgraphs with $r < n - 1$. The short proof is due to Peck[1984].

Theorem 4.1.3.   *Let the complete graph K_n of order n have a decomposition $G_1, G_2, \ldots, G_r$ into complete bipartite subgraphs. Then $r \geq n - 1$.*

Proof. Let G_i' be the spanning subgraph of K_n with the same set of edges as G_i $(i = 1, 2, \ldots, r)$. The conditions of the theorem imply that we may write

$$J - I = \sum_{i=1}^{r} A_i', \qquad (4.2)$$

where A_i' is the adjacency matrix of G_i'. Let A_i be the adjacency matrix of G_i $(i = 1, 2, \ldots, r)$. Then A_i is a principal submatrix of A_i' and contains all the nonzero entries of A_i'. The matrix A_i, and hence the matrix A_i', is of rank 2 because in the special form (4.1) we have B equal to a matrix of all 1's. We now replace all of the 1's in A_i corresponding to the 1's in B^T with 0's. The resulting matrix A_i'' is clearly of rank 1. Furthermore the matrix

$$Q_i = A_i' - 2A_i''$$

is skew-symmetric. We may now write (4.2) in the form

$$I + Q = J - 2\sum_{i=1}^{r} A_i''$$

where Q is a skew-symmetric matrix of order n. A real skew-symmetric matrix has pure imaginary eigenvalues so that we may conclude that $I + Q$ is nonsingular. But the rank of a sum of matrices does not exceed the sum of the ranks and hence it follows that $1 + r \geq n$. □

Another proof of Theorem 4.1.3 is given by Tverberg[1982]. A third proof is indicated in the exercises.

Exercises

1. Let B be the m by n reduced adjacency matrix of a bipartite graph G. Prove that G is connected if and only if there do not exist permutation matrices P and Q such that

$$PBQ = \begin{bmatrix} B_1 & O \\ O & B_2 \end{bmatrix}$$

   where B_1 is a p by q matrix for some nonnegative integers p and q satisfying $1 \leq p + q \leq m + n - 1$.
2. Let A be a tournament matrix of order n. Prove that the rank of A is at least equal to $n - 1$. (Hint: Consider the matrix N of size n by $n + 1$ obtained from A by adjoining a column of 1's and show that only the zero vector is in the left null space of N.)
3. Let $G_1, G_2, \ldots, G_r$ be a decomposition of the complete graph K_n into complete bipartite graphs. For each G_i direct the edges from the vertices in one set of its bipartition to the vertices in the other set of its bipartition. The result is a tournament of order n and hence a tournament matrix A of order n. Obtain from the decomposition $G_1, G_2, \ldots, G_r$ a factorization $A = CD$ of A into an

n by r (0,1)-matrix C and an r by n (0,1)-matrix D. Now use Exercise 2 to obtain an alternative proof of Theorem 4.1.3 (de Caen and Hoffman[1989]).

References

R.A. Brualdi, F. Harary and Z. Miller[1980], Bigraphs versus digraphs via matrices, *J. Graph Theory*, 4, pp. 51–73.

D. de Caen and D.G. Hoffman[1989], Impossibility of decomposing the complete graph on n points into $n-1$ isomorphic complete bipartite graphs, *SIAM J. Discrete Math.*, 2, pp. 48–50.

R.L. Graham and H.O. Pollak[1971], On the addressing problem for loop switching, *Bell System Tech. J.*, 50, pp. 2495–2519.

     [1972], On embedding graphs in squashed cubes, *Lecture Notes in Math*, vol. 303, Springer-Verlag, New York, pp. 99–110.

G.W. Peck[1984], A new proof of a theorem of Graham and Pollak, *Discrete Math.*, 49, pp. 327–328.

H. Tverberg[1982], On the decomposition of K_n into complete bipartite graphs, *J. Graph Theory*, 6, pp. 493–494.

4.2 Fully Indecomposable Matrices

In this section we deal primarily with m by n (0,1)-matrices. The definitions and results apply to arbitrary matrices upon replacing each nonzero element with a 1.

Let A be an m by n (0,1)-matrix. The *term rank* $\rho = \rho(A)$ of A is defined in section 1.2 to be the maximal number of 1's of A with no two of the 1's on a line. A *line cover* of A is a collection of lines of A which together contain all the 1's of A. By Theorem 1.2.1 the minimal number of lines in a line cover is equal to the term rank of A. A line cover with the minimal number of lines is called a *minimum line cover* of A. We denote the set of minimum line covers of A by $\mathcal{L} = \mathcal{L}(A)$. An *essential line* of A is a line which belongs to every minimum line cover. An essential line may be either an *essential row* or an *essential column*.

Since the term rank of A is ρ, we may permute the lines of A so that there are 1's in the first ρ positions on the main diagonal. The permuted A assumes the form

$$\begin{bmatrix} A' & A_{12} \\ A_{21} & O \end{bmatrix} \tag{4.3}$$

where A' is a ρ by ρ matrix with 1's everywhere on its main diagonal. Without loss of generality we assume that A has the form (4.3). For each $i = 1, 2, \ldots, \rho$ each minimum line cover of A contains either row i or column i, but not both. Thus row i and column i cannot both be essential lines of A, $(i = 1, 2, \ldots, \rho)$. Let r be the number of essential rows of A and let s be the number of essential columns. Let $t = \rho - r - s$. Then r, s and t

are nonnegative integers summing to ρ. We now simultaneously permute the first ρ rows and the first ρ columns of A so that the resulting matrix assumes the form

$$
\begin{array}{c}
 \\
r \\
s \\
t
\end{array}
\begin{array}{c}
\begin{array}{cccc} r & s & & t \end{array} \\
\left[
\begin{array}{cccc}
A_1 & X & Y & Z \\
* & A_2 & * & * \\
* & S & A_3 & * \\
* & T & * & O
\end{array}
\right]
\end{array},
\tag{4.4}
$$

where rows $1, 2, \ldots, r$ are the essential rows and columns $r+1, r+2, \ldots, r+s$ are the essential columns of the permuted A. For each $i = r + s + 1, r + s + 2, \ldots, \rho$ there is a minimum line cover of the permuted A in (4.4) which does not contain row i (and thus contains column i) and a minimum line cover which does not contain column i (and thus contains row i). It now follows that all the submatrices marked with a $*$ in (4.4) are zero matrices. We summarize these conclusions in the following theorem (Dulmage and Mendelsohn[1958] and Brualdi[1966]).

Theorem 4.2.1.   *Let A be an m by n $(0,1)$-matrix with term rank equal to ρ. Then there is a permutation matrix P of order m and a permutation matrix Q of order n such that*

$$
PAQ =
\left[
\begin{array}{cccc}
A_1 & X & Y & Z \\
O & A_2 & O & O \\
O & S & A_3 & O \\
O & T & O & O
\end{array}
\right].
\tag{4.5}
$$

The matrices A_1, A_2 and A_3 are square, possibly vacuous, matrices with 1's everywhere on their main diagonals, and the sum of their orders is ρ. The essential rows of the matrix in (4.5) are those rows which meet A_1, and the essential columns are those columns which meet A_2.

It follows from the description of the matrix (4.5) given in the statement of Theorem 4.2.1 that the only minimum line cover of the matrix

$$
\begin{bmatrix} A_1 & Z \end{bmatrix}
$$

is the line cover of all rows. Also the only minimum line cover of the matrix

$$
\begin{bmatrix} A_2 \\ T \end{bmatrix}
$$

is the line cover of all columns. The matrix A_3 in (4.5), if not vacuous, has at least two minimum line covers. These are the line cover of all rows and the line cover of all columns. Whether or not there are other minimum line covers depends on the additional structure of A_3. This leads us to a study of matrices whose only minimum line covers are the all rows cover and the all columns cover. Such matrices are necessarily square.

We now assume that A is a $(0,1)$-matrix of order n. The matrix A is *partly decomposable* provided there exists an integer k with $1 \leq k \leq n-1$ such that A has a k by $n-k$ zero submatrix. This means that we may permute the lines of A to obtain a matrix of the form

$$\begin{bmatrix} B & O \\ D & C \end{bmatrix}$$

where the zero matrix O is of size k by $n-k$. The matrices B and C are square matrices of orders k and $n-k$, respectively. The partly decomposable matrix A has a line cover consisting of $n-k \geq 1$ rows and $k \geq 1$ columns. This line cover may or may not be a minimum line cover since A may have a p by q zero submatrix with $p+q > n$. But it follows that *the square matrix A is partly decomposable if and only if it has a minimum line cover other than the all rows cover and the all columns cover.* The square matrix A is *fully indecomposable*[1] provided it is not partly decomposable. The property of full indecomposability is not affected by arbitrary line permutations. If $n = 1$, then A is fully indecomposable if and only if A is not the zero matrix of order 1. Each line of a fully indecomposable matrix of order $n \geq 2$ has at least two 1's. It follows from Theorem 1.2.1 that the fully indecomposable matrix A of order n has term rank equal to n. But even more is true. If $n \geq 2$ and we choose any entry of A and delete its row and column from A, the resulting matrix B of order $n-1$ has term rank $\rho(B)$ equal to $n-1$. For, if $\rho(B) < n-1$, then by Theorem 1.2.1 B and hence A would have a p by q zero submatrix for some positive integers p and q with $p+q = (n-1)+1 = n$. Conversely, if A is a $(0,1)$-matrix of order n and every submatrix of order $n-1$ has term rank equal to $n-1$, then A is fully indecomposable. This characterization of fully indecomposable matrices (Marcus and Minc[1963] and Brualdi[1966]) is formulated in the next theorem. A collection of n elements of A (or the positions of those elements) is called a *diagonal* of A provided no two of the elements belong to the same row or column of A. A *nonzero diagonal* of A is a diagonal not containing any 0's. If G is the bipartite graph whose reduced adjacency matrix is A, then the nonzero diagonals of A are in one-to-one correspondence with the perfect matchings of G.

---

[1] One may wonder why the adverbs "partly" and "fully" are being used here. The reason is that the terminology "decomposable" and "indecomposable" is sometimes used in place of "reducible" and "irreducible." A reducible matrix of order n also has a k by $n-k$ zero submatrix for some integer k, but the row indices and column indices of the zero submatrix are disjoint sets. Thus reducibility is a more severe restriction on a matrix than that of part decomposability.

Theorem 4.2.2.   *Let A be a* $(0, 1)$*-matrix of order* $n \geq 2$*. Then A is fully indecomposable if and only if every* 1 *of A belongs to a nonzero diagonal and every* 0 *of A belongs to a diagonal all of whose other elements equal* 1*.*

There is a close connection between fully indecomposable matrices and the irreducible matrices of Chapter 3 (see Brualdi[1979] and Brualdi and Hedrick[1979]).

Theorem 4.2.3.   *Let A be a* $(0, 1)$*-matrix of order* n*. Let* $A^{\#}$ *be the matrix obtained from A by replacing each entry on the main diagonal with a* 1*. Then A is irreducible if and only if* $A^{\#}$ *is fully indecomposable.*

Proof. We know that $A^{\#}$ is fully indecomposable if and only if it does not have a k by $n - k$ zero submatrix for any integer k with $1 \leq k \leq n - 1$. It is a consequence of the definition of irreducibility that A is irreducible if and only if each k by $n - k$ zero submatrix of A with $1 \leq k \leq n - 1$ contains a 0 from the main diagonal of A. Since $A^{\#}$ is obtained from A by replacing the 0's that occur on the main diagonal with 1's, the theorem follows. $\square$

The conclusion of Theorem 4.2.3 can also be formulated as: *The square matrix A is irreducible if and only if* $A + I$ *is fully indecomposable.* The following characterization of fully indecomposable matrices is given in Brualdi, Parter and Schneider[1966].

Corollary 4.2.4.   *Let A be a* $(0, 1)$*-matrix of order* n*. Then A is fully indecomposable if and only if there exist permutation matrices P and Q of order* n *such that PAQ has all* 1*'s on its main diagonal and PAQ is irreducible.*

Proof. Assume that A is fully indecomposable. The term rank of A equals n and thus there exist permutation matrices P and Q of order n such that all diagonal elements of PAQ equal 1. The matrix PAQ is also fully indecomposable and it follows from Theorem 4.2.3 that PAQ is irreducible. The converse proposition is derived in a very similar way. $\square$

We now continue with the study of matrices for which the all rows cover and the all columns cover are minimum line covers (but for which there may be other minimum line covers). By Theorem 1.2.1 this is equivalent to the study of (0,1)-matrices of order n with term rank equal to n. First we prove the following preliminary result.

Lemma 4.2.5.   *Let A be a* $(0, 1)$*-matrix of order n of the form*

$$\begin{bmatrix} X & Z \\ O & Y \end{bmatrix}$$

where X and Y are square matrices of orders k and l, respectively. Then there does not exist a nonzero diagonal of A which contains a 1 *of Z.*

Proof. Consider a 1 of Z lying in row i and column j of A. Let B be the matrix of order $n-1$ obtained from A by deleting row i and column j. Then B has a line cover consisting of $k-1$ rows and $l-1$ columns where $(k-1)+(l-1) = n-2$. By Theorem 1.2.1 $\rho(B) \leq n-2$, and the conclusion follows.   □

The following theorem is contained in Dulmage and Mendelsohn[1958] and Brualdi[1966].

Theorem 4.2.6.   *Let A be a $(0,1)$-matrix of order n with term rank $\rho(A)$ equal to n. Then there exist permutation matrices P and Q of order n and an integer $t \geq 1$ such that PAQ has the form*

$$\begin{bmatrix} B_1 & B_{12} & \cdots & B_{1t} \\ O & B_2 & \cdots & B_{2t} \\ \vdots & \vdots & \ddots & \vdots \\ O & O & \cdots & B_t \end{bmatrix} \tag{4.6}$$

where $B_1, B_2, \ldots, B_t$ are square fully indecomposable matrices. The matrices $B_1, B_2, \ldots, B_t$ that occur as diagonal blocks in (4.6) are uniquely determined to within arbitrary permutations of their lines, but their ordering in (4.6) is not necessarily unique.

Proof. Because $\rho(A) = n$, we may permute the lines of A so that the resulting matrix B has only 1's on its main diagonal. We now apply Theorem 3.2.4 to B. According to that theorem there exists an integer $t \geq 1$ such that the lines of B can be *simultaneously* permuted to obtain a matrix of the form

$$\begin{bmatrix} A_1 & A_{12} & \cdots & A_{1t} \\ O & A_2 & \cdots & A_{2t} \\ \vdots & \vdots & \ddots & \vdots \\ O & O & \cdots & A_t \end{bmatrix} \tag{4.7}$$

where $A_1, A_2, \ldots, A_t$ are square irreducible matrices which are uniquely determined to within simultaneous permutations of their lines. Each of the matrices A_i in (4.7) has only 1's on its main diagonal. By Theorem 4.2.3 the matrices $A_1, A_2, \ldots, A_t$ are fully indecomposable. It follows from Theorem 4.2.2 that each 1 of the matrix (4.7) which belongs to one of the matrices $A_1, A_2, \ldots, A_t$ is part of a nonzero diagonal, and from Lemma 4.2.5 that any other 1 does not belong to a nonzero diagonal. Hence the nonzero diagonals of the matrix (4.7) are the same as those of the matrix

$$B' = \begin{bmatrix} A_1 & O & \cdots & O \\ O & A_2 & \cdots & O \\ \vdots & \vdots & \ddots & \vdots \\ O & O & \cdots & A_t \end{bmatrix}, \tag{4.8}$$

and every 1 of B' belongs to a nonzero diagonal.

Suppose that the lines of A could also be permuted to give

$$C = \begin{bmatrix} C_1 & C_{12} & \cdots & C_{1s} \\ O & C_2 & \cdots & C_{2s} \\ \vdots & \vdots & \ddots & \vdots \\ O & O & \cdots & C_s \end{bmatrix} \tag{4.9}$$

where $C_1, C_2, \ldots, C_s$ are fully indecomposable matrices. Arguing as above the nonzero diagonals of C are the same as those of

$$C' = \begin{bmatrix} C_1 & O & \cdots & O \\ O & C_2 & \cdots & O \\ \vdots & \vdots & \ddots & \vdots \\ O & O & \cdots & C_s \end{bmatrix} \tag{4.10}$$

and every 1 of C belongs to a nonzero diagonal. Thus B' and C' are both obtained by replacing with 0's all 1's of A that do not belong to a nonzero diagonal and then permuting the lines of A. Therefore C' can be obtained from B' by permuting its lines. Since the matrices $A_1, A_2, \ldots, A_t$ and $C_1, C_2, \ldots, C_s$ are fully indecomposable, we conclude that $s = t$ and that there exists a permutation $i_1, i_2, \ldots, i_t$ of $1, 2, \ldots, t$ such that C_{i_j} can be obtained from A_j by line permutations for each $j = 1, 2, \ldots, t$. The ordering of the diagonal blocks in (4.6) is not unique, if, for instance, the matrix A is a direct sum of two fully indecomposable matrices of different orders. □

Let A be a (0,1)-matrix of order n with $\rho(A) = n$. The matrices $B_1, B_2, \ldots, B_t$ that occur as diagonal blocks in (4.6) are called the *fully indecomposable components* of A. By Theorem 4.2.6 the fully indecomposable components of A are uniquely determined to within permutations of their lines. As demonstrated in the proof of Theorem 4.2.6, the fully indecomposable components of A are the irreducible components of any matrix obtained from A by permuting lines so that there are no 0's on the main diagonal. Notice that the matrix A is fully indecomposable if and only if it has exactly one fully indecomposable component.

A (0,1)-matrix A of order n has *total support* provided each of its 1's belongs to a nonzero diagonal. If $A = O$, then A has total support. If

$A \neq O$, then it follows from Lemma 4.2.5 and Theorem 4.2.6 that A has *total support if and only if there are permutation matrices P and Q of order n such that PAQ is a direct sum of fully indecomposable matrices.*

Fully indecomposable matrices can be characterized within the set of matrices with total support by using bipartite graphs.

Theorem 4.2.7.   *Let A be a nonzero $(0,1)$-matrix of order n with total support, and let G be the bipartite graph whose reduced adjacency matrix is A. Then A is fully indecomposable if and only if G is connected.*

Proof. If A is not fully indecomposable, then there are permutation matrices P and Q such that PAQ is a direct sum of two or more fully indecomposable matrices, and G is not connected.

Conversely, suppose that G is not connected. Then there are permutation matrices R and S such that

$$RAS = \begin{bmatrix} A' & O \\ O & A'' \end{bmatrix}$$

where A' is a p by q matrix for some nonnegative integers p and q with $1 \leq p + q \leq 2n - 1$. Without loss of generality we assume that $p \leq q$. Then A has a line cover consisting of p rows and $n - q$ columns where $p + (n - q) = n - (q - p)$. Since A has total support and $A \neq O$, we have $\rho(A) = n$ and it follows that $p = q$. But then A has a zero submatrix of size p by $n - p$ and A is not fully indecomposable.   □

We close this section by applying Theorem 3.3.8 to obtain an inductive structure of Hartfiel[1975] for fully indecomposable matrices.

Theorem 4.2.8.   *Let A be a fully indecomposable $(0,1)$-matrix of order $n \geq 2$. Then there exist permutation matrices P and Q of order n and an integer $m \geq 2$ such that PAQ has the form*

$$\begin{bmatrix} A_1 & O & \cdots & O & E_1 \\ E_2 & A_2 & \cdots & O & O \\ \vdots & \vdots & \ddots & \vdots & \vdots \\ O & O & \cdots & A_{m-1} & O \\ O & O & \cdots & E_m & A_m \end{bmatrix} \qquad (4.11)$$

where $A_1, A_2, \ldots, A_m$ are fully indecomposable matrices and the matrices $E_1, E_2, \ldots, E_m$ each contain at least one 1.

Proof. By Corollary 4.2.4 we may permute the lines of A so that the resulting matrix B has all 1's on its main diagonal and is irreducible. By Theorem 3.3.8 the lines of B can be simultaneously permuted to obtain the form (4.11) where

$m \geq 2$, the A_i are irreducible matrices with all 1's on their main diagonals, and the E_i each contain at least one 1. By Theorem 4.2.3 the matrices A_i are fully indecomposable and the theorem follows. $\square$

Exercises

1. Let A be an m by n $(0,1)$-matrix with $m < n$, and assume that each row of A is an essential row. Prove that there exist a positive integer p and permutation matrices P and Q such that

$$PAQ = \begin{bmatrix} B_{11} & B_{12} & B_{13} & \cdots & B_{p1} & B_{10} \\ B_{21} & B_{22} & B_{23} & \cdots & B_{p2} & O \\ O & B_{32} & B_{33} & \cdots & B_{3p} & O \\ O & O & B_{43} & \cdots & B_{4p} & O \\ \vdots & \vdots & \vdots & & \vdots & \vdots \\ O & O & O & \cdots & B_{pp} & O \end{bmatrix}$$

   where $B_{11}, B_{22}, \ldots, B_{pp}$ are square matrices with 1's everywhere on their main diagonals and $B_{10}, B_{21}, B_{32}, \ldots, B_{p,p-1}$ have no zero rows (Brualdi[1966]).
2. Let $n \geq 2$. Prove that the permanent of a fully indecomposable $(0,1)$-matrix of order n is at least 2 and characterize those fully indecomposable matrices whose permanent equals 2.
3. Prove that the product of two fully indecomposable $(0,1)$-matrices of the same order is a fully indecomposable matrix (Lewin[1971]).
4. Let A be a fully indecomposable $(0,1)$-matrix of order $n \geq 2$. Prove that A^{n-1} is a positive matrix and then deduce that A is a primitive matrix with exponent at most equal to $n - 1$ (Lewin[1971]).
5. Find an example of a fully indecomposable $(0,1)$-matrix of order n whose exponent equals $n - 1$.
6. Let A be a fully indecomposable $(0,1)$-matrix of order n. Prove that there is a doubly stochastic matrix D of order n such that the matrix obtained from D by replacing each positive element with a 1 equals A.

References

R.A. Brualdi[1966], Term rank of the direct product of matrices, *Canad. J. Math.*, 18, pp. 126–138.

[1979], Matrices permutation equivalent to irreducible matrices and applications, *Linear and Multilin. Alg.*, 7, pp. 1–12.

R.A. Brualdi, F. Harary and Z. Miller[1980], Bigraphs versus digraphs via matrices, *J. Graph Theory*, 4, pp. 51–73.

R.A. Brualdi and M.B. Hedrick[1979], A unified treatment of nearly reducible and nearly decomposable matrices, *Linear Alg. Applics.*, 24, pp. 51–73.

R.A. Brualdi, S.V. Parter and H. Schneider[1966], The diagonal equivalence of a nonnegative matrix to a stochastic matrix, *J. Math. Anal. Applics.*, 16, pp. 31–50.

A.L. Dulmage and N.S. Mendelsohn[1958], Coverings of bipartite graphs, *Canad. J. Math.*, 10, pp. 517–534.

[1959], A structure theory of bipartite graphs of finite exterior dimension, *Trans. Roy. Soc. Canad.*, Third Ser. Sec. III 53, pp. 1–3.

[1967], Graphs and Matrices, *Graph Theory and Theoretical Physics* (F. Harary, ed.), Academic Press, New York, pp. 167–227.

R.P. Gupta[1967], On basis diagraphs, *J. Comb. Theory*, 3, pp. 16–24.

D.J. Hartfiel[1975], A canonical form for fully indecomposable (0,1)-matrices, *Canad. Math. Bull*, 18, pp. 223–227.

M. Lewin[1971], On nonnegative matrices, *Pacific. J. Math.*, 36, pp. 753–759.

M. Marcus and H. Minc[1963], Disjoint pairs of sets and incidence matrices, *Illinois J. Math*, 7, pp. 137–147.

E.J. Roberts[1970], The fully indecomposable matrix and its associated bipartite graph—an investigation of combinatorial and structural properties, *NASA Tech. Memorandum* TM X-58037.

R. Sinkhorn and P. Knopp[1969], Problems involving diagonal products in nonnegative matrices, *Trans. Amer. Math. Soc.*, 136, pp. 67–75.

4.3 Nearly Decomposable Matrices

We continue to frame our discussion in terms of (0,1)-matrices with the understanding that the 1's can be replaced by arbitrary nonzero numbers.

Let A be a fully indecomposable (0,1)-matrix. The matrix A is called *nearly decomposable* provided whenever a 1 of A is replaced with a 0, the resulting matrix is partly decomposable. Thus the nearly decomposable matrices are the "minimal" fully indecomposable matrices. Two examples of nearly decomposable matrices are

$$A_1 = \begin{bmatrix} 0 & 1 & 1 \\ 1 & 0 & 1 \\ 1 & 1 & 0 \end{bmatrix} \quad A_2 = \begin{bmatrix} 1 & 0 & 0 & 1 \\ 0 & 1 & 0 & 1 \\ 0 & 0 & 1 & 1 \\ 1 & 1 & 1 & 0 \end{bmatrix}. \tag{4.12}$$

The relationship between fully indecomposable matrices and irreducible matrices as described in Theorem 4.2.3 and Corollary 4.2.4 only partially extends to nearly decomposable matrices and nearly reducible matrices.

Theorem 4.3.1.   *Let A be a $(0,1)$-matrix of order n. If each element on the main diagonal of A is 0 and $A + I$ is nearly decomposable, then A is nearly reducible. If A is nearly reducible, then each element on the main diagonal of A is 0 and $A + I$ is fully indecomposable, but $A + I$ need not be nearly decomposable.*

Proof. Assume that A has all 0's on its main diagonal. By Theorem 4.2.3 A is irreducible if and only if the matrix $B = A + I$ is fully indecomposable. Suppose that $A + I$ is nearly decomposable. Let A' be a matrix obtained from A by replacing a 1 with a 0. If A' is irreducible, then by Theorem 4.2.3 $A' + I$ is fully indecomposable and we contradict the near decomposability of $A + I$. Hence A is nearly reducible.

Now suppose that A is nearly reducible. Then each element on the main diagonal of A is 0, and the matrix $B = A + I$ is fully indecomposable. By Theorem 4.2.3 the replacement of an off-diagonal 1 of B with a 0 results in a partly decomposable matrix. The nearly decomposable matrix A_2 in (4.12) shows that it may be possible to replace a 1 on the main diagonal of B with a 0 and obtain a fully indecomposable matrix.   □

The fact that $A + I$ need not be nearly decomposable if A is nearly reducible prevents in general theorems about nearly reducible matrices and nearly decomposable matrices from being directly obtainable from one another. We can, however, use the inductive structure of nearly reducible matrices given in Theorem 3.3.4 to obtain an inductive structure for nearly decomposable matrices. First we prove two lemmas.

Lemma 4.3.2.   *Let B be a $(0,1)$-matrix having the form*

$$
\begin{bmatrix}
1 & 0 & 0 & \cdots & 0 & 0 & \\
1 & 1 & 0 & \cdots & 0 & 0 & \\
0 & 1 & 1 & \cdots & 0 & 0 & \\
\vdots & \vdots & \vdots & \ddots & \vdots & \vdots & F_1 \\
0 & 0 & 0 & \cdots & 1 & 0 & \\
0 & 0 & 0 & \cdots & 1 & 1 & \\
& & F_2 & & & & B_1
\end{bmatrix}
\tag{4.13}
$$

where B_1 is a fully indecomposable matrix, F_1 has a 1 in its first row and F_2 has a 1 in its last column. Then B is a fully indecomposable matrix.

Proof. By Theorem 4.2.7 it suffices to show that B has total support and that the bipartite graph G whose reduced adjacency matrix is B is connected. It is a direct consequence of Theorem 4.2.2 that each 1 of B belongs to a nonzero diagonal. By Theorem 4.2.7 the bipartite graph G_1 whose reduced adjacency matrix is B_1 is connected. The bipartite graph G is obtained from G_1 by attaching a chain whose endpoints are two vertices of G_1 (and possibly some additional edges). Hence G is connected as well.
                              □

Lemma 4.3.3.   *Assume that in Lemma 4.3.2 the matrix B is nearly decomposable. Then the matrix B_1 in (4.13) is nearly decomposable and F_1 and F_2 each contain exactly one 1. Let the unique 1 in F_1 belong to column j of F_1, and let the unique 1 in F_2 belong to row i of F_2. If the order of B_1 is at least 2, then element in position (i, j) of B_1 is a 0.*

Proof. If the replacement of some 1 of B_1 with a 0 results in a fully indecomposable matrix, then by Lemma 4.3.2 the replacement of that 1 in B with a 0 also results in a fully indecomposable matrix. It follows that

B_1 is nearly decomposable. Lemma 4.3.2 also implies that F_1 and F_2 each contain exactly one 1.

Now assume that the order of B_1 is at least 2. Let G_1 be the connected bipartite graph whose reduced adjacency matrix is B_1. Suppose that the element in the position (i,j) of B_1 equals 1. Let the matrices B' and B_1' be obtained from B and B_1, respectively, by replacing this 1 with a 0. Since the order of B_1 is at least 2, B_1' is not a zero matrix. Each nonzero diagonal of B_1 extends to a nonzero diagonal of B by including the leading 1's on the main diagonal of B which are displayed in (4.13). Since B_1 is fully indecomposable, B_1 has a nonzero diagonal which includes the 1 in position (i,j). Consider such a nonzero diagonal of B_1. Removing the 1 in position (i,j) and including the 1's of F_1 and F_2 as well as the 1's below the main diagonal of B which are displayed in (4.13) results in a nonzero diagonal of B. It follows from these considerations that the matrix B' has total support. The bipartite graph G' whose reduced adjacency matrix is B' is connected since it is obtained from the connected graph G_1 by replacing an edge with a chain joining its two endpoints. We now apply Theorem 4.2.7 to B' and conclude that B' is fully indecomposable, contradicting the near decomposability assumption of B. Hence the element in the position (i,j) of B_1 equals 0.   □

The following inductive structure for a nearly decomposable matrix is due to Hartfiel[1970]. It is a simplification of an inductive structure obtained by Sinkhorn and Knopp[1969].

Theorem 4.3.4.   *Let A be a nearly decomposable $(0,1)$-matrix of order $n \geq 2$. Then there exist permutation matrices P and Q of order n and an integer m with $1 \leq m \leq n-1$ such that PAQ has the form (4.13) where B_1 is a nearly decomposable matrix of order m. The matrix F_1 contains a unique 1 and it belongs to its first row and column j for some j with $1 \leq j \leq m$. The matrix F_2 contains a unique 1 and it belongs to its last column and row i for some i with $1 \leq i \leq m$. If $m \geq 2$, then $m \neq 2$ and the element in position (i,j) of B_1 is 0.*

Proof. The matrix A is fully indecomposable and thus has term rank equal to n. We permute the lines of A and obtain a nearly decomposable matrix B all of whose diagonal elements equal 1. By Theorem 4.3.1 the matrix $B - I$ is nearly reducible. By Theorem 3.3.4 there is a permutation matrix R of order n such that

$$R(B-I)R^T = RBR^T - I = \begin{bmatrix} 0 & 0 & 0 & \cdots & 0 & 0 & \\ 1 & 0 & 0 & \cdots & 0 & 0 & \\ 0 & 1 & 0 & \cdots & 0 & 0 & F_1 \\ \vdots & \vdots & \vdots & \ddots & \vdots & \vdots & \\ 0 & 0 & 0 & \cdots & 1 & 0 & \\ & & F_2 & & & & A_1 \end{bmatrix}$$

where A_1 is a nearly reducible matrix of order m for some integer m with $1 \leq m \leq n - 1$. The matrix F_1 contains a single 1 and it belongs to its first row and column j where $1 \leq j \leq m$. The matrix F_2 contains a single 1 and it belongs to its last column and row i where $1 \leq i \leq m$. The element in position (i, j) of A_1 is 0. Hence RBR^T has the form (4.13) with $B_1 = A_1 + I$. By Theorem 4.3.1, B_1 is fully indecomposable. Since B is nearly decomposable, we may apply Lemma 4.3.3 and conclude that B_1 is nearly decomposable and that the element in position (i, j) of B_1 is 0 if $m \geq 2$. Finally we note that if $m \geq 2$, then $m \neq 2$, since no nearly decomposable matrix of order 2 contains a 0. $\square$

We remark that a matrix of the form (4.13) satisfying the conclusions of Theorem 4.3.4 need not be nearly decomposable. The matrix

$$B_1 = \begin{bmatrix} 1 & 1 & 0 & 0 \\ 1 & 0 & 1 & 1 \\ 0 & 1 & 1 & 0 \\ 0 & 1 & 0 & 1 \end{bmatrix}$$

is a nearly decomposable matrix (an easy way to see this is to notice that each 1 belongs to a line which contains exactly two 1's). However the matrix

$$B = \begin{bmatrix} 1 & 1 & 0 & 0 & 0 \\ 0 & 1 & 1 & 0 & 0 \\ 0 & 1 & 0 & 1 & 1 \\ 1 & 0 & 1 & 1 & 0 \\ 0 & 0 & 1 & 0 & 1 \end{bmatrix}$$

is not nearly decomposable. This is because replacing the 1's in positions (3,2) and (4,3) positions with 0's results in a fully indecomposable matrix.

The matrix B_1 that occurs in the inductive structure of nearly decomposable matrices given in Theorem 4.3.4 can be any nearly decomposable matrix except for the 2 by 2 matrix of all 1's (Hartfiel[1971]). The nearly decomposable matrix of order 1, whose unique entry is a 1, occurs when the matrix of all 1's of order 2 is written in the form (4.13). Now let B be any nearly decomposable matrix of order $n \geq 3$. Without loss of generality we may assume that B has the form (4.13) and the conclusions of Theorem 4.3.4 are satisfied. Let A be the matrix

$$\begin{bmatrix} 1 & E_1 \\ E_2 & B \end{bmatrix}$$

where E_1 is a 1 by n (0,1)-matrix with a single 1 and this 1 belongs to the same column in which F_1 has its 1, and E_2 is an n by 1 (0,1)-matrix with a single 1 and this 1 belongs to the same row in which F_1 has its 1. By Lemma

4.3.2 the matrix A is fully indecomposable. The near decomposability of B implies the near decomposability of A.

It was pointed out in section 3.3 that an irreducible principal submatrix of a nearly reducible matrix is nearly reducible. Since a nearly decomposable matrix remains nearly decomposable under arbitrary line permutations, one might suspect that a fully indecomposable submatrix B of a nearly decomposable matrix A is nearly decomposable. This turns out to be false. However, if the submatrix of A which is complementary to B has a nonzero diagonal, then B is nearly decomposable. These two properties of nearly decomposable matrices can be found in Brualdi and Hedrick[1979].

A nearly decomposable $(0,1)$-matrix of order 1 has exactly one 1. A nearly decomposable $(0,1)$-matrix of order 2 has exactly four. Minc[1972] determined the largest number of 1's that a nearly decomposable $(0,1)$-matrix of order n can have. Combining Theorem 4.3.1 with Theorem 3.3.6 we see that $3n - 2$ is an upper bound for the number of 1's in a nearly decomposable matrix of order n. This bound cannot be attained for $n \geq 3$.

Theorem 4.3.5.   *Let A be a nearly decomposable $(0,1)$-matrix of order $n \geq 3$. Then the number of 1's in A is between $2n$ and $3(n-1)$. The number of 1's in A is $2n$ if and only if there are permutation matrices P and Q of order n such that PAQ equals*

$$\begin{bmatrix} 1 & 1 & 0 & \cdots & 0 & 0 \\ 0 & 1 & 1 & \cdots & 0 & 0 \\ 0 & 0 & 1 & \cdots & 0 & 0 \\ \vdots & \vdots & \vdots & \ddots & \vdots & \vdots \\ 0 & 0 & 0 & \cdots & 1 & 1 \\ 1 & 0 & 0 & \cdots & 0 & 1 \end{bmatrix} . \tag{4.14}$$

The number of 1's in A equals $3(n-1)$ if and only if there are permutation matrices P and Q of order n such that PAQ equals the matrix

$$S_n = \begin{bmatrix} 1 & 0 & \cdots & 0 & 1 \\ 0 & 1 & \cdots & 0 & 1 \\ \vdots & \vdots & \ddots & \vdots & \vdots \\ 0 & 0 & \cdots & 1 & 1 \\ 1 & 1 & \cdots & 1 & 0 \end{bmatrix} . \tag{4.15}$$

Proof. The matrices in (4.14) and (4.15) are nearly decomposable, since each of its 1's is contained in a line which has exactly two 1's. Since $n \geq 3$, each line of the nearly decomposable matrix A has at least two 1's. Hence A has at least $2n$ 1's. Assume that A has exactly $2n$ 1's. We permute the lines of A so that there are only 1's on the main diagonal. The permuted A contains exactly one off-diagonal 1 in each line and hence equals $I + R$ for some permutation matrix R of order n. Since A is fully indecomposable, the permutation of $\{1, 2, \ldots, n\}$ derived from R is a cycle of length n. It follows that (4.14) holds for some permutation matrices P and Q of order n.

We next investigate the largest number of 1's in a nearly decomposable matrix A of order $n \geq 3$. We verify the conclusions of the theorem by induction on n. Each nearly decomposable matrix of order 3 is a permuted form of the matrix S_n in (4.15) and thus has exactly 6 1's. [We remark that if $n = 3$, the matrix in (4.14) is a permuted form of the matrix in (4.15).] Now assume that $n > 3$. We use the notation $\sigma(X)$ for the number of 1's in a (0,1)-matrix X. Let P and Q be permutation matrices such that PAQ satisfies the conclusions of Theorem 4.3.4. Then

$$\sigma(A) \leq 2(n - m) + 1 + \sigma(B_1) \tag{4.16}$$

where B_1 is a nearly decomposable matrix of order m. If $m = 1$, then $\sigma(B_1) = 1$ and by (4.16), $\sigma(A) = 2n$ which is strictly less than $3(n-1)$ for $n > 3$. The case $m = 2$ cannot occur in Theorem 4.3.4, so we now assume that $m \geq 3$. By the induction hypothesis, $\sigma(B_1) \leq 3(m - 1)$. Using this inequality in (4.16) we obtain

$$\sigma(A) \leq 2n + m - 2 \leq 2n + (n - 1) - 2 = 3(n - 1).$$

Suppose that $\sigma(A) = 3(n-1)$. Then we must have $m = n - 1$ and $\sigma(B_1) = 3(n - 2)$. By the inductive assumption, the lines of B_1 can be permuted to obtain the matrix S_{n-1}. Thus the lines of A can be permuted to obtain

$$\begin{bmatrix} 1 & F_1 \\ F_2 & S_{n-1} \end{bmatrix} \tag{4.17}$$

where the matrices F_1 and F_2 each contain exactly one 1. Let the 1 of F_2 occur in row i of F_2, and let the 1 of F_1 occur in column j of F_1. If $i = j = n - 1$, then (4.17) is the same as (4.15). We now assume that it is not the case that $i = j = n - 1$. It now follows from Theorem 4.3.4 that $i \neq j$ and that $i \neq n - 1$ and $j \neq n - 1$. We apply additional line permutations to A and assume that $i = 1$ and $j = 2$. The matrix in (4.17) can be repartitioned to give

$$\begin{bmatrix} 1 & 0 & 1 & 0 & \cdots & 0 & 0 \\ 1 & 1 & 0 & 0 & \cdots & 0 & 1 \\ 1 & 0 & & & & & \\ 0 & 0 & & & & & \\ \vdots & \vdots & & & S_{n-2} & & \\ 0 & 0 & & & & & \\ 0 & 1 & & & & & \end{bmatrix} . \tag{4.18}$$

If $n > 4$, the matrix obtained from (4.18) by replacing the 1 in position $(2, n)$ with a 0 is fully indecomposable by Lemma 4.3.2. Hence we must have $n = 4$. But in this case (4.18) is a permuted form of (4.15).   □

Lovász and Plummer[1977] call a bipartite graph *elementary* provided it is connected and each edge is contained in a perfect matching. A *minimal elementary bipartite graph* is one such that the removal of any edge results in a bipartite graph which is not elementary. By Theorem 4.2.7 a bipartite graph is elementary if and only if its reduced adjacency matrix is fully indecomposable. The reduced adjacency matrix of a minimal elementary bipartite graph is a nearly decomposable matrix. Estimates for the number of lines in a nearly decomposable matrix which have exactly two 1's follow from their investigations.

Exercises

1. For each integer $n \geq 3$ give an example of a nearly reducible matrix A of order n such that $A + I$ is not nearly decomposable.
2. Let $n \geq 3$ and k be integers with $2n \leq k \leq 3(n-1)$. Prove that there exists a nearly decomposable (0,1)-matrix of order n with exactly k 1's (Brualdi and Hedrick[1979].
3. Give an example to show that a fully indecomposable submatrix of a nearly decomposable matrix need not be nearly decomposable (Brualdi and Hedrick [1979]).
4. Let A be a nearly decomposable (0,1)-matrix which is partitioned as

$$A = \left[\begin{array}{cc} A_1 & B_1 \\ B_2 & A_2 \end{array} \right]$$

where A_1 is a square matrix with 1's everywhere on its main diagonal and A_2 is a fully indecomposable matrix. Prove that A_2 is a nearly decomposable matrix (Brualdi and Hedrick[1979]).

References

R.A. Brualdi and M.B. Hedrick[1979], A unified treatment of nearly reducible and nearly decomposable matrices, *Linear Alg. Applics.*, 24, pp. 51–73.
D.J. Hartfiel[1970], A simplified form for nearly reducible and nearly decomposable matrices, *Proc. Amer. Math. Soc.*, 24, pp. 388–393.
    [1971], On constructing nearly decomposable matrices, *Proc. Amer. Math. Soc.*, 27, pp. 222–228.
L. Lovász and M.D. Plummer[1977], On minimal elementary bipartite graphs, *J. Combin. Theory, Ser. A*, 23, pp. 127–138.
H. Minc[1969], Nearly decomposable matrices, *Linear Alg. Applics.*, 5, pp. 181–187.
R. Sinkhorn and P. Knopp[1969], Problems involving diagonal products in non-negative matrices, *Trans. Amer. Math. Soc.*, 136, pp. 67–75.

4.4 Decomposition Theorems

Let A be an m by n matrix, and let $\mathcal{P}$ denote a class of matrices. By a *decomposition theorem* we mean a theorem which asserts that there is an expression for A of the form

$$A = P_1 + P_2 + \cdots + P_k + X \qquad (4.19)$$

where the matrices $P_1, P_2, \ldots, P_k$ are in the class $\mathcal{P}$. We may require X to be restricted in some way, perhaps equal to a zero matrix. The purpose of the theorem may be to maximize k in (4.19) or to minimize k in the event that X is required to be a zero matrix, or the purpose may be to maximize or to minimize some other quantity that can be associated with the decomposition (4.19).

The König theorem (Theorem 1.2.1) can be viewed as a decomposition theorem. Recall that an m by n (0,1)-matrix P is a *subpermutation matrix of rank r* (an *r-subpermutation matrix*) provided P has exactly r 1's and no two 1's of P are on the same line. Let A be an m by n (0,1)-matrix. Then the König theorem asserts that *A can be expressed in the form*

$$A = P + X$$

where P is an r-subpermutation matrix and X is a $(0,1)$-matrix if and only if A does not have a line cover consisting of fewer than r lines. The theorem of Vizing (Theorem 2.6.2) is a decomposition theorem for symmetric (0,1)-matrices in which the P_i are required to be symmetric permutation matrices and X is required to be a zero matrix.

Let A be an m by n (0,1)-matrix having no zero lines. We define the *co-term rank* of A to be the minimal number of 1's in A with the property that each line of A contains at least one of these 1's. We denote this basic invariant by $\rho^*(A)$ and derive the following basic relationship.

Theorem 4.4.1. *Let A be an m by n $(0,1)$-matrix having no zero lines. Then the co-term rank $\rho^*(A)$ equals*

$$\max\{r + s\} \qquad (4.20)$$

where the maximum is taken over all r by s (possibly vacuous) zero submatrices of A with $0 \le r \le m$ and $0 \le s \le n$.

Proof. Suppose that r and s are nonnegative integers for which A has an r by s zero submatrix. We permute the lines of A to bring A to the form

$$
\begin{array}{c}
m - r \\
r
\end{array}
\begin{array}{cc}
n-s \quad s \\
\left[\begin{array}{cc} A_3 & A_1 \\ A_2 & O \end{array} \right] .
\end{array}
$$

Clearly, $\rho^*(A) \geq r + s$. We now assume that r and s are integers for which the maximum occurs in (4.20) and verify the reverse inequality. The first $m - r$ rows and the first $n - s$ columns of the permuted A form a line cover with the fewest number of lines. Hence by Theorem 1.2.1

$$\rho(A) = m - r + n - s = m + n - (r + s). \qquad (4.21)$$

In addition it follows from Theorem 1.2.1 that $\rho(A_1) = m - r$ and $\rho(A_2) = n - s$. We select $m - r$ 1's of A_1 with no two of the 1's in the same line and then $s - (m - r)$ additional 1's one from each of the remaining columns of A_1. We also select $n - s$ 1's of A_2 with no two of the 1's from the same line and then $r - (n - s)$ additional 1's one from each of the remaining rows of A_2. We obtain in this way a total $r + s$ 1's of A with the property that each line of A contains at least one of these 1's. Therefore $\rho^*(A) \leq r + s$ and hence we have equality. $\square$

We remark that for an m by n (0,1)-matrix A with no zero lines, it follows from equation (4.21) above that the two basic invariants are related by the equation

$$\rho(A) + \rho^*(A) = m + n. \qquad (4.22)$$

Stated as a decomposition theorem, Theorem 4.4.1 asserts: The m by n (0,1)-matrix A with no zero lines can be expressed in the form

$$A = Q + Y$$

where Q is a (0,1)-matrix with no zero lines and with at most t 1's if and only if A does not have an r by s zero submatrix with $r + s > t$.

Every (0,1)-matrix A has a decomposition (4.19) in which $P_1, P_2, \ldots, P_k$ are subpermutation matrices and $X = O$. We may, for instance, choose the P_i to have rank 1 and the integer k to be the number of 1's of A. It is natural to ask for the smallest k for which a decomposition of A into subpermutation matrices exists. Now assume that A has no zero lines. Then A has a decomposition (4.19) in which the matrices $P_1, P_2, \ldots, P_k$ have no zero lines and $X = O$. We may, for instance, choose $k = 1$ and $P_1 = A$. It is also natural to ask for the largest k for which A admits a decomposition into matrices with no zero lines. Both of these questions can be answered by appealing to the following theorem of Gupta[1967, 1974, 1978]. This theorem is stated in terms of nonnegative integral matrices, but its proof is more conveniently expressed in terms of bipartite multigraphs.

Theorem 4.4.2.   *Let A be an m by n nonnegative integral matrix with row sums $r_1, r_2, \ldots, r_m$ and column sums $s_1, s_2, \ldots, s_n$. Let k be a positive integer. Then A has a decomposition of the form*

$$A = P_1 + P_2 + \cdots + P_k \qquad (4.23)$$

where $P_1, P_2, \ldots, P_k$ are nonnegative integral matrices satisfying the following two properties:

(i) *The number of P_t's which have a positive element in row i equals* $\min\{k, r_i\}, (i = 1, 2, \ldots, m)$.

(ii) *The number of P_t's which have a positive element in column j equals* $\min\{k, s_j\}, (j = 1, 2, \ldots, n)$.

Proof. The matrix $A = [a_{ij}], (i = 1, 2, \ldots, m; j = 1, 2, \ldots, n)$ is the reduced adjacency matrix of a bipartite multigraph G with bipartition $\{X, Y\}$ where $X = \{x_1, x_2, \ldots, x_m\}$ and $Y = \{y_1, y_2, \ldots, y_n\}$. The edge $\{x_i, y_j\}$ has multiplicity a_{ij}. The degree of the vertex x_i equals r_i and the degree of the vertex y_j equals s_j. Let σ be a function which assigns to each edge of G an integer k from the set $\{1, 2, \ldots, k\}$. As in the proof of Theorem 2.6.2 we think of σ as assigning a *color* to each edge of G from a set $\{1, 2, \ldots, k\}$ of k colors. Adjacent edges of G need not be assigned different colors by σ, nor are the a_{ij} edges of the form $\{x_i, y_j\}$ required to have identical colors.

For each integer $r = 1, 2, \ldots, k$, we define a bipartite multigraph $G_r(\sigma)$ with bipartition $\{X, Y\}$. The multiplicity of the edge $\{x_i, y_j\}$ in $G_r(\sigma)$ equals the number of edges of G of the form $\{x_i, y_j\}$ which are assigned color r by $\sigma, (i = 1, 2, \ldots, m; j = 1, 2, \ldots, n)$. Let $P_r = A_r(\sigma)$ be the reduced adjacency matrix of $G_r(\sigma)$. Then $A = P_1 + P_2 + \cdots + P_r$. The properties (i) and (ii) in the theorem hold for this decomposition of A if and only if σ assigns $\min\{k, r_i\}$ distinct colors to the edges of G incident to vertex $x_i, (i = 1, 2, \ldots, m)$ and $\min\{k, s_j\}$ distinct colors to the edges incident to vertex $y_j, (j = 1, 2, \ldots, n)$.

For each vertex z of G let $f(z, \sigma)$ be the number of distinct colors assigned by σ to the edges of G which are incident to z. We have

$$f(z, \sigma) \leq \min\{k, \text{degree of } z\}, (z \in X \cup Y). \tag{4.24}$$

We now assume that σ has been chosen so that

$$\sum_{z \in X \cup Y} f(z, \sigma)$$

is as large as possible. We show that for this choice of σ equality holds throughout (4.24). Assume that this were not the case. Without loss of generality we may assume that for a vertex $z = x_{i_0}$ we have

$$f(x_{i_0}, \sigma) < \min\{k, r_{i_0}\}. \tag{4.25}$$

It follows that some color p is assigned to two or more edges incident to x_{i_0} while some color q is assigned to no edge incident to x_{i_0}. Starting at vertex x_{i_0} we determine a walk

$$x_{i_0} \rightarrow y_{i_0} \rightarrow x_{i_1} \rightarrow y_{i_1} \rightarrow x_{i_2} \rightarrow \cdots \tag{4.26}$$

in which the edges $\{x_{i_0}, y_{i_0}\}, \{x_{i_1}, y_{i_1}\}, \cdots$ are assigned color p and the edges $\{y_{i_0}, x_{i_1}\}, \{y_{i_1}, x_{i_2}\}, \cdots$ are assigned color q. The walk (4.26) continues until one of the following occurs:

(a) a vertex is reached which is incident to another edge of the same assigned color as the incoming edge.

(b) a vertex is reached which is not incident with any edge of one of the two colors p and q.

Since G is a bipartite multigraph and since there is no edge incident to x_{i_0} which is assigned color q, the vertex x_{i_0} occurs exactly once on the walk (4.26). We now reassign colors to the edges that occur on the walk (4.26) by interchanging the two colors p and q. Let τ be the coloring of the edges of G thus obtained. It follows that

$$f(x_{i_0}, \tau) = f(x_{i_0}, \sigma) + 1,$$

and

$$f(z, \tau) \geq f(z, \sigma)$$

for all vertices z of G. But then

$$\sum_{z \in X \cup Y} f(z, \tau) > \sum_{z \in X \cup Y} f(z, \sigma),$$

contradicting our choice of σ. Hence equality holds throughout (4.24). □

There are two special cases of Theorem 4.4.2 which are of particular interest. The first of these is another theorem of König[1936].

Theorem 4.4.3. *Let A be an m by n nonnegative integral matrix with maximal line sum equal to k. Then A has a decomposition of the form*

$$A = P_1 + P_2 + \cdots + P_k \tag{4.27}$$

where $P_1, P_2, \ldots, P_k$ are m by n subpermutation matrices.

Proof. Let $r_1, r_2, \ldots, r_m$ be the row sums of A and let $s_1, s_2, \ldots, s_n$ be the column sums. We apply Theorem 4.4.2 to A with k equal to the maximum line sum of A. We obtain a decomposition (4.23) in which the P_i are nonnegative integral matrices satisfying (i) and (ii) in Theorem 4.4.2. For the chosen k, we have $\min\{k, r_i\} = r_i, (i = 1, 2, \ldots, m)$ and $\min\{k, s_j\} = s_j, (j = 1, 2, \ldots, n)$. It follows that $P_1, P_2, \ldots, P_k$ are subpermutation matrices. □

Corollary 4.4.4. *Let A be an m by n nonnegative integral matrix with maximum line sum equal to k and let r be an integer with $0 < r < k$. Then A has a decomposition $A = A_1 + A_2$ where A_1 is a nonnegative integral matrix* *with maximum line sum equal to r, and A_2 is a nonnegative integral matrix with maximum line sum equal to $k - r$.*

Proof. By Theorem 4.4.3 there is a decomposition (4.27) of A into sub-permutation matrices. We now choose $A_1 = P_1 + P_2 + \cdots + P_r$ and $A_2 = A - A_1$.   □

Corollary 4.4.5.   *Let A be an m by n nonnegative integral matrix each of whose line sums equals k or $k-1$. Let r be an integer with $1 \leq r < k$. Then A has a decomposition $A = A_1 + A_2$ into nonnegative integral matrices A_1 and A_2 where each line sum of A_1 equals r or $r - 1$ and each line sum of A_2 equals $k - r$ or $k - r + 1$.*

Proof. We again use the decomposition (4.27). Since the line sums of A equal k or $k - 1$, at most one of $P_1, P_2, \ldots, P_k$ does not have a 1 in any specified line. The result now follows as in the preceding corollary.   □

There are analogues of the preceding two corollaries for symmetric matrices. The following theorem of Thomassen[1981] is a slight extension of a theorem of Tutte[1978].

Theorem 4.4.6.   *Let A be a symmetric nonnegative integral matrix of order n each of whose line sums equals k or $k - 1$. Let r be an integer with $1 \leq r < k$. Then A has a decomposition $A = A_1 + A_2$ into symmetric nonnegative integral matrices A_1 and A_2 where each line sum of A_1 equals r or $r - 1$.*

Proof. It suffices to prove the theorem in the case that $r = k - 1$. Let p be the number of rows of A which sum to k. We assume that $p > 0$, for otherwise we may choose $A_1 = A$. We simultaneously permute the lines of $A = [a_{ij}]$ and assume that A has the form

$$\begin{bmatrix} C & B \\ B^T & D \end{bmatrix},$$

where C is a matrix of order p and D is a matrix of order $n - p$, and the first p rows of A are the rows which sum to k. Suppose that $a_{ij} \neq 0$ for some i and j with $1 \leq i, j \leq p$. Then we may subtract 1 from a_{ij} and, in the case $i \neq j$, 1 from a_{ji} and argue on the resulting matrix. Thus we may assume that $C = O$. Then the matrix B has all row sums equal to k and all column sums at most equal to k. It follows from Theorem 1.2.1 (or from Theorem 4.4.3) that B has a decomposition $B = P + B_1$ where P is a subpermutation matrix of rank p. We now let

$$A_2 = \begin{bmatrix} O & P \\ P^T & D \end{bmatrix},$$

and $A_1 = A - A_2$.   □

The following corollary is due to Lovász[1970].

Corollary 4.4.7. *Let B be a symmetric nonnegative integral matrix of order n with maximum line sum at most equal to $r + s - 1$ where r and s are positive integers. Then there is a decomposition $A = A_1 + A_2$ where A_1 and A_2 are symmetric nonnegative integral matrices with maximum line sum at most equal to r and s, respectively.*

Proof. We increase some of the diagonal elements of B and obtain a symmetric nonnegative integral matrix A each of whose line sums equals $r + s - 1$ and apply Theorem 4.4.6 with $k = r + s - 1$.   □

The following theorem was discovered by Gupta[1967, 1974, 1978] and Fulkerson[1968].

Theorem 4.4.8. *Let A be a nonnegative integral matrix with minimum line sum equal to a positive integer k. Then A has a decomposition of the form*

$$A = P_1 + P_2 + \cdots + P_k$$

where $P_1, P_2, \ldots, P_k$ are m by n nonnegative integral matrices each of whose line sums is positive.

Proof. As in the proof of Theorem 4.4.3 we apply Theorem 4.4.2 but this time with k equal to the minimum line sum of A. In order for (i) and (ii) of Theorem 4.4.2 to be satisfied each of the line sums of the matrices $P_1, P_2 \ldots, P_k$ in (4.23) must be positive.   □

We now turn to a decomposition theorem of a different type. Theorem 4.1.3 asserts that in a decomposition of the complete graph K_n of order n into complete bipartite subgraphs, the number of complete bipartite graphs is at least $n - 1$. This theorem, as was done in its proof, can be viewed as a decomposition theorem for the matrix $J - I$ of order n. We now prove a general theorem of Graham and Pollak[1971, 1973] which gives a lower bound for the number of bipartite graphs in a decomposition of a multigraph.

Theorem 4.4.9. *Let G be a multigraph of order n and let $G_1, G_2, \ldots G_r$ be a decomposition of G into complete bipartite subgraphs. Let $A = [a_{ij}]$, $(i, j = 1, 2, \ldots, n)$ be the adjacency matrix of G and let n_+ be the number of positive eigenvalues of A and let n_- be the number of negative eigenvalues. Then $r \geq \max\{n_+, n_-\}$.*

Proof. Let the set of vertices of G be $V = \{a_1, a_2, \ldots, a_n\}$. A complete bipartite subgraph of G is obtained by specifying two nonempty subsets X and Y of V for which $\{x, y\}$ is an edge of G for each x in X and each y in Y.

The pair $\{X, Y\}$ is the bipartition of the subgraph and since G has no loops, the sets X and Y are disjoint. Let $\{X_i, Y_i\}$ be the bipartition of the complete bipartite graph G_i in the decomposition of G ($i = 1, 2, \ldots, r$).

Let $z_1, z_2, \ldots, z_n$ be n indeterminates and let $z = (z_1, z_2, \ldots, z_n)^T$. We consider the quadratic form

$$q(z) = z^T A z = 2 \sum_{1 \leq i < j \leq n} a_{ij} z_i z_j.$$

With each of the bipartite graphs G_i in the decomposition of G we associate the quadratic form

$$q_i(z) = q_i(z_1, z_2, \ldots, z_n) = \left(\sum_{\{k : a_k \in X_i\}} z_k \right) \left(\sum_{\{l : a_l \in Y_i\}} z_l \right).$$

Since $G_1, G_2, \ldots, G_r$ is a decomposition of G we have

$$q(z) = z^T A z = 2 \sum_{i=1}^{r} q_i(z). \tag{4.28}$$

We apply the elementary algebraic identity

$$ab = \frac{1}{4}((a + b)^2 - (a - b)^2)$$

to $q_i(z)$ and obtain from (4.28)

$$q(z) = z^T A z = \frac{1}{2} \left(\sum_{i=1}^{r} l_i'(z)^2 - \sum_{i=1}^{r} l_i''(z)^2 \right), \tag{4.29}$$

where the $l_i'(z)$ and the $l_i''(z)$ are linear forms in $z_1, z_2, \ldots, z_n$. The linear forms $l_1'(z), l_2'(z), \ldots, l_r'(z)$ vanish on a subspace W of dimension at least $n - r$ of real n-space. Hence the quadratic form $q(z)$ is negative semi-definite on W. Let E^+ be the linear space of dimension n_+ spanned by the eigenvectors of A corresponding to its positive eigenvalues. Then $q(z)$ is positive definite on E^+. It follows that

$$(n - r) + n_+ = \dim W + \dim E^+ \leq n$$

and hence $r \geq n_+$. One concludes in a similar way that $r \geq n_-$.   □

Alon, Brualdi and Shader[1991] have extended Theorem 4.4.9 by proving that every bipartite decomposition of the graph G has an additional special property. We state this theorem without proof below. Given a coloring of a graph G, we say that a subgraph of G is *multicolored* provided that no two of its edges have the same color. A bipartite decomposition of G corresponds to an edge coloring in which two edges receive the same color if and only if they belong to the same bipartite graph of the decomposition.

Theorem 4.4.10.   *Let G be a graph of order n and suppose the adjacency matrix of G has n_+ positive eigenvalues and n_- negative eigenvalues. Then in any decomposition of G into complete bipartite graphs there is a multicolored forest with at least $\max\{n_+, n_-\}$ edges. In any decomposition of K_n into complete bipartite graphs there is a multicolored spanning tree.*

Let G be the complete graph K_n of order n. The adjacency matrix $A = J - I$ has $n - 1$ negative eigenvalues. Hence Theorem 4.1.3, which asserts that the complete graph of order n cannot be decomposed into fewer than $n - 1$ complete bipartite subgraphs, is a special case of Theorem 4.4.9.

If n is even, the complete graph K_n can be decomposed into $n - 1$ complete bipartite subgraphs, each of which is isomorphic to $K_{1,n/2}$. If n is odd, K_n cannot have a decomposition into $n - 1$ isomorphic complete bipartite subgraphs $K_{1,m}$ for any positive integer m. The following theorem of de Caen and Hoffman[1989], stated without proof, asserts that other decompositions of K_n into $n - 1$ isomorphic complete bipartite subgraphs are impossible.

Theorem 4.4.11.   *Let n be a positive integer. If r and s are integers with $r, s \geq 2$, then there does not exist a decomposition of K_n into complete bipartite subgraphs each of which is isomorphic to the complete bipartite graph $K_{r,s}$.*

Theorem 4.4.9 has implications for an addressing problem in graphs. We refer to the papers of Graham and Pollak cited above and to Winkler[1983] and van Lint[1985].

Let $K_{n,n}^*$ be the bipartite graph of order $2n$ which is obtained from the complete bipartite graph $K_{n,n}$ by removing the edges of a perfect matching. The reduced adjacency matrix of $K_{n,n}^*$ is the matrix $J - I$ of order n. The graph $K_{n,n}^*$ can be decomposed into n complete bipartite subgraphs each of which is isomorphic to $K_{1,n-1}$. The following theorem of de Caen and Gregory [1987] asserts that there are no decompositions with fewer than n complete bipartite graphs.

Theorem 4.4.12.   *Let $n \geq 2$. Let the bipartite graph $K_{n,n}^*$ of order $2n$ have a decomposition $G_1, G_2, \ldots, G_r$ into complete bipartite subgraphs. Then $r \geq n$. If $r = n$, then there exist positive integers p and q such that $pq = n - 1$ and each G_i is isomorphic to $K_{p,q}$.*

Proof. Let $\{X, Y\}$ be a bipartition of $K_{n,n}^*$ where $X = \{x_1, x_2, \ldots, x_n\}$ and $Y = \{y_1, y_2, \ldots, y_n\}$. Each of the bipartite subgraphs G_i has a bipartition $\{X_i, Y_i\}$. Let G_i' be the spanning subgraph of $K_{n,n}^*$ with the same set of edges as G_i and let A_i be the reduced adjacency matrix of $G'_i, (i = 1, 2, \ldots, r)$. The hypothesis of the theorem implies that

$$J - I = A_1 + A_2 + \cdots + A_r. \tag{4.30}$$

Let $\widehat{X}_i$ be the (0,1)-matrix of size n by 1 whose kth component equals 1 if and only if x_k is in $X_i, (k = 1, 2, \ldots, n)$. Let $\widehat{Y}_i$ be the (0,1)-matrix of size 1 by n whose kth component is 1 if and only if y_k is in $Y_i, (k = 1, 2, \ldots, n)$. We have $A_i = \widehat{X}_i \widehat{Y}_i, (i = 1, 2, \ldots, r)$. We define an n by r (0,1)-matrix by

$$\widehat{X} = \begin{bmatrix} \widehat{X}_1 & \widehat{X}_2 & \cdots & \widehat{X}_r \end{bmatrix}$$

and we define an r by n (0,1)-matrix by

$$\widehat{Y} = \begin{bmatrix} \widehat{Y}_1 \\ \widehat{Y}_2 \\ \vdots \\ \widehat{Y}_n \end{bmatrix}.$$

By (4.30) we have

$$J - I = \widehat{X}\widehat{Y}. \tag{4.31}$$

From equation (4.31) we conclude that a decomposition of $K_{n,n}^*$ into r complete bipartite subgraphs is equivalent to a factorization of $J - I$ into two (0,1)-matrices of sizes n by r and r by n, respectively. The matrix $J - I$ has rank equal to n, and the ranks of $\widehat{X}$ and $\widehat{Y}$ cannot exceed r. Hence it follows from (4.31) that

$$n = \operatorname{rank}(J - I) \leq r.$$

We now assume that $r = n$. Since the elements on the main diagonal of $J - I$ equal 0, we have $\widehat{Y}_i^T \widehat{X}_i = 0, (i = 1, 2, \ldots, n)$. Let i and j be distinct integers between 1 and n. Let U be the n by $n - 1$ matrix obtained from $\widehat{X}$ by deleting columns i and j and appending a column of 1's as a new first column. Let V be the $n - 1$ by n matrix obtained from $\widehat{Y}$ by deleting rows i and j and appending a row of 1's as a new first row. Then

$$UV = I + \widehat{X}_i \widehat{Y}_i + \widehat{X}_j \widehat{Y}_j$$

is a singular matrix. Let

$$U' = \begin{bmatrix} \widehat{X}_i & \widehat{X}_j \end{bmatrix}$$

and

$$V' = \left[\begin{array}{c} \widehat{Y}_i \\ \widehat{Y}_j \end{array} \right].$$

Using these equations and taking determinants we obtain

$$0 = \det(UV) = \det(I + U'V') = \det(I_2 + V'U') = 1 - (\widehat{Y}_i\widehat{X}_j)(\widehat{Y}_j\widehat{X}_i),$$

where I_2 denotes the identity matrix of order 2. Hence $\widehat{Y}_i\widehat{X}_j = \widehat{Y}_j\widehat{X}_i = 1$ for all i and j with $i \neq j$. Thus in the case $r = n$ the equation (4.31) implies that

$$\widehat{Y}\widehat{X} = J - I. \tag{4.32}$$

The two equations (4.31) and (4.32) imply that both X and Y commute with J, and hence $\widehat{X}$ and $\widehat{Y}$ each have constant line sums. There exist integers p and q such that

$$XJ = JX = pJ \qquad \text{and} \qquad YJ = JY = qJ.$$

Now

$$(n - 1)J = (J - I)J = (\widehat{X}\widehat{Y})J = \widehat{X}(\widehat{Y}J) = \widehat{X}(qJ) = pqJ.$$

Thus $pq = n - 1$ and the theorem now follows.   □

In Chapter 6 we shall obtain additional decomposition theorems.

Exercises

1. Let A be an m by n (0,1)-matrix with no zero lines. Prove that $\rho(A) \leq \rho^*(A)$. Investigate the case of equality.
2. Determine the largest co-term rank possible for an m by n (0,1)-matrix with no zero lines and characterize those matrices for which equality holds.
3. Let D be a digraph with no isolated vertices. A *matching* of D is a collection of pairwise vertex-disjoint directed chains and cycles of D. A *cover* of D is a collection of arcs which meet all vertices of D. Let $\chi(D)$ denote the minimum number of matchings into which the arcs of D can be partitioned, and let $\kappa(D)$ denote the maximum number of covers of D into which the arcs of D can be partitioned. Prove that $\chi(D)$ equals the smallest number s such that both the indegree and outdegree of each vertex of D are at most s, and that $\kappa(D)$ equals the largest number t such that both the indegree and outdegree of each vertex of D are at least t (Gupta[1978]).
4. Let A be a nonnegative integral matrix of order n each of whose line sums equals k. Theorem 4.4.4 asserts that A can be written as a sum of permutation matrices of order n. Prove this special case of Theorem 4.4.4 using Theorem 1.2.1. In Exercises 5 and 6 we refer to this special case of Theorem 4.4.4 as the *regular* case.

5. Prove Theorem 4.4.4 from the regular case as follows: Let A be an m by n nonnegative integral matrix with maximal line sum equal to k. Assume that $m \geq n$. Extend A to a matrix B of order m by including $m - n$ additional columns of 0's. Now increase the elements of B by integer values in order to obtain a nonnegative integral matrix B' of order m each of whose line sums equals k. Apply the regular case of Theorem 4.4.4 to B' and deduce that A is the sum of k subpermutation matrices. Give an example to show that for a given choice of B' not every decomposition of A as a sum of k subpermutation matrices arises in this way (Brualdi and Csima[1991]).

6. Prove Theorem 4.4.4 from the regular case as follows: Let A be an m by n nonnegative integral matrix with maximal line sum equal to k. Let the row sums of A be $r_1, r_2, \ldots, r_m$ and let the column sums of A be $s_1, s_2, \ldots, s_n$. Let A' be the matrix of order $m + n$ defined by

$$\begin{bmatrix} D_1 & A \\ A^T & D_2 \end{bmatrix},$$

where D_1 is the diagonal matrix of order m with diagonal elements $k - r_1, k - r_2, \ldots, k - r_m$ and D_2 is the diagonal matrix of order n with diagonal elements $k - s_1, k - s_2, \ldots, k - s_n$. Apply the regular case of Theorem 4.4.4 to A' and deduce that A is the sum of k subpermutation matrices. Show that every decomposition of A as a sum of k subpermutation matrices arises in this way (Brualdi and Csima[1991]).

7. Let G be the graph of order n which is obtained from the complete graph K_n by removing an edge. Determine the smallest number of complete bipartite graphs into which the edges of G can be partitioned.

8. Let m and n be positive integers with $m \leq n$. Let G be the graph of order n obtained from the complete graph K_n by removing the edges of a complete graph K_m. Prove that the smallest number of complete bipartite graphs into which the edges of G can be partitioned equals $n - m$ (Jones, Lundgren, Pullman and Rees[1988]).

References

N. Alon, R.A. Brualdi and B.L. Shader[1991], Multicolored trees in bipartite decompositions of graphs, *J. Combin. Theory, Ser. B*, to be published.

R.A. Brualdi and J. Csima[1991], Butterfly embedding proof of a theorem of König, *Amer. Math. Monthly*, to be published.

D. de Caen and D. Gregory[1987], On the decomposition of a directed graph into complete bipartite subgraphs, *Ars Combinatoria*, 23B, pp. 139–146.

D. de Caen and D.G. Hoffman[1989], Impossibility of decomposing the complete graph on n points into $n - 1$ isomorphic complete bipartite graphs, *SIAM J. Disc. Math*, 2, pp. 48–50.

D.R. Fulkerson[1970], Blocking Polyhedra, *Graph Theory and Its Applications* (B. Harris, ed.), Academic Press, New York, pp. 93–111.

R.L. Graham and H.O. Pollak[1971], On the addressing problem for loop switching, *Bell System Tech. J.*, 50, pp. 2495–2519.

[1973], On embedding graphs in squashed cubes, *Lecture Notes in Math.*, vol. 303, Springer-Verlag, New York, pp. 99–110.

R.P. Gupta[1967], A decomposition theorem for bipartite graphs, *Theorie des Graphes Rome I.C.C.* (P. Rosenstiehl, ed.), Dunod, Paris, pp. 135–138.

[1974], On decompositions of a multigraph with spanning subgraphs, *Bull. Amer. Math. Soc.*, 80, pp. 500–502.

[1978], An edge-coloration theorem for bipartite graphs with applications, *Discrete Math.*, 23, pp. 229–233.

K. Jones, J.R. Lundgren, N.J. Pullman and R. Rees[1988], A note on the covering numbers of $K_n - K_m$ and complete t-partite graphs, *Congressus Num.*, 66, pp. 181–184.

D. König[1936], *Theorie der endlichen und unendlichen Graphen*, Leipzig. Reprinted [1950], Chelsea, New York.

J.H. van Lint[1985], $(0, 1, *)$ distance problems in Combinatorics, *Surveys in Combinatorics* (I. Anderson, ed.), London Math. Soc. Lecture Notes 103, Cambridge University Press, pp. 113–135.

L. Lovász[1970], Subgraphs with prescribed valencies, *J. Combin. Theory*, 8, pp. 391–416.

J. Orlin[1977], Contentment in graph theory: covering graphs with cliques, *Indag. Math.*, 39, pp. 406–424.

C. Thomassen[1981], A remark on the factor theorems of Lovász and Tutte, *J. Graph Theory*, 5, pp. 441–442.

W.T. Tutte[1978], The subgraph problem, *Discrete Math.*, 3, pp. 289–295.

4.5 Diagonal Structure of a Matrix

Let $A = [a_{ij}], (i, j = 1, 2, \ldots, n)$ be a (0,1)-matrix of order n. A nonzero diagonal of A, as defined in section 4.2, is a collection of n 1's of A with no two of the 1's on a line. More formally, a *nonzero diagonal* of A is a set

$$D = \{(1, j_1), (2, j_2), \ldots, (n, j_n)\} \tag{4.33}$$

of n positions of A for which $(j_1, j_2, \ldots, j_n)$ is a permutation of the set $\{1, 2, \ldots, n\}$ and $a_{1j_1} = a_{2j_2} = \cdots = a_{n,j_n} = 1$. Let $\mathcal{D} = \mathcal{D}(A)$ be the set of all nonzero diagonals of A. The cardinality of the set $\mathcal{D}$ equals the permanent of A. In this section we are concerned with some basic properties of the *diagonal structure* $\mathcal{D}$ of A. Let X be the set of positions of A which contain 1's. The pair $(X, \mathcal{D})$ is called the *diagonal hypergraph* of A.[2]

Those positions of A containing 1's which do not belong to any nonzero diagonal are of no importance for the diagonal structure of A. Thus throughout this section we assume that each position in X is contained in a nonzero diagonal, that is, the matrix A has total support. Moreover, we implicitly assume that A is not a zero matrix.

Let B be another (0,1)-matrix of order n with total support, and let Y be the set of positions of B which contain 1's. An *isomorphism of the diagonal hypergraphs* $(X, \mathcal{D}(A))$ and $(Y, \mathcal{D}(B))$ is a bijection $\phi : X \to Y$

---

[2] In the terminology of hypergraphs (see Berge[1973]), the elements of X are *vertices* and the elements of $\mathcal{D}$ are *hyperedges*. Thus a graph is a hypergraph in which all hyperedges have cardinality equal to two. All the hyperedges of $\mathcal{D}$ have cardinality equal to n. In general, a hypergraph may have edges of different cardinalities.

with the property that for each $D \subseteq X$, D is a nonzero diagonal of A if and only if $\phi(D)$ is a nonzero diagonal of B.

A collection of 1's of the matrix A with the property that no two of the 1's belong to the same nonzero diagonal of A is called *strongly stable*.[3] More formally, a set

$$S = \{(i_1, k_1), (i_2, k_2), \ldots, (i_t, k_t)\}$$

is a *strongly stable set* of A provided $a_{i_1 k_1} = a_{i_2 k_2} = \cdots = a_{i_t k_t} = 1$ and each nonzero diagonal D of A has at most one element in common with S. The collection of strongly stable sets of A is denoted by $\mathcal{S} = \mathcal{S}(A)$, and the pair $(X, \mathcal{S})$ is the *strongly stable hypergraph* of A. An isomorphism of the strongly stable hypergraphs of two matrices is defined very much like an isomorphism of diagonal hypergraphs.

By its very definition the strongly stable hypergraph of a matrix is determined by its diagonal hypergraph. It follows from the following theorem of Brualdi and Ross[1981] that the diagonal hypergraph is determined by the strongly stable hypergraph.

Theorem 4.5.1.   *Let A and B be $(0,1)$-matrices of order n with total support. Let X be the set of positions of A which contain 1's and let Y be the set of positions of B that contain 1's. Let $\phi : X \to Y$ be a bijection. Then ϕ is an isomorphism of the diagonal hypergraphs $(X, \mathcal{D}(A))$ and $(Y, \mathcal{D}(B))$ if and only if ϕ is an isomorphism of the strongly stable hypergraphs $(X, \mathcal{S}(A))$ and $(Y, \mathcal{S}(B))$.*

Proof. First assume that ϕ is an isomorphism of $(X, \mathcal{D}(A))$ and $(Y, \mathcal{D}(B))$. Let F be a subset of X and let D be a nonzero diagonal of A. Then $|D \cap F| \leq 1$ if and only if $|\phi(D) \cap \phi(F)| \leq 1$. It follows that F is a strongly stable set of A if and only if $\phi(A)$ is a strongly stable set of B. Hence ϕ is an isomorphism of $(X, \mathcal{S}(A))$ and $(Y, \mathcal{S}(B))$.

Now assume that ϕ is an isomorphism of $(X, \mathcal{S}(A))$ and $(Y, \mathcal{S}(B))$. Let D be a nonzero diagonal of A. Suppose that $\phi(D)$ is not a nonzero diagonal of B. Since $\phi(D)$ is a set of n positions of B containing 1's, there are distinct positions p_1 and p_2 in $\phi(D)$ which belong to the same line of B. The set $\{p_1, p_2\}$ is a strongly stable set of B, but since $\{\phi^{-1}(p_1), \phi^{-1}(p_2)\}$ is a subset of D, $\{\phi^{-1}(p_1), \phi^{-1}(p_2)\}$ is not a strongly stable set of A. This contradicts the assumption that ϕ is an isomorphism of the strongly stable hypergraphs of A and B. Hence $\phi(D)$ is a nonzero diagonal of B. In a similar way one proves that if D' is a nonzero diagonal of B then $\phi^{-1}(D')$ is a nonzero diagonal of A. Thus ϕ is an isomorphism of the diagonal hypergraphs of A and B. $\square$

---

[3] This terminology comes from the theory of hypergraphs. A *strongly stable set* of a hypergraph is a subset of its vertices containing at most one vertex of each hyperedge.

The strongly stable sets of the $(0,1)$-matrix A have been defined in terms of the nonzero diagonals of A. The next theorem of Brualdi[1979] contains a simple instrinsic characterization of strongly stable sets.

Theorem 4.5.2.   *Let A be a $(0,1)$-matrix of order n with total support. Let X be the set of positions of A which contain 1's, and let S be a subset of X. Then S is a strongly stable set of A if and only if there exist nonnegative integers p and q with $p + q = n - 1$ and a p by q zero submatrix B of A such that S is a subset of the positions of the submatrix of A which is complementary to B.*

Proof. First assume that there exists a zero submatrix B satisfying the properties stated in the theorem. If $p = 0$ or $q = 0$, then the submatrix complementary to B is a line and S is a strongly stable set. Now suppose that $p > 0$ and $q > 0$. Let (i_1, j_1) and (i_2, j_2) be two positions of S which belong to different lines of the complementary submatrix of B. The submatrix of A of order $n - 2$ obtained by deleting rows i_1 and i_2 and columns j_1 and j_2 contains the p by q zero submatrix B. Since $p + q = n - 1$, it follows from Theorem 1.2.1 that there does not exist a nonzero diagonal of A containing both of the positions (i_1, j_1) and (i_2, j_2). Therefore S is a strongly stable set of A.

Now assume that S is a strongly stable set of positions of A, and let $|S| = m$. We first assume that A is fully indecomposable, and prove by induction on m that there exists a p by q zero submatrix B of A with $p + q = n - 1$ such that S is a subset of the positions of the submatrix of A which is complementary to B. If the positions of S all belong to one line, we may find B with $p = 0$ or $q = 0$. If $m = 2$ the existence of B is a consequence of Theorem 1.2.1. We now proceed under the added assumption that $m > 2$ and that S contains positions from at least two different rows and at least two different columns. If there is a position (k, l) in S which is in the same row as another position in S and in the same column as a third position in S, then the conclusions hold by applying the induction hypothesis to $S - \{(k, l)\}$. Hence we further assume that S satisfies the condition:

(*) For each position (k, l) in S, (k, l) is either the only position of S in row k or the only position of S in column l.

We now distinguish two cases.

Case 1. There exist distinct positions s_1 and s_2 in S which belong to the same line of A. Without loss of generality we assume that the line containing both s_1 and s_2 is a row. Let $S_1 = S - \{s_1\}$ and let $S_2 = S - \{s_2\}$. We apply the induction hypothesis to S_1 and S_2 and obtain for $k = 1$ and $k = 2$ a p_k by q_k zero submatrix B_k with $p_k + q_k = n - 1$ such that S_k is a subset of the positions in the submatrix of A complementary to B_k. If the column containing s_1 does not meet B_1 or the column containing s_2 does not meet B_2, then the conclusion holds. Hence we assume that the column containing s_1 meets B_1 and the column containing s_2 meets B_2. It now follows from (*) that the column containing s_1 contains no other position in S, $(k = 1, 2)$ and that each of the integers p_1, q_1, p_2 and q_2 is positive. Let B_1 and B_2 lie in exactly v common rows of A and exactly u common columns of A. Then A has zero submatrices B_3 and B_4 of sizes v by $q_1 + q_2 - u$ and $p_1 + p_2 - v$ by u, respectively. We have

$$(v + (q_1 + q_2 - u)) + ((p_1 + p_2 - v) + u) = 2(n - 1). \qquad (4.34)$$

Hence either $v + (q_1 + q_2 - u) \geq n - 1$ or $u + (p_1 + p_2 - v) \geq n - 1$. Suppose that $v + (q_1 + q_2 - u) \geq n$. Since $q_1 > 0$ and $q_2 > 0$, we have $q_1 + q_2 - u > 0$. Since all positions in $S - \{s_1, s_2\}$ lie in the $n - (q_1 + q_2 - u)$ columns of A complementary to those of B_3 and since $|S| > 2$, we have $q_1 + q_2 - u < n$. Thus B_3 is a nonvacuous zero submatrix of A and we contradict the full indecomposable assumption of A. Hence we have $v + (q_1 + q_2 - u) \leq n - 1$ and it follows from (4.34) that $(p_1 + p_2 - v) + u \geq n - 1$. Since S is contained in the set of positions of the submatrix complementary to the $(p_1 + p_2 - v)$ by u zero submatrix B_4 of A, the conclusion holds in this case.

Case 2. Each line of A contains at most one position in S. We permute the lines of A and assume that $S = \{(1, 1), (2, 2), \ldots, (m, m)\}$. For $i = 1, 2$ and 3, let $S_i = S - \{(i, i)\}$. By the induction assumption there exists a p_i by q_i zero submatrix B_i of A with $p_i + q_i = n - 1$ such that the positions of S_i are positions of the submatrix of A complementary to B_i. Since $|S_i| \geq 2$, p_i and q_i are both positive, $(i = 1, 2, 3)$. If for some i, neither row i nor column i meets B_i, then S is a subset of the positions of the submatrix complementary to B_i. Hence we assume that row i or column i meets B_i, $(i = 1, 2, 3)$. If for some i and j with $i \neq j$ row i does not meet B_i and row j does not meet B_j, or column i does not meet B_i and column j does not meet B_j, then an argument like that in Case 1 completes the proof. Without loss of generality, we assume that row 1 does not meet B_1 and column 2 does not meet B_2. But now if row 3 does not meet B_3 we apply the argument of Case 1 to B_1 and B_3, and if column 3 does not meet B_3 we apply the argument of Case 1 to B_2 and B_3. Hence the conclusion follows by induction if A is fully indecomposable. A strongly stable set of A can contain positions from only one fully indecomposable component of A, and the conclusion now holds in general.   □

Let A be a $(0,1)$-matrix of order n with total support, and let X be the set of positions of A that contain 1's. A *linear set* of A [or of the diagonal hypergraph $(X, \mathcal{D}(A))$] is the set of positions occupied by 1's in a line of A. According as the line is a row or column, we speak of a *row-linear set* and a *column-linear set*. Let $P = [p_{ij}]$, $(i, j = 1, 2, \ldots, n)$ and $Q = [q_{ij}]$, $(i, j = 1, 2, \ldots, n)$ be permutation matrices of order n, and let σ

and τ be the permutations of $\{1, 2, \ldots, n\}$ defined by $\sigma(i) = j$ if $p_{ij} = 1$ and $\tau(i) = j$ if $q_{ij} = 1$ $(i, j = 1, 2, \ldots, n)$. Let Y be the set of positions occupied by 1's in PAQ. The bijection ψ from the set X to the set Y defined by $\psi((i, j)) = (\sigma(i), \tau(j))$ is an isomorphism of the diagonal hypergraphs $(X, \mathcal{D}(A))$ and $(Y, \mathcal{D}(PAQ))$. The isomorphism ψ is the *isomorphism induced by the permutation matrices P and Q*. The bijection θ from the set X to the set Z of positions occupied by 1's in the transposed matrix A^T defined by $\theta((i, j)) = (j, i)$ is an isomorphism of the diagonal hypergraphs $(X, \mathcal{D}(A))$ and $(Z, \mathcal{D}(A^T))$. The isomorphism θ is the *isomorphism induced by transposition*. Isomorphisms induced by permutation matrices or by transposition map the linear sets of one diagonal hypergraph onto the linear sets of another diagonal hypergraph.

The following theorem is from Brualdi and Ross[1981].

Theorem 4.5.3.   *Let A and B be $(0, 1)$-matrices of order n with B fully indecomposable, and let θ be an isomorphism of the diagonal hypergraphs $(X, \mathcal{D}(A))$ and $(Y, \mathcal{D}(B))$. Then each linear set of A is mapped by θ onto a linear set of B if and only if there are permutation matrices P and Q of order n such that one of the following holds:*

(i) $B = PAQ$ and θ is induced by P and Q;

(ii) $B = PA^T Q$ and $\theta = \phi\rho$ where ϕ is an isomorphism induced by transposition and ρ is an isomorphism induced by P and Q.

Proof. If (i) or (ii) holds, then it is evident that each linear set of A is mapped by θ onto a linear set of B. We now prove the converse statement. Suppose that e row-linear sets of A are mapped onto row-linear sets of B and $(n - e)$ row-linear sets are mapped onto column-linear sets. Since the row-linear sets of A are pairwise disjoint, B has an e by $n - e$ zero submatrix. Since B is fully indecomposable, we have $e = 0$ or $e = n$. If $e = n$ the column-linear sets of A sets are mapped by θ onto the column-linear sets of B, and (i) holds. If $e = 0$ the column-linear sets of A are mapped onto the row linear sets of B, and (ii) holds. $\square$

If the assumptions of Theorem 4.5.3 hold and each linear set of A is mapped onto a linear set of B, then Theorem 4.5.3 implies that A is fully indecomposable. More generally, the number of fully indecomposable components is invariant under a diagonal hypergraph isomorphism (Brualdi and Ross[1979]).

Theorem 4.5.4.   *Let A and B be $(0, 1)$-matrices of order n with total support, and let $A_1, A_2, \ldots, A_r$ and $B_1, B_2, \ldots, B_s$ be the fully indecomposable components of A and B, respectively. Then the diagonal hypergraph $(X, \mathcal{D}(A))$ of A is isomorphic to the diagonal hypergraph of B if and only if $r = s$ and there is a permutation σ of $\{1, 2, \ldots, r\}$ such that the diagonal hypergraph of A_i is isomorphic to the diagonal hypergraph of $B_{\sigma(i)}$ for each $i = 1, 2, \ldots, r$.*

Proof. Without loss of generality we assume that $A = A_1 \oplus A_2 \oplus \cdots \oplus A_r$ and that $B = B_1 \oplus B_2 \oplus \cdots \oplus B_s$. First assume that ψ is an isomorphism of $(X, \mathcal{D}(\mathcal{A}))$ and $(Y, \mathcal{D}(\mathcal{B}))$. If $r = s = 1$, then the conclusions hold. Without loss of generality we now assume that $r > 1$. Then $A = A_1 \oplus A'$ where $A' = A_2 \oplus \cdots \oplus A_r$ is a matrix of order $k < n$. The set X can be partitioned into sets X_1 and X' where X_1 is the set of positions of A_1 that contain 1's. Each nonzero diagonal D of A can be partitioned into sets D_1 and D' where $D_1 \subseteq X_1$ and $D' \subseteq X'$. Let $E = E_1 \cup E'$ be any nonzero diagonal of A. Then for each nonzero diagonal D of A, $D_1 \cup E'$ and $E_1 \cup D'$ are nonzero diagonals of A. Since A has total support, each position in X_1 belongs to a nonzero diagonal D_1 of A_1. Since $D_1 \cup E'$ is a nonzero diagonal of A, $\psi(D_1) \cup \psi(E')$ is a nonzero diagonal of B. It now follows that $\psi(X_1)$ is contained in rows and columns of B which are complementary to those of $\psi(E')$. In a similar way we conclude that $\psi(X')$ is contained in the rows and columns complementary to those of $\psi(E_1)$. Thus ψ induces an isomorphism of the diagonal hypergraphs of A_1 and $B_{i_1} \oplus \cdots \oplus B_{i_m}$ for some positive integer m and some $i_1, \ldots, i_m$ with $1 \leq i_1 < \cdots < i_m \leq s$. It follows that $r \leq s$, and in a similar way that $s \leq r$. Hence $r = s$ and there exists a permutation σ of $\{1, 2, \ldots, r\}$ such that the diagonal hypergraph of A_i is isomorphic to the diagonal hypergraph of $B_{\sigma(i)}$ for each $i = 1, 2, \ldots, r$.

The converse is immediate.   □

Let A and B be (0,1)-matrices of order n with total support. The isomorphisms of the diagonal hypergraphs of A and B which map the linear sets of A onto the linear sets of B can be characterized by using Theorems 4.5.3 and 4.5.4. Informally described, such isomorphisms are obtained by replacing some of the fully indecomposable components of A with their transposes and permuting the lines of A. We note that in general Theorem 4.5.5 implies that for the further investigation of isomorphisms of diagonal hypergraphs, it suffices to consider only fully indecomposable matrices.

In order to have an isomorphism of diagonal hypergraphs which is different from those described in (i) and (ii) of Theorem 4.5.4, there must be a linear set of one of the matrices which is mapped onto a nonlinear set of the other. That such isomorphisms exist is demonstrated in the following example.

Let A and B be the fully indecomposable matrices of order 5 defined by

$$
A = \begin{bmatrix} 0 & 1 & 1 & 0 & 0 \\ 1 & 1 & 1 & 0 & 0 \\ 1 & 0 & 1 & 1 & 1 \\ 0 & 0 & 1 & 0 & 1 \\ 0 & 0 & 0 & 1 & 1 \end{bmatrix}, \quad B = \begin{bmatrix} 0 & 1 & 1 & 0 & 0 \\ 1 & 1 & 1 & 0 & 0 \\ 1 & 0 & 1 & 1 & 0 \\ 0 & 0 & 1 & 0 & 1 \\ 0 & 0 & 1 & 1 & 1 \end{bmatrix}. \tag{4.35}
$$

Notice that B has a line sum equal to five while no line sum of A equals five. Thus there do not exist permutation matrices P and Q such that

$PAQ = B$ or $PA^T Q = B$. Yet the diagonal hypergraphs of A and B are isomorphic. If we label the positions of A and B with the elements of the set $\{a, b, \ldots, l, m\}$, then an isomorphism of the diagonal hypergraphs of A and B is defined schematically by

$$
\begin{bmatrix}
0 & j & a & 0 & 0 \\
k & l & b & 0 & 0 \\
m & 0 & c & d & e \\
0 & 0 & g & 0 & f \\
0 & 0 & 0 & h & i
\end{bmatrix},
\quad
\begin{bmatrix}
0 & j & a & 0 & 0 \\
k & l & b & 0 & 0 \\
m & 0 & c & g & 0 \\
0 & 0 & d & 0 & h \\
0 & 0 & e & f & i
\end{bmatrix}.
\tag{4.36}
$$

The nonzero diagonals of A and of B are:

$$\{a, f, h, l, m\}, \{b, f, h, j, m\}, \{c, f, h, j, k\}, \{e, g, h, j, k\}, \{d, g, i, j, k\}.$$

The set of positions $\{a, b, c, d, e\}$ is a linear set of B but not of A.

We now generalize the idea used in the construction of the previous example.

Let $C = [c_{ij}], (i, j = 1, 2, \ldots, m)$ and $D = [d_{kl}], (k, l = 1, 2, \ldots, n)$ be matrices of orders m and n, respectively, such that $c_{mm} = d_{11}$. The matrix $C \star D$ of order $m + n - 1$ is defined schematically by

$$
A = C \star D =
\begin{bmatrix}
C & & O_{m-1,n-1} \\
& x & \\
O_{n-1,m-1} & & D
\end{bmatrix}
\quad (x = c_{mm} = d_{11}).
$$

In the matrix $C \star D$ the matrices C and D "overlap" in one position with common value x and there is an $m - 1$ by $n - 1$ zero submatrix in the upper right corner and an $n - 1$ by $m - 1$ zero submatrix in the lower left corner. The matrices $C^T \star D$ and $C \star D^T$ are said to be obtained from $A = C \star D$ by a *partial transposition* on C and D, respectively. We remark that if the order of C is 1, then $A = D$ and $A^T = C \star D^T$. Thus transposition of a matrix is a special instance of partial transposition.

Suppose that the matrix $A = C \star D$ has total support. Then the matrix $B = C^T \star D$ also has total support. Let the diagonal hypergraphs of A and B be $(X, \mathcal{D}(A))$ and $(Y, \mathcal{D}(B))$, respectively. Then the mapping $\theta : X \to Y$ defined by

$$
\theta(i, j) =
\begin{cases}
(j, i) & \text{if } (i, j) \text{ is a position in } C, \\
(i, j) & \text{otherwise,}
\end{cases}
$$

is an isomorphism of the diagonal hypergraphs of A and B. The mapping θ is called the *isomorphism of the diagonal hypergraphs of A and B induced by partial transposition on C*. In an analogous way we define the isomorphism induced by partial transposition on D.

The binary operation $\star$ is an associative operation and we may write without ambiguity $A = A_1 \star A_2 \star \cdots \star A_k$ whenever $A_1, A_2, \ldots, A_k$ are matrices such that the element in the lower right corner of A_i equals the entry in the upper left corner of $A_{i+1}, (i = 1, 2, \ldots, k-1)$. Let $B = A_1' \star A_2' \star \cdots \star A_k'$ where for each $i = 1, 2, \ldots, k$ we have $A_i' = A_i$ or A_i^T. Then it follows by an inductive argument that the diagonal hypergraph of A is isomorphic to the diagonal hypergraph of B under a composition of isomorphisms induced by partial transposition. For instance, if $k = 3$ and $B = A_1 \star A_2^T \star A_3$, then a partial transposition of $A = A_1 \star A_2 \star A_3$ on $A_1 \star A_2$ followed by a partial transposition on A_1^T results in B. It has been *conjectured* by Brualdi and Ross[1981] that if A and B are matrices with total support and θ is an isomorphism of the diagonal hypergraphs of A and B, then θ is a composition of isomorphisms induced by permutation matrices and partial transpositions. If each linear set of A is mapped by θ onto a linear set of B, then the validity of the conjecture is a consequence of Theorem 4.5.4. The conjecture has also been verified in Brualdi and Ross[1981] in another circumstance which we briefly describe.

Let A be a $(0,1)$-matrix of order n with total support, and let S be a subset of the positions of A occupied by 1's. Then S is called a *linearizable set* of A provided there is a $(0,1)$-matrix B of order n with total support and an isomorphism θ of the diagonal hypergraphs of A and B such that $\theta(S)$ is a subset of a linear set of B. A linearizable set S of A contains at most n elements, and it is a consequence of Theorem 4.5.1 that S is a strongly stable set of A. For the matrix A of order $n = 5$ in (4.35), the set $S = \{a, b, c, d, e\}$ of positions indicated in (4.36) is a linearizable set of A with 5 elements. Linearizable sets are characterized in Brualdi and Ross[1981] where the following theorem is also proved.

Theorem 4.5.5.   *Let A be a $(0,1)$-matrix of order n with total support such that A has a linearizable set S of n elements. Let B be a $(0,1)$-matrix of order n with total support such that there is an isomorphism of the diagonal hypergraphs of A and B for which $\theta(S)$ is a linear set of B. Then θ is a composition of isomorphisms induced by permutation matrices and partial transpositions.*

Ross[1980] has obtained the same conclusion under a weakening of the hypothesis of the theorem.

Exercises

1. Let A be a fully indecomposable $(0,1)$-matrix of order n. Prove that the size of a strongly stable set of the diagonal hypergraph of A does not exceed $\sigma(A) - 2n + 2$, where $\sigma(A)$ denotes the number of 1's of A (Brualdi[1979]).
2. Prove that the mapping induced by partial transposition is an isomorphism of diagonal hypergraphs.

3. Let A be a (0,1)-matrix of order n and let X be the set of positions of A which contain 1's. Let $C = C(A)$ denote the collection C of subsets of X for which C is the set of edges of a cycle of the bipartite graph whose reduced adjacency matrix is A. We call (X, C) the *cycle hypergraph* of A. Now assume that A and B are (0,1)-matrices of order n with total support. Let θ be an isomorphism of the diagonal hypergraphs of A and B. Prove that θ is an isomorphism of the cycle hypergraphs of A and B. Conclude that a linearizable set of A does not contain a member of $C(A)$ (Brualdi and Ross[1981]).

4. (Continuation of Exercise 3) Show that the converse of Exercise 3 is false by exhibiting an isomorphism θ of the cycle hypergraphs of

$$A = B = \begin{bmatrix} 1 & 1 & 0 \\ 1 & 1 & 1 \\ 1 & 1 & 1 \end{bmatrix}$$

which is not a diagonal hypergraph isomorphism (Brualdi and Ross[1981]).

5. (Continuation of Exercises 3 and 4) Prove that the converse of Exercise 3 is true if θ maps some nonzero diagonal of A to a nonzero diagonal of B (Brualdi and Ross[1981]).

References

C. Berge[1973], *Graphs and Hypergraphs*, North-Holland, Amsterdam.

R.A. Brualdi[1979], The diagonal hypergraph of a matrix (bipartite graph), *Discrete Math.*, 27, pp. 127–147.

[1980], On the diagonal hypergraph of a matrix, *Annals of Discrete Math.*, 8, pp. 261–264.

R.A. Brualdi and J.A. Ross[1981], Matrices with isomorphic diagonal hypergraphs, *Discrete Math.*, 33, pp. 123–138.

J.A. Ross[1980], *Some problems in combinatorial matrix theory*, Ph.D. thesis, University of Wisconsin, Madison.

5

Some Special Graphs

5.1 Regular Graphs

We begin our study of special graphs with two lemmas on nonnegative matrices. We again let e_n denote the column vector of n 1's.

Lemma 5.1.1. *Let A be a nonnegative real matrix of order n and let all of the line sums of A equal k. Then k is an eigenvalue of A corresponding to the eigenvector e_n and the modulus of every other eigenvalue of A does not exceed k. Furthermore, if $n > 1$ then the eigenvalue k is of multiplicity one if and only if A is irreducible.*

Proof. The equation $Ae_n = ke_n$ implies at once that k is an eigenvalue of A corresponding to the eigenvector e_n. By Theorem 3.6.2 no other eigenvalue can have larger modulus. If A is reducible, then all of the line sums of each irreducible component of A also equal k and it follows that the multiplicity of the eigenvalue k is at least two. If A is irreducible, then it follows from the Perron–Frobenius theory (see, e.g., Horn and Johnson[1985]) of nonnegative matrices that the multiplicity of k as an eigenvalue of A equals one. □

Lemma 5.1.2. *Let A be a nonnegative real matrix of order n. Then there exists a polynomial $p(x)$ such that*

$$J = p(A) \tag{5.1}$$

if and only if A is irreducible and all of the line sums of A are equal.

Proof. Suppose that (5.1) is valid. Then $AJ = JA$ and it follows that all of the line sums of A are equal. If A is reducible, then all of the positive integral powers of A have certain fixed positions occupied by zeros and this contradicts (5.1).

Conversely, suppose that A is irreducible and that all of the line sums of A are equal to k. Then by Lemma 5.1.1 we know that k is a simple eigenvalue of A. We may write the minimum polynomial of A in the form

$$m(\lambda) = (\lambda - k)q(\lambda)$$

and this implies that

$$Aq(A) = kq(A).$$

Thus each nonzero column of $q(A)$ is an eigenvector of A corresponding to the eigenvalue k. But the eigenspace associated with the eigenvalue k has dimension one and hence each column of $q(A)$ is a suitable multiple of e_n. The same argument applies to the transposed situation

$$A^T q(A)^T = kq(A)^T,$$

and we may conclude that each column of $q(A)^T$ is also a multiple of e_n. Hence each row of $q(A)$ is a multiple of e_n. But this means that $q(A)$ is a multiple of J. We cannot have $q(A) = O$ because $m(\lambda)$ is the minimum polynomial of A. Thus J is a polynomial in A.   □

We may apply the preceding lemma directly to the adjacency matrix of a graph and obtain the following theorem of Hoffman[1963].

Theorem 5.1.3. *Let A be the adjacency matrix of a graph G of order $n > 1$. Then there exists a polynomial $p(x)$ such that*

$$J = p(A) \tag{5.2}$$

if and only if G is a regular connected graph.

Corollary 5.1.4. *Let G be a regular connected graph of order $n > 1$ and let the distinct eigenvalues of G be denoted by $k > \lambda_1 > \cdots > \lambda_{t-1}$. Then if*

$$q(\lambda) = \prod_{i=1}^{t-1} (\lambda - \lambda_i),$$

we have

$$J = \left(\frac{n}{q(k)} \right) q(A).$$

The polynomial

$$p(\lambda) = \left(\frac{n}{q(k)} \right) q(\lambda)$$

is the unique polynomial of lowest degree such that $p(A) = J$.

Proof. Since A is symmetric we know that the zeros of the minimum polynomial of A are distinct. Then by the proof of Lemma 5.1.2 we have that $q(A) = cJ$ for some nonzero constant c. The eigenvalues of $q(A)$ are $q(k)$ and $q(\lambda_i)$ $(i = 1, 2, \ldots, t - 1)$ and all of these are zero with the exception of $q(k)$. But the only nonzero eigenvalue of cJ is cn and hence $c = q(k)/n$.

Let $p(\lambda)$ be a polynomial such that $p(A) = J$. The eigenvalues of $p(A)$ are $p(k)$ and $p(\lambda_i)$ $(i = 1, 2, \ldots, t - 1)$. Since e_n is an eigenvector of $p(A)$ and of J corresponding to the eigenvalues $p(k)$ and n, respectively, we have $p(\lambda_i) = 0$ for $i = 1, 2, \ldots, t - 1$. $\qquad\square$

The polynomial

$$p(\lambda) = \left(\frac{n}{q(k)}\right) q(\lambda)$$

in Corollary 5.1.4 is called the *Hoffman polynomial* of the regular connected graph G.

We illustrate the preceding discussion by showing that the only connected graph G of order n with exactly two distinct eigenvalues is the complete graph K_n. Let A be the adjacency matrix of such a graph with eigenvalues $\lambda_1 > \lambda_2$. Then we have

$$A^2 - (\lambda_1 + \lambda_2)A + \lambda_1\lambda_2 I = O.$$

Since A is symmetric and of trace zero, it follows that G is regular of degree $-\lambda_1\lambda_2$. Thus the Hoffman polynomial of G is of degree 1 and this implies that $J = A + I$.

In the following section we study in some detail regular connected graphs with exactly three distinct eigenvalues, that is, graphs whose Hoffman polynomial is of degree 2.

Exercises

1. Let G be a graph of order n which is regular of degree k. Prove that the sum of the squares of its eigenvalues equals kn.
2. Determine the spectrum and Hoffman polynomial of the complete bipartite graph $K_{m,m}$.
3. Determine the spectrum and Hoffman polynomial of the complete multipartite graph $K_{m,m,\ldots,m}(k\ m's)$. (This graph has km vertices partitioned into k parts of size m and there is an edge joining two vertices if and only if they belong to different parts.)

References

N. Biggs[1974], *Algebraic Graph Theory*, Cambridge Tracts in Mathematics No. 67, Cambridge University Press, Cambridge.

A.J. Hoffman[1963], On the polynomial of a graph, *Amer. Math. Monthly*, 70, pp. 30–36.

A.J. Hoffman and M.H. McAndrew[1965], The polynomial of a directed graph, *Proc. Amer. Math. Soc.*, 16, pp. 303–309.

R.A. Horn and C.R. Johnson[1985], *Matrix Analysis*, Cambridge University Press, Cambridge.

A.J. Schwenk and R.J. Wilson[1978], On the eigenvalues of a graph, *Selected Topics in Graph Theory* (L.W. Beineke and R.J. Wilson, eds.), Academic Press, New York, pp. 307–336.

5.2 Strongly Regular Graphs

Throughout this section G denotes a graph of order $n, (n \geq 3)$ with vertices $a_1, a_2, \ldots, a_n$ and we let A denote the adjacency matrix of G.

A *strongly regular graph* on the parameters (n, k, λ, μ) is a graph G of order $n, (n \geq 3)$ which is regular of degree k and satisfies the following additional requirements:

(i) If a and b are any two distinct vertices of G which are joined by an edge, then there are exactly λ further vertices of G which are joined to both a and b.

(ii) If a and b are any two distinct vertices of G which are not joined by an edge, then there are exactly μ further vertices of G which are joined to both a and b.

We exclude from consideration the complete graph K_n and its complement, the void graph, so that neither property (i) nor (ii) is vacuous. Strongly regular graphs were introduced by Bose[1963] and have subsequently been investigated by many authors. We mention, in particular, the studies of Seidel[1968,1969,1974,1976] and the book by Brouwer, Cohen and Neumaier[1989].

We begin with some simple examples of strongly regular graphs.

The 4-cycle and the 5-cycle are strongly regular graphs on the parameters

$$(4, 2, 0, 2) \qquad \text{and} \qquad (5, 2, 0, 1),$$

respectively. No other n-cycle qualifies as a strongly regular graph.

The *Petersen graph* in Figure 5.1 is a strongly regular graph on the parameters

$$(10, 3, 0, 1).$$

The graph with two connected components each of which is a 3-cycle is a strongly regular graph on the parameters

$$(6, 2, 1, 0).$$

The complete bipartite graph $K_{m,m}, (m \geq 2)$ is a strongly regular graph on the parameters

$$(2m, m, 0, m).$$

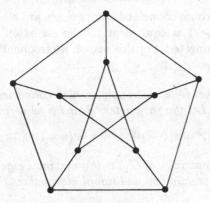

Figure 5.1

Let G be a strongly regular graph on the parameters (n, k, λ, μ) and let A be its adjacency matrix. We know that the entry in the (i, j) position of A^2 equals the number of walks of length 2 with a_i and a_j as endpoints. This number equals k, λ or μ according as these vertices are equal, adjacent or nonadjacent. Hence we have

$$A^2 = kI + \lambda A + \mu(J - I - A), \tag{5.3}$$

or, equivalently,

$$A^2 - (\lambda - \mu)A - (k - \mu)I = \mu J. \tag{5.4}$$

We introduce another parameter, namely,

$$l = n - k - 1. \tag{5.5}$$

The integer l is the degree of the complement $\bar{G}$ of G. An elementary calculation involving (5.3) tells us that

$$(J - I - A)^2 = lI + (l - k + \mu - 1)(J - I - A) + (l - k + \lambda + 1)A.$$

Hence it follows that if G is a strongly regular graph, then its complement $\bar{G}$ is also a strongly regular graph on the parameters

$$(\bar{n} = n, \bar{k} = l, \bar{\lambda} = l - k + \mu - 1, \bar{\mu} = l - k + \lambda + 1).$$

If we multiply the equation (5.3) by the column vector e_n then we obtain

$$k^2 = k + \lambda k + \mu(n - 1 - k),$$

and we write this relation in the form

$$l\mu = k(k - \lambda - 1). \tag{5.6}$$

In case $\mu = 0$, then by (5.6) we have $\lambda = k - 1$. This means that every vertex of G belongs to a complete graph K_{k+1}, and thus G is a disconnected graph whose connected components are all of the form K_{k+1}. The requirement $\mu \geq 1$ is equivalent to the assertion that the strongly regular graph G is connected. For this reason we frequently require strongly regular graphs to have $\mu \geq 1$.

Theorem 5.2.1. *Let G be a strongly regular connected graph on the parameters (n, k, λ, μ). Let the parameters d and δ be defined by*

$$d = (\lambda - \mu)^2 + 4(k - \mu), \qquad \delta = (k + l)(\lambda - \mu) + 2k. \qquad (5.7)$$

Then the adjacency matrix A of G has the maximal eigenvalue k of multiplicity 1, and A has exactly two additional eigenvalues

$$\rho = \frac{1}{2}(\lambda - \mu + \sqrt{d}) \geq 0, \qquad \sigma = \frac{1}{2}(\lambda - \mu - \sqrt{d}) \leq -1 \qquad (5.8)$$

of multiplicities

$$r = \frac{1}{2}\left(k + l - \frac{\delta}{\sqrt{d}}\right), \qquad s = \frac{1}{2}\left(k + l + \frac{\delta}{\sqrt{d}}\right), \qquad (5.9)$$

respectively.

Proof. Since G is a connected graph and is not a complete graph, A has at least three distinct eigenvalues. The first assertion in the theorem follows from Lemma 5.1.1. We next multiply (5.4) by $A - kI$ and this implies

$$(A - kI)(A^2 - (\lambda - \mu)A - (k - \mu)I) = O.$$

Thus the quantities ρ and σ displayed in (5.8) are eigenvalues of A.

If $d = 0$ then $\lambda = \mu = k$. But since G is regular of degree k we must have $\lambda \leq k - 1$ so that $d \neq 0$ and $\rho > \sigma$. Notice that the parameters λ and μ are expressible in terms of the quantities $k > \rho > \sigma$:

$$\lambda = k + \rho + \sigma + \rho\sigma, \qquad \mu = k + \rho\sigma.$$

We know that $\mu \leq k$ so that $\rho \geq 0$ and $\sigma \leq 0$. But $\sigma = 0$ implies that $\lambda = k + \rho$ and this contradicts $\lambda \leq k - 1$. Hence we have $\rho \geq 0$ and $\sigma < 0$.

We now turn to the complement $\bar{G}$ of G. An elementary calculation tells us that for $\bar{G}$ we have

$$\bar{d} = d, \qquad \bar{\rho} = -\sigma - 1, \qquad \bar{\sigma} = -\rho - 1.$$

But again for $\bar{G}$ we have $\bar{\rho} \geq 0$ so that we may conclude that $\rho \geq 0$ and $\sigma \leq -1$, as required.

Let r and s denote the multiplicities of ρ and σ, respectively, as eigenvalues of A. Then we have

$$r + s = n - 1,$$

and since A has trace zero, we have

$$k + r\rho + s\sigma = 0.$$

We solve these equations for r and s and this gives (5.9).   □

The eigenvalue multiplicities r and s are nonnegative integers, and this fact in conjunction with (5.9) places severe restrictions on the parameter sets for strongly regular graphs.

Theorem 5.2.2.   *Let G be a strongly regular connected graph on the parameters (n, k, λ, μ).*

(i) *If $\delta = 0$, then*

$$\lambda = \mu - 1, \qquad k = l = 2\mu = r = s = (n-1)/2.$$

(ii) *If $\delta \neq 0$, then $\sqrt{d}$ is an integer and the eigenvalues ρ and σ are also integers. Furthermore if n is even, then $\sqrt{d} \mid \delta$ whereas $2\sqrt{d} \nmid \delta$, and if n is odd, then $2\sqrt{d} \mid \delta$.*

Proof. If $\delta = 0$, then $k + l = 2k/(\mu - \lambda) > k$ and thus $0 < \mu - \lambda < 2$. Therefore we have $\lambda = \mu - 1$. The remaining equations of (1) now follow from (5.6) and (5.9).

If $\delta \neq 0$, then the conclusion (2) follows directly from (5.8) and (5.9).   □

Strongly regular graphs of the form (1) in Theorem 5.2.2 are called *conference graphs*. They arise in a wide variety of mathematical investigations (see Cameron and van Lint[1975], Goethals and Seidel[1967, 1970 and van Lint and Seidel[1966]). They have the same parameter sets as their complements and have been constructed for orders n equal to a prime power congruent to 1 (modulo 4). Let F be a finite field on n elements, where n is a prime power congruent to 1 (modulo 4). Then we may construct a graph G of order n whose vertices are the elements of F. Two vertices a and b are adjacent in G if and only if $a - b$ is a nonzero square in F. Notice that -1 is a square in F so that G is undirected. The resulting graph is a strongly regular graph on the parameters

$$(n, k = (n-1)/2, \lambda = (n-5)/4, \mu = (n-1)/4).$$

These special conference graphs are called *Paley graphs*.

We now apply the preceding theory to a proof of the *friendship theorem* of Erdös, Rényi and Sós[1966]. In other terms the theorem says that in a finite society in which each pair of members has exactly one common friend, there is someone who is a friend to everyone else. Our account follows Cameron[1978].

Theorem 5.2.3.   *Let G be a graph of order n and suppose that for any two distinct vertices a and b there is a unique vertex c which is joined to* *both a and b. Then n is odd and G consists of a number of triangles with a common vertex.*

Proof. Let G be a graph fulfilling the hypothesis of the theorem. Let a and b be nonadjacent vertices of G. Then there is a unique vertex c which is adjacent to both a and b. There are also unique vertices $d \neq b$ adjacent to both a and c and $e \neq a$ adjacent to both b and c. If x is any vertex different from c and d which is adjacent to a then there exists a unique vertex y different from c and e which is adjacent to both x and b. A similar statement holds with a and b interchanged. Hence the degrees of the vertices a and b are equal.

Now suppose that G is not a regular graph. Let a and b be vertices of unequal degrees, and let c be the unique vertex which is adjacent to both a and b. The preceding paragraph implies that a and b are adjacent.

We may suppose by interchanging a and b if necessary that the degrees of a and c are unequal. Let d be any further vertex. Then d is adjacent to at least one of a and b because a and b are of unequal degrees. Similarly, d is adjacent to at least one of a and c. But d is not adjacent to both b and c because a is already adjacent to both b and c. Hence d is adjacent to a. It follows that G consists of a number of triangles with a common vertex a.

Hence we may assume that G is regular of degree k. By the hypothesis of the theorem we then have a strongly regular graph with $\lambda = \mu = 1$. By Theorem 5.2.1 it follows that $s - r = \delta/\sqrt{d} = k/\sqrt{k-1}$ is an integer. But then $(k-1)|k^2$ and it follows easily that the only possibilities are $k = 0$ and $k = 2$. These yield the cases of a single vertex and a triangle.   □

We look next at some further examples of strongly regular graphs. The *triangular graph* $T(m)$ is defined as the line graph of the complete graph K_m, $(m \geq 4)$. Thus the vertices of $T(m)$ may be identified as the 2-subsets of $\{1, 2, \ldots, m\}$, and two vertices are adjacent in $T(m)$ provided the corresponding 2-subsets have a nonempty intersection. An inspection of the structure of $T(m)$ reveals that $T(m)$ is a strongly regular graph on the parameters

$$(n = m(m-1)/2, k = 2(m-2), \lambda = m-2, \mu = 4).$$

The following classification theorem is due to Chang[1959, 1960] and Hoffman[1960].

Theorem 5.2.4. *Let G be a strongly regular graph on the parameters $(m(m-1)/2, 2(m-2), m-2, 4)$, $(m \geq 4)$. If $m \neq 8$, then G is isomorphic to the triangular graph $T(m)$. If $m = 8$, then G is isomorphic to one of four graphs, one of which is $T(8)$.*

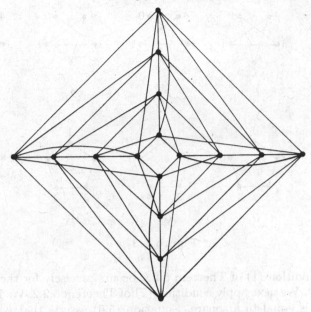

Figure 5.2

The *lattice graph* $L_2(m)$ is defined as the line graph of the complete bipartite graph $K_{m,m}$, $(m \geq 2)$. These are strongly regular graphs on the parameters

$$(n = m^2, k = 2(m-2), \lambda = m-2, \mu = 2).$$

The following classification theorem is due to Shrikhande[1959].

Theorem 5.2.5.   *Let G be a strongly regular graph on the parameters $(m^2, 2(m-2), m-2, 2)$, $(m \geq 2)$. If $m \neq 4$, then G is isomorphic to the lattice graph $L_2(m)$. If $m = 4$, then G is isomorphic to $L_2(4)$ or to the graph in Figure 5.2.*

A *Moore graph* (of diameter 2) is a strongly regular graph with $\lambda = 0$ and $\mu = 1$. These graphs contain no triangles and for any two nonadjacent vertices there is a unique vertex adjacent to both. Hoffman and Singleton[1960] showed that the parameter sets of Moore graphs are severely restricted.

Theorem 5.2.6.   *The only possible parameter sets (n, k, λ, μ) of a Moore graph are*

$$(5, 2, 0, 1), (10, 3, 0, 1), (50, 7, 0, 1) \text{ and } (3250, 57, 0, 1).$$

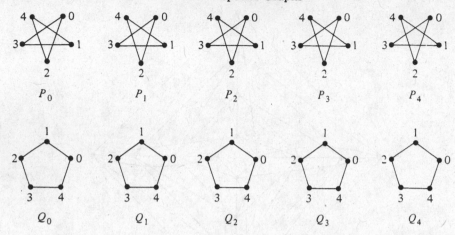

Figure 5.3

Proof. Condition (1) of Theorem 5.2.2 occurs precisely for the parameters (5,2,0,1). We next apply condition (2) of Theorem 5.2.2. We have that $d = 4k - 3$ is equal to a square. Equation (5.6) asserts that $k + l = k^2$ and hence $\delta = k(2 - k)$. Thus we have $k(2 - k) \equiv 0 \pmod{\sqrt{d}}$. We also have $4k - 3 \equiv 0 \pmod{\sqrt{d}}$. Multiplying the first of these congruences by 4 and the second by k and then adding we obtain $5k \equiv 0 \pmod{\sqrt{d}}$. This and $4k - 3 \equiv 0 \pmod{\sqrt{d}}$ now imply that $15 \equiv 0 \pmod{\sqrt{d}}$. Thus the only possibilities for $\sqrt{d}$ are 1,3,5 and 15. The first case is an excluded degeneracy, and the other three values yield the last three parameter sets displayed in the theorem.   □

The first of the parameter sets in Theorem 5.2.6 is satisfied by the pentagon, the second by the Petersen graph and the third by the *Hoffman–Singleton graph*. They are the unique strongly regular graphs on these parameter sets. The existence of a strongly regular graph corresponding to the last of the parameter sets is unknown. Aschbacher[1971] has shown that its automorphism group cannot be too large.

The Hoffman–Singleton graph may be represented by the ten cycles of order 5 labeled as shown in Figure 5.3, where vertex i of P_j is joined to vertex $i + jk \pmod 5$ of Q_k (Bondy and Murty[1976]).

We remark that Moore graphs may be defined under certain more general conditions so that their diameter is allowed to exceed 2 (see Cameron[1978] and Cameron and van Lint[1975]). But in this case Bannai and Ito[1973] and Damerell[1973] have shown that the only additional graphs introduced consist of a single cycle.

A *generalized Moore graph* is a strongly regular graph with $\mu = 1$. The parameter λ is allowed to take on any value in such a graph, but none has yet been found with $\lambda \geq 1$.

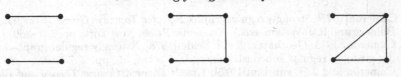

Figure 5.4

Exercises

1. Prove that a regular connected graph with three distinct eigenvalues is strongly regular.
2. Let G be a connected graph of order n which is regular of degree k. Assume that G satisfies requirements (i) and (ii) for a strongly regular graph but with the words *exactly* λ and *exactly* μ replaced by *at most* λ and *at most* μ, respectively. Prove that

$$n \le k + 1 + k(k - 1 - \lambda)/\mu$$

   with equality if and only if G is strongly regular on the parameters (n, k, λ, μ) (Seidel[1979]).
3. A $(0, 1, -1)$-matrix C of order $n+1$ all of whose main diagonal elements equal 0 is a *conference matrix* provided $CC^T = nI$. Prove that there exists a symmetric conference matrix of order $n + 1$ if and only if there exists a conference graph of order n.
4. Construct the conference matrices of orders 6 and 10 corresponding to the Paley graphs of orders 5 and 9.
5. Prove Theorem 5.2.4 when $m > 8$.
6. Let G be a regular connected graph of order n with at most 4 distinct eigenvalues. Prove that a graph H of order n is cospectral with G if and only if H is a connected regular graph having the same set of distinct eigenvalues as G (Cvetković, Doob and Sachs[1982]).
7. Let G be a graph with no vertex of degree 0 which is not a complete multipartite graph. Prove that G contains one of the three graphs in Figure 5.4 as an induced subgraph.
8. Let G be a graph with no vertex of degree 0. Assume that G has exactly one positive eigenvalue. Use Exercise 7 and the interlacing inequalities for the eigenvalues of symmetric matrices to prove that G is a complete multipartite graph (Smith[1970]).

References

M. Aschbacher[1971], The non-existence of rank three permutation groups of degree 3250 and subdegree 57, *J. Algebra*, 19, pp. 538–540.

E. Bannai and T. Ito[1973], On finite Moore graphs, *J. Fac. Sci. Univ. Tokyo*, 20, pp. 191–208.

J.A. Bondy and U.S.R. Murty[1976], *Graph Theory with Applications*, North-Holland, New York.

R.C. Bose[1963], Strongly regular graphs, partial geometries, and partially balanced designs, *Pacific J. Math.*, 13, pp. 389–419.

A. Brouwer, A. Cohen and A. Neumaier[1989], *Distance Regular Graphs*, Springer-Verlag, Berlin.

P.J. Cameron[1978], Strongly regular graphs, *Selected Topics in Graph Theory* (L.W. Beineke and R.J. Wilson, eds.), Academic Press, New York, pp. 337–360.

P.J. Cameron, J.-M. Goethals and J.J. Seidel[1978], Strongly regular graphs having strongly regular subconstituents, *J. Algebra*, 55, pp. 257–280.

P.J. Cameron and J.H. van Lint[1975], *Graph Theory, Coding Theory and Block Designs*, London Math. Soc. Lecture Note Series No. 19, Cambridge University Press, Cambridge.

D.M. Cvetković, M. Doob, and H. Sachs[1982], *Spectra of Graphs—Theory and Application*, 2nd ed., Deutscher Verlag der Wissenschaften, Berlin, Academic Press, New York.

D.M. Cvetković, M. Doob. I. Gutman and A. Torgašev[1988], *Recent Results in the Theory of Graph Spectra*, Annals of Discrete Math. No. 36, North-Holland, Amsterdam.

R.H. Damerell[1973], On Moore graphs, *Proc. Cambridge Phil. Soc.*, 74, pp. 227–236.

P. Erdös, A. Rényi and V. T. Sós[1966], On a problem of graph theory, *Studia Sci. Math. Hungar.*, 1, pp. 215–235.

J.-M. Goethals and J.J. Seidel[1967], Orthogonal matrices with zero diagonal, *Canad. J. Math.*, 19, pp. 1001–1010.

  [1970], Strongly regular graphs derived from combinatorial designs, *Canad. J. Math.*, 22, pp. 597–614.

W. Haemers[1979], *Eigenvalue Techniques in Design and Graph Theory*, Mathematisch Centrum, Amsterdam.

D.G. Higman[1971], Partial geometries, generalized quadrangles and strongly regular graphs, *Atti di Conv. Geometria Combinatoria e sue Applicazione* (A. Barlotti, ed.), Perugia, pp. 263–293.

A.J. Hoffman[1960], On the uniqueness of the triangular association scheme, *Ann. Math. Statist.*, 31, pp. 492–497.

X.L. Hubaut[1975], Strongly regular graphs, *Discrete Math.*, 13, pp. 357–381.

Chang Li-Chien[1959], The uniqueness and non-uniqueness of the triangular association scheme, *Sci. Record. Peking Math.* (New Ser.), 3, pp. 604–613.

  [1960], Associations of partially balanced designs with parameters $v = 28$, $n_1 = 12$, $n_2 = 15$, and $p_{11}^2 = 4$, *Sci. Record. Peking Math.* (New Ser.), 4, pp. 12–18.

J.H. van Lint and J.J. Seidel[1966], Equilateral point sets in elliptic geometry, *Nederl. Akad. Wetensch. Proc.* Ser. A, 69(=*Indag. Math.*, 28), pp. 335–348.

J.J. Seidel[1968], Strongly regular graphs with (-1,1,0) adjacency matrix having eigenvalue 3, *Linear Alg. Applics.*, 1, pp. 281–298.

  [1969], Strongly regular graphs, *Recent Progress in Combinatorics*, (W.T. Tutte, ed.), Academic Press, New York, pp. 185–198.

  [1974], Graphs and two-graphs, Proceedings of the Fifth Southeastern Conference on Combinatorics, Graph Theory and Computing, Congressus Numerantium X, *Utilitas Math.*, Winnipeg, pp. 125–143.

  [1976], A survey of two-graphs, *Teorie Combinatorie*, Tomo I (B. Segre, ed.), Accademia Nazionale dei Lincei, Rome, pp. 481–511.

  [1979], Strongly regular graphs, *Surveys in Combinatorics, Proc. 7th British Combinatorial Conference*, London Math. Soc. Lecture Note Ser. 38 (B. Bollobás, ed.), Cambridge University Press, Cambridge.

S.S. Shrikhande[1959], The uniqueness of the L_2 association scheme, *Ann. Math. Statist.*, 30, pp. 781–798.

J.H. Smith[1970], Some properties of the spectrum of a graph, *Combinatorial Structures and Their Applications* (R. Guy, N. Sauer and J. Schönheim, eds.), Gordon and Breach, New York, pp. 403–406.

5.3 Polynomial Digraphs

We may directly generalize the proofs of Theorem 5.1.3 and Corollary 5.1.4 and obtain the following theorem of Hoffman and McAndrew[1965].

Theorem 5.3.1. *Let A be the adjacency matrix of a digraph D of order $n > 1$. Then there exists a polynomial $p(x)$ such that*

$$J = p(A) \qquad (5.10)$$

if and only if D is a regular strongly connected digraph. Let D be a strongly connected digraph which is regular of degree k and let $m(\lambda)$ be the minimum polynomial of A. If

$$q(\lambda) = \frac{m(\lambda)}{\lambda - k}$$

then the polynomial

$$p(\lambda) = \left(\frac{n}{q(k)} \right) q(\lambda)$$

is the unique polynomial of lowest degree such that $p(A) = J$.

By Lemma 5.1.1 the modulus of each eigenvalue of A is at most equal to k. The roots of the polynomial $p(\lambda)$ are eigenvalues of A and it follows that $|p(\lambda)|$ is a monotone increasing function if λ is real and $\lambda \geq k$. We have $p(k) = n$ and we therefore conclude that the degree k of regularity of D equals the greatest real root of the equation $p(\lambda) = n$. Extending our definition in section 5.1 to digraphs, we call the polynomial $p(\lambda)$ in Theorem 5.3.1 the *Hoffman polynomial* of the regular strongly connected digraph D.

Let A be the adjacency matrix of a digraph D. We say that A is *regular of degree k* provided D is regular of degree k. Similarly, two adjacency matrices are called *isomorphic* provided their corresponding digraphs are isomorphic.

We now consider the special polynomials $p(\lambda) = (\lambda^m + d)/c$, where c is a positive integer and d is a nonnegative integer.

Theorem 5.3.2. *Let m and c be positive integers and let d be a nonnegative integer. Let A be a $(0,1)$-matrix of order n satisfying the equation*

$$A^m = -dI + cJ. \qquad (5.11)$$

Then there exists a positive integer k such that A is regular of degree k and $k^m = -d + cn$. If $d = 0$ then the trace of A is also equal to k.

Proof. The regularity of A is a consequence of Theorem 5.3.1. We now multiply (5.11) by J and this gives $k^m = -d + cn$. Now assume that $d = 0$.

Let the characteristic roots of A be $\lambda_1, \lambda_2, \ldots, \lambda_n$. The characteristic roots of cJ are cn of multiplicity one and 0 of multiplicity $n-1$. Hence we may write $\lambda_1 = k, \lambda_2 = 0, \ldots, \lambda_n = 0$ and the trace of A is k.   □

We note that if $d = 0$ in Theorem 5.3.2 it is essential that the digraph D associated with A have loops because otherwise the configuration is impossible. In terms of D equation (5.11) asserts that for each pair of distinct vertices a and b of D there are exactly c walks of length m from a to b and each vertex is on exactly $c - d$ closed walks of length m. In the case $c = 1$ and $d = 0$, a $(0,1)$-matrix A satisfying $A^m = J$ is a primitive matrix of exponent m, and the polynomial $p(\lambda) = \lambda^m$ is the Hoffman polynomial of D.

We next turn to showing that if $d = 0$ then the condition $k^m = cn$ in Theorem 5.3.2 is sufficient for there to exist a $(0,1)$-matrix A of order n satisfying $A^m = cJ$. A *g-circulant matrix* is a matrix of order n in which each row other than the first is obtained from the preceding row by shifting the elements cyclically g columns to the right. Let $A = [a_{ij}], (i, j = 1, 2, \ldots, n)$ be a g-circulant. Then

$$a_{ij} = a_{i+1,j+g}$$

in which the subscripts are computed modulo n. A 1-circulant matrix is more commonly called a *circulant matrix*. A detailed study of circulant matrices can be found in Ablow and Brenner[1963] and in the book by Davis[1979].

Let $a_0, a_1, \ldots, a_{n-1}$ be the first row of the g-circulant matrix A of order n. The *Hall polynomial* of A is defined to be the polynomial

$$\theta_A(x) = \sum_{i=0}^{n-1} a_i x^i.$$

The following lemma is a direct consequence of the definitions involved.

Lemma 5.3.3.   *Let A be a g-circulant matrix of order n and let B be an h-circulant matrix of order n. Then the product AB is a gh-circulant matrix of order n and we have*

$$\theta_{AB}(x) \equiv \theta_A(x^h)\theta_B(x) \quad (\mod x^n - 1).$$

Corollary 5.3.4.   *Let A be a g-circulant matrix of order n. Then for each positive integer m A^m is a g^m-circulant matrix and we have*

$$\theta_{A^m}(x) \equiv \theta_A(x)\theta_A(x^g)\cdots\theta_A(x^{g^{m-1}}) \quad (\mod x^n - 1). \tag{5.12}$$

Proof. We apply Lemma 5.3.3 and use induction on m.   □

The g-circulant solutions of the equation $A^m = -dI + cJ$ are characterized by their Hall polynomials in the following result of Lam[1977].

Lemma 5.3.5. *Let A be a g-circulant matrix. Then $A^m = cJ$ if and only if*

$$\theta_A(x)\theta_A(x^g)\cdots\theta_A(x^{g^{m-1}}) \equiv c(1+x+\cdots+x^{n-1}) \quad (\text{mod } x^n - 1). \quad (5.13)$$

If $d \neq 0$, then $A^m = dI + cJ$ if and only if

$$\theta_A(x)\theta_A(x^g)\cdots\theta_A(x^{g^{m-1}}) \equiv d+c(1+x+\cdots+x^{n-1}) \quad (\text{mod } x^n - 1) \quad (5.14)$$

and

$$g^m \equiv 1 \quad (\text{mod } n). \qquad (5.15)$$

Proof. By Corollary 5.3.4 A^m is a g^m-circulant and its Hall polynomial is given by equation (5.12). If $A^m = cJ$, then the first row of A^m is $(c, c, \ldots, c)$ and this implies (5.13). Conversely, if (5.13) is satisfied, then the first row of A^m is $(c, c, \ldots, c)$, and it follows that $A^m = cJ$.

Suppose that $d \neq 0$ and $A^m = dI + cJ$. It follows as above that (5.14) holds. Moreover, since $dI + cJ$ is a circulant, we have $g^m \equiv 1 \pmod{n}$ and (5.15) holds. Next suppose that $d \neq 0$ and (5.14) and (5.15) are satisfied. Then the first row of A^m equals $(d + c, c, \ldots, c)$. By (5.15) A^m is a 1-circulant and hence we have $A^m = dI + cJ$. $\qquad\square$

The following theorem is from Lam[1977].

Theorem 5.3.6. *Suppose that $k^m = cn$. Then the k-circulant matrix A of order n whose first row consists of k 1's followed by $n - k$ 0's satisfies*

$$A^m = cJ.$$

Proof. The Hall polynomial of the matrix A defined in the theorem satisfies

$$\theta_A(x) = 1 + x + \cdots + x^{k-1}.$$

It follows by induction on k that

$$\theta_A(x)\theta_A(x^k)\cdots\theta_A(x^{k^{m-1}}) = 1 + x + x^2 + \cdots + x^{k^m - 1}.$$

Since $k^m = cn$, we have

$$\theta_A(x)\theta_A(x^k)\cdots\theta_A(x^{k^{m-1}}) = 1 + x + x^2 + \cdots + x^{cn-1}$$
$$\equiv c(1 + x + \cdots + x^{n-1}) \quad (\text{mod } x^n - 1).$$

We now apply Lemma 5.3.5 and obtain the desired conclusion. $\qquad\square$

If $d \neq 0$, then the condition $k^m = -d + cn$ is not in general sufficient to guarantee the existence of a $(0,1)$-matrix A of order n satisfying $A^m = -dI + cJ$. Let $d = -1$. If $k^m = 1 + cn$, then Lemma 5.3.5 also implies that the k-circulant matrix A constructed in Theorem 5.3.6 satisfies $A^m = I + cJ$.

The following theorem of Lam and van Lint[1978] completely settles the existence question in the case $c = d = 1$.

Theorem 5.3.7.   *Let m be a positive integer. There exists a $(0,1)$-matrix A of order n satisfying the equation*

$$A^m = -I + J$$

if and only if m is odd and $n = k^m + 1$ for some positive integer k.

Proof. Suppose that A is a $(0,1)$-matrix satisfying $A^m = -I + J$. Then A has trace equal to zero and by Theorem 5.3.2 A is a regular matrix of degree k with $n = k^m + 1$. First assume that $m = 2$. The eigenvalues of $J - I$ are $n - 1 = k^2$ with multiplicity 1 and -1 with multiplicity $n - 1$. Hence the eigenvalues of A are k with multiplicity 1 and $\pm i$ with equal multiplicities. This contradicts the fact that A has zero trace. If m is even, then $(A^{m/2})^2 = -I + J$ where $A^{m/2}$ is also a $(0,1)$-matrix. We conclude that m is an odd integer.

We now suppose that m is an odd integer and $n = k^m + 1$. Let $g = -k$ and let A be the g-circulant matrix of order n whose first row consists of 0 followed by k 1's and $(n - 1 - k)$ 0's. The Hall polynomial of A satisfies

$$\theta_A(x) = x + x^2 + \cdots + x^k.$$

We have

$$\theta_A(x)\theta_A(x^g)\cdots\theta_A(x^{g^{m-1}}) \equiv -1 + (1 + x + \cdots + x^{n-1}) \pmod{x^n - 1},$$

and

$$g^m = (-k)^m = -k^m = -n + 1 \equiv 1 \pmod{n}.$$

We now deduce from Lemma 5.3.5 that $A^m = -I + J$.   □

Although the matrix equation $A^m = -dI + cJ$ has a simple form some difficult questions emerge. A complete characterization of those integers m, d and c for which there exists a $(0,1)$-matrix solution is very much unsettled. In those instances where a solution is known to exist, virtually nothing is known about the number of nonisomorphic solutions for general n. If $n = k^2$ the number of regular $(0,1)$-matrices of order n satisfying $A^2 = J$ is unknown (Hoffman[1967]).

If in the equation $A^m = dI + cJ$, we do not regard d and c as prescribed, then different questions emerge. In these circumstances we seek $(0,1)$-matrices A of order n for which A^m has all elements on its main diagonal equal and all off-diagonal elements equal. A trivial solution is the matrix $A = J$, and in this case $d = 0$ and $c = n^{m-1}$. Ma and Waterhouse[1987] have completely settled the existence of nontrivial solutions in the case $d = 0$.

Theorem 5.3.8.   *Let $m \geq 2$ and $n \geq 2$ be integers. There exists a $(0,1)$-matrix A of order n with $A \neq J$ such that A^m has all of its elements equal if and only if n is divisible by the square of some prime number. Let g be the product of the distinct prime divisors of n. If n is divisible by the square of some prime number, then there exists a g-circulant matrix $A \neq J$ for which all elements of A^m are equal.*

Let A be a $(0,1)$-matrix of order n. We now consider the more general matrix equation

$$A^2 = E + cJ, \tag{5.16}$$

where E is a diagonal matrix and c is a positive integer. This corresponds to the case of a digraph D of order n with exactly c walks of length 2 between every pair of *distinct* vertices.

It is at once evident that the matrix A of (5.16) need no longer be regular. The following matrices with $c = 1$ provide counterexamples:

$$\begin{bmatrix} 1 & 1 & \cdots & 1 \\ 1 & & & \\ \vdots & & O & \\ 1 & & & \end{bmatrix} \quad (n \geq 2), \tag{5.17}$$

where O is the zero matrix of order $n-1$, and

$$\begin{bmatrix} 0 & 1 & \cdots & 1 \\ 1 & & & \\ \vdots & & Q & \\ 1 & & & \end{bmatrix} \quad (n \geq 4), \tag{5.18}$$

where Q is a symmetric permutation matrix of order $n-1$.

In this connection Ryser[1970] has established the following.

Theorem 5.3.9.   *Let A be a $(0,1)$-matrix of order $n > 1$ that satisfies the matrix equation*

$$A^2 = E + cJ,$$

where E is a diagonal matrix and c is a positive integer. Then there exists an integer k such that A is regular of degree k except for the $(0,1)$-matrices of order n with $c = 1$ isomorphic to (5.17) or (5.18) and the $(0,1)$-matrix of order 5 with $c = 2$ isomorphic to

$$\begin{bmatrix} 0 & 1 & 1 & 1 & 1 \\ 1 & 1 & 1 & 0 & 0 \\ 1 & 0 & 0 & 1 & 1 \\ 1 & 1 & 1 & 0 & 0 \\ 1 & 0 & 0 & 1 & 1 \end{bmatrix}.$$

Furthermore, if A is regular of degree k, then

$$A^2 = dI + cJ,$$

where

$$k^2 = d + cn$$

and

$$-c < d \leq k - c.$$

Bridges[1971] has extended the preceding result and found all of the non-regular solutions of

$$A^2 - aA = E + cJ.$$

In addition Bridges[1972] and Bridges and Mena[1981] have completely settled the regularity question for matrix equations of the form $A^r = E + cJ$.

Theorem 5.3.10. *Let A be a (0, 1)-matrix of order n that satisfies the equation*

$$A^r = E + cJ,$$

where E is a diagonal matrix and r and c are positive integers. Then A is regular provided $n > 3$ and $r > 3$.

Additional information on regular solutions of the various types of matrix equations described here can be found in Chao and Wang[1987], King and Wang[1985], Knuth[1970], Lam[1975] and Wang[1980, 1981, 1982].

Exercises

1. Determine the Hoffman polynomial of the strongly connected digraph of order n each of whose vertices has indegree and outdegree equal to 1 (a directed cycle of length n).
2. Determine the Hoffman polynomial of the digraph obtained from the complete bipartite graph $K_{m,m}$ by replacing each edge by two oppositely directed arcs.
3. Prove Lemma 5.3.3.
4. Suppose that $k^m = 1 + cn$. Prove that the k-circulant matrix A in Theorem 5.3.6 satisfies the equation $A^m = I + cJ$.
5. Construct the digraph of the solution A of order $n = 9$ of the equation $A^3 = -I + J$ given in the proof of Theorem 5.3.7.

References

C.M. Ablow and J.L. Brenner[1963], Roots and canonical forms for circulant matrices, *Trans. Amer. Math. Soc.*, 107, pp. 360–376.

W.G. Bridges[1971], The polynomial of a non-regular digraph, *Pacific J. Math.*, 38, pp. 325–341.

[1972], The regularity of x^3-graphs, *J. Combin. Theory, Ser. B*, 12, pp. 174–176.

W.G. Bridges and R.A. Mena[1981], x^k-digraphs, *J. Combin. Theory, Ser. B*, 30, pp. 136–143.

C.Y. Chao and T. Wang[1987], On the matrix equation $A^2 = J$, *J. Math. Res. and Exposition*, 2, pp. 207–215.

P.J. Davis[1979], *Circulant Matrices*, Wiley, New York.

A.J. Hoffman[1967], Research problem 2-11, *J. Combin. Theory*, 2, p. 393.

A.J. Hoffman and M.H. McAndrew[1965], The polynomial of a directed graph, *Proc. Amer. Math. Soc.*, 16, pp. 303–309.

F. King and K. Wang[1985], On the g-circulant solutions to the matrix equation $A^m = \lambda J$, II, *J. Combin. Theory, Ser. A*, 38, pp. 182–186.

D.E. Knuth[1970], Notes on central groupoids, *J. Combin. Theory*, 8, pp. 376–390.

C.W.H. Lam[1975], A generalization of cyclic difference sets I, *J. Combin. Theory, Ser. A*, 19, pp. 51–65.

[1975], A generalization of cyclic difference sets II, *J. Combin. Theory, Ser. A*, 19, pp. 177–191.

[1977], On some solutions of $A^k = dI + \lambda J$, *J. Combin. Theory, Ser. A*, 23, pp. 140–147.

C.W.H. Lam and J.H. van Lint[1978], Directed graphs with unique paths of fixed length, *J. Combin. Theory, Ser. B*, 24, pp. 331–337.

S.L. Ma and W.C. Waterhouse[1987], The g-circulant solutions of $A^m = \lambda J$, *Linear Alg. Applics.*, 85, pp. 211–220.

H.J. Ryser[1970], A generalization of the matrix equation $A^2 = J$, *Linear Alg. Applics.*, 3, pp. 451–460.

K. Wang[1980], On the matrix equation $A^m = \lambda J$, *J. Combin. Theory, Ser. A*, 29, pp. 134–141.

[1981], A generalization of group difference sets and the matrix equation $A^m = dI + \lambda J$, *Aequationes Math.*, 23, pp. 212–222.

[1982], On the g-circulant solutions to the matrix equation $A^m = \lambda J$, *J. Combin. Theory, Ser. A*, 33, pp. 287–296.

6

Existence Theorems

6.1 Network Flows

Let D be a digraph of order n whose set of vertices is the n-set V. Let E be the set of arcs of D and let $c : E \to Z^+$ be a function which assigns to each arc $\alpha = (x, y)$ a nonnegative integer

$$c(\alpha) = c(x, y).$$

The integer $c(x, y)$ is called the *capacity* of the arc (x, y) and the function c is a *capacity function* for the digraph D. In this chapter, loops (arcs joining a vertex to itself) are of no significance and thus we implicitly assume throughout that D has no loops.

Let s and t be two distinguished vertices of D, which we call the *source* and *sink*, respectively, of D. The quadruple

$$N = \langle D, c, s, t \rangle$$

is called a *capacity-constrained network*. We could replace D with a general digraph in which the arc (x, y) of D has multiplicity $c(x, y)$. However, it is more convenient and suggestive to continue with a digraph in which $c(x, y)$ represents the capacity of the arc (x, y).

A *flow from s to t* in the network N is a function $f : E \to Z^+$ from the set of arcs of D to the nonnegative integers which satisfies the constraints

$$0 \le f(x, y) \le c(x, y) \text{ for each arc } (x, y) \text{ of } D, \tag{6.1}$$

and

$$\sum_y f(x, y) - \sum_z f(z, x) = 0 \text{ for each vertex } x \neq s, t. \tag{6.2}$$

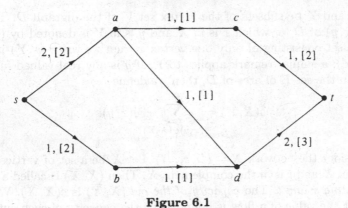

Figure 6.1

In (6.2) the first summation is over all vertices y such that (x,y) is an arc of D and the second summation is over all vertices z such that (z,x) is an arc of D. Since

$$\sum_{x \in V} \left(\sum_y f(x,y) - \sum_z f(z,x) \right) = \sum_{x \in V} \sum_y f(x,y) - \sum_{x \in V} \sum_z f(z,x) = 0,$$

it follows from (6.2) that

$$\sum_y f(s,y) - \sum_z f(z,s) = \sum_z f(z,t) - \sum_y f(t,y). \qquad (6.3)$$

Equations (6.2) are interpreted to mean that the net flow out of a vertex different from the source s and sink t is zero, while equation (6.3) means that the net flow out of the source s equals the net flow into the sink t. We let $v = v(f)$ be the common value of the two quantities in (6.3) and call v the *value of the flow* f. In a capacity-constrained network it is usual to allow the capacity and flow functions (and other imposed constraints) to take on any nonnegative real values. For combinatorial applications integer values are required. The theorems to follow remain valid if real values are permitted for both the capacity and flow values. With our restriction to integer values, there are only finitely many flows.

For an example, let N be the capacity-constrained network illustrated in Figure 6.1 where the numbers in brackets denote capacities of arcs and the other numbers denote the function values of a flow f. The value of this flow f is $v = 3$. No flow from s to t in N can have value greater than 3, since the value of a flow cannot exceed the amount of flow from the set $X = \{s, a, b\}$ of vertices to the set $Y = \{t, c, d\}$ of vertices and this latter quantity is bounded by the sum of the capacities of those arcs whose initial vertices lie in X and whose terminal vertices lie in Y. We now investigate in general the maximum value of a flow in a capacity-constrained network.

Let X and Y be subsets of the vertex set V of the digraph D. The set of arcs (x,y) of D for which x is in X and y is in Y is denoted by (X,Y). If X is a set consisting of only one vertex x then we write (x,Y) instead of $(\{x\},Y)$; a similar remark applies to Y. If g is any real-valued function defined on the set E of arcs of D, then we define

$$g(X,Y) = \sum_{(x,y)\in(X,Y)} g(x,y).$$

Now consider the network $N = \langle D,c,s,t\rangle$. Let X be a set of vertices such that s is in X and t is in the complement $\overline{X}$. Then $(X,\overline{X})$ is called a *cut in N separating s and t*. The *capacity of the cut* $(X,\overline{X})$ is $c(X,\overline{X})$. We first show that the value of a flow is bounded by the capacity of each cut.

Lemma 6.1.1.   *Let f be a flow with value v in the network $\langle D,c,s,t\rangle$, and let $(X,\overline{X})$ be a cut separating s and t. Then*

$$v = f(X,\overline{X}) - f(\overline{X},X) \le c(X,\overline{X}).$$

Proof. It follows from (6.2) and the definition of v as given by (6.3) that

$$f(x,V) - f(V,x) = \begin{cases} 0 & \text{if } x \ne s,t \\ v & \text{if } x = s \\ -v & \text{if } x = t \end{cases}.$$

Since s is in X and t is in $\overline{X}$, we have

$$\begin{aligned}
v &= \textstyle\sum_{x\in X}(f(x,V) - f(V,x)) = f(X,V) - f(V,X) \\
&= f(X, X\cup\overline{X}) - f(X\cup\overline{X}, X) \\
&= f(X,X) + f(X,\overline{X}) - f(X,X) - f(\overline{X},X) \\
&= f(X,\overline{X}) - f(\overline{X},X).
\end{aligned}$$

From (6.1) we conclude that $f(X,\overline{X}) \le c(X,\overline{X})$ and $f(\overline{X},X) \ge 0$, and the conclusion follows.   □

We now state and prove the fundamental *maxflow-mincut theorem* of Ford and Fulkerson[1956,1962] which asserts that there is a cut for which equality holds in Lemma 6.1.1.

Theorem 6.1.2.   *In a capacity-constrained network $N = \langle D,c,s,t\rangle$ the maximum value of a flow from the source s to the sink t equals the minimum capacity of a cut separating s and t.*

Proof. Let f be a flow from s to t in N whose value v is largest. By Lemma 6.1.1 it suffices to define a cut $(X, \overline{X})$ separating s and t whose capacity equals v. The set X is defined recursively as follows:

(i) $s \in X$;

(ii) if $x \in X$ and y is a vertex for which (x, y) is an arc and $f(x, y) < c(x, y)$, then $y \in X$;

(iii) if $x \in X$ and y is a vertex for which (y, x) is an arc and $f(y, x) > 0$, then $y \in X$.

We first show that $(X, \overline{X})$ is a cut separating s and t by verifying that $t \notin X$. Assume, to the contrary, that $t \in X$. It follows from the definition of X that there is a sequence $x_0 = s, x_1, \ldots, x_m = t$ of vertices such that for each $i = 0, 1, \ldots, m - 1$ either (x_i, x_{i+1}) or (x_{i+1}, x_i) is an arc, and

$$(x_i, x_{i+1}) \text{ is an arc and } c(x_i, x_{i+1}) - f(x_i, x_{i+1}) > 0 \qquad (6.4)$$

$$\text{or } (x_{i+1}, x_i) \text{ is an arc and } f(x_{i+1}, x_i) > 0. \qquad (6.5)$$

Let a be the positive integer equal to the minimum of the numbers occurring in (6.4) and (6.5). Let g be the function defined on the arcs of the digraph D which has the same values as f except that

$$g(x_i, x_{i+1}) = f(x_i, x_{i+1}) + a \text{ if } (x_i, x_{i+1}) \text{ is an arc,}$$

and

$$g(x_{i+1}, x_i) = f(x_{i+1}, x_i) - a \text{ if } (x_{i+1}, x_i) \text{ is an arc.}$$

Then g is a flow in N from s to t with value $v + a > v$ contradicting the choice of f. Hence $t \in \overline{X}$ and $(X, \overline{X})$ is a cut separating s and t.

From the definition of X it follows that

$$f(x, y) = c(x, y) \text{ if } (x, y) \in (X, \overline{X})$$

and

$$f(y, x) = 0 \text{ if } (y, x) \in (\overline{X}, X).$$

Hence $f(X, \overline{X}) = c(X, \overline{X})$ and $f(\overline{X}, X) = 0$. Applying Lemma 6.1.1 we conclude that $v = c(X, \overline{X})$.   □

Let $l : E \to Z^+$ be an integer-valued function defined on the set E of arcs of the digraph D such that for each arc (x, y),

$$0 \le l(x, y) \le c(x, y).$$

Suppose that in the definition of a flow f we replace (6.1) with

$$l(x, y) \le f(x, y) \le c(x, y) \text{ for each arc } (x, y) \text{ of } D. \qquad (6.6)$$

Thus l determines lower bounds on the flows of arcs. The proof of Theorem 6.1.2 can be adapted to yield: *the maximum value of a flow equals the minimum value of $c(X, \overline{X}) - l(\overline{X}, X)$ taken over all cuts separating s and t provided there is a flow satisfying* (6.2) *and* (6.6).

We now discuss the existence of flows satisfying (6.6) which satisfy (6.2) for *all* vertices x. More general results can be found in Ford and Fulkerson[1962].

A function $f : E \to Z^+$ defined on the set E of arcs of the digraph D with vertex set V which satisfies (6.6) and

$$f(x, V) - f(V, x) = 0 \text{ for all } x \in V \tag{6.7}$$

is called a *circulation on D with constraints* (6.6). The following fundamental theorem of Hoffman[1960] establishes necessary and sufficient conditions for the existence of circulations.

Theorem 6.1.3.   *There exists a circulation on the digraph D with constraints* (6.6) *if and only if for every subset X of the vertex set V,*

$$c(X, \overline{X}) \geq l(\overline{X}, X).$$

Proof. We define a network $N^* = (D^*, c^*, s, t)$ as follows. The digraph D^* is obtained from D by adjoining two new vertices s and t and all the arcs (s, x) and (x, t) with $x \in V$. If (x, y) is an arc of D, then $c^*(x, y) = c(x, y) - l(x, y)$. If $x \in V$, then $c^*(s, x) = l(V, x)$ and $c^*(x, t) = l(x, V)$.

From the rules

$$f^*(x, y) = f(x, y) - l(x, y) \text{ if } (x, y) \text{ is an arc of } D,$$
$$f^*(s, x) = l(V, x) \text{ if } x \in V,$$
$$f^*(x, t) = l(x, V) \text{ if } x \in V,$$

we see that there is a circulation f on D with constraints (6.6) if and only if there is a flow f^* in N^* with value equal to $l(V, V)$. The subsets X of V and the cuts $(X^*, \overline{X}^*)$ separating s and t are in one-to-one correspondence by the rules

$$X^* = X \cup \{s\}, \overline{X}^* = \overline{X} \cup \{t\}.$$

Moreover,

$$
\begin{aligned}
c^*(X^*, \overline{X}^*) &= c^*(X \cup \{s\}, \overline{X} \cup \{t\}) \\
&= c^*(X, \overline{X}) + c^*(s, \overline{X}) + c^*(X, t) \\
&= c(X, \overline{X}) - l(X, \overline{X}) + l(V, \overline{X}) + l(X, V) \\
&= c(X, \overline{X}) + l(\overline{X}, \overline{X}) + l(X, V) \\
&= c(X, \overline{X}) + l(V, V) - l(\overline{X}, X).
\end{aligned}
$$

Hence by Theorem 6.1.2 there is a flow f^* in N^* with value $l(V, V)$ if and only if $c(X, \overline{X}) \geq l(\overline{X}, X)$ for all $X \subseteq V$.   □

Theorem 6.1.3 can be used to obtain conditions for the existence of a flow f in a network N satisfying the lower and upper bound constraints (6.6). We add new arcs (s, t) and (t, s) with infinite capacity (any capacity larger than the capacity of each cut suffices) and apply Theorem 6.1.3. The resulting necessary and sufficient condition is that $c(X, \overline{X}) \geq l(\overline{X}, X)$ for all subsets X of vertices for which $\{s, t\} \subseteq X$ or $\{s, t\} \subseteq \overline{X}$.

The final flow theorem that we present concerns a network with multiple sources and multiple sinks in which an upper bound is placed on the net flow out of each of the source vertices and a lower bound is placed on the net flow into each of the sink vertices.

Let D be a digraph of order n with vertex set V. Let c be a nonnegative integer-valued capacity function defined on the set E of arcs of D. Suppose that S and T are disjoint subsets of the vertex set V, and let $W = V - (S \cup T)$. Let $a : S \rightarrow Z^+$ be a nonnegative integer-valued function defined on the vertices in S, and let $b : T \rightarrow Z^+$ be a nonnegative integer-valued function defined on the vertices in T. For $s \in S$, a(s) can be regarded as the *supply* at the source vertex s. For $t \in T$, b(t) is the *demand* at the sink vertex t. We call

$$N = \langle D, c, S, a, T, b \rangle$$

a *capacity-constrained, supply-demand network*, and we are interested in when a flow exists that satisfies the demands at the vertices in T without exceeding the supplies at the vertices in S. Let $f : E \rightarrow Z^+$ be a function which assigns to each arc in E a nonnegative integer. Then f is a *supply-demand flow* in N provided

$$f(s, V) - f(V, s) \leq a(s), (s \in S), \tag{6.8}$$

$$f(V, t) - f(t, V) \geq b(t), (t \in T), \tag{6.9}$$

$$f(x, V) - f(V, x) = 0, (x \in W), \tag{6.10}$$

and

$$0 \leq f(x, y) \leq c(x, y), ((x, y) \in E). \tag{6.11}$$

The following theorem is from Gale[1957].

Theorem 6.1.4. *In the capacity-constrained, supply-demand network $N = \langle D, c, S, a, T, b \rangle$ there exists a supply-demand flow if and only if*

$$b(T \cap \overline{X}) - a(S \cap \overline{X}) \leq c(X, \overline{X}) \tag{6.12}$$

for each subset X of the vertex set V.

Proof. We remark that if $X = \emptyset$, then (6.12) asserts that $\sum_{s \in S} a(s) \geq \sum_{t \in T} b(t)$, that is, the total demand does not exceed the total supply.

First suppose that there is a flow f satisfying (6.8)–(6.11). Let $X \subseteq V$. Then summing the constraints (6.8)–(6.10) over all vertices in $\overline{X}$ and using (6.11) we obtain

$$b(T \cap \overline{X}) - a(S \cap \overline{X}) \leq f(V, \overline{X}) - f(\overline{X}, V)$$
$$= f(X, \overline{X}) - f(\overline{X}, X) \leq c(X, \overline{X}).$$

Hence (6.12) holds for all $X \subseteq V$.

Now suppose that (6.12) holds for all $X \subseteq V$. We define a capacity-constrained network $N^* = \langle D^*, c^*, s^*, t^* \rangle$ by adjoining to D a new source vertex s^* and a new sink vertex t^* and all arcs of the form (s^*, s) with $s \in S$ and all arcs of the form (t, t^*) with $t \in T$. If (x, y) is an arc of D we define $c^*(x, y) = c(x, y)$. If $s \in S$, then $c^*(s^*, s) = a(s)$. If $t \in T$, then $c^*(t, t^*) = b(t)$. Let $(X^*, \overline{X}^*)$ be any cut of N^* which separates s^* and t^*, and let $X = X^* - \{s^*\}$ and $\overline{X} = \overline{X}^* - \{t^*\}$. Then

$$c^*(X^*, \overline{X}^*) - c^*(T, t^*)$$
$$= c^*(X, t^*) + c^*(s^*, \overline{X}) + c^*(X, \overline{X}) - c^*(T, t^*)$$
$$= (T \cap X) + a(S \cap \overline{X}) + c(X, \overline{X}) - b(T)$$
$$= -b(T \cap \overline{X}) + a(S \cap \overline{X}) + c(X, \overline{X}).$$

Hence by (6.12)

$$c^*(X^*, \overline{X}^*) \geq c^*(T, t^*)$$

for all cuts $(X^*, \overline{X}^*)$ of N^* which separate s^* and t^*. It follows that the minimum capacity of a cut separating s^* and t^* equals $c^*(T, t^*)$. By Theorem 6.1.2 there is a flow f^* in N^* with value equal to $c^*(T, t^*)$. Since $c^*(t, t^*) = b(t)$ for all t in T we have $f^*(t, t^*) = b(t), (t \in T)$. Let f be the restriction of f^* to the arcs of D. Then f satisfies (6.10) and (6.11). Moreover, for s in S

$$a(s) \geq f^*(s^*, s) = f^*(s, V) - f^*(V, s) = f(s, V) - f(V, s),$$

and for t in T

$$b(t) = f^*(t, t^*) = f^*(V, t) - f^*(t, V) = f(V, t) - f(t, V).$$

Thus f also satisfies (6.8) and (6.9). $\square$

A reformulation of Theorem 6.1.4 asserts that a supply-demand flow exists in N if and only if for each set of sink vertices there is a flow that satisfies the combined demand of those sink vertices without exceeding the supply at each source vertex. A precise statement is given below.

Corollary 6.1.5. *There exists a supply-demand flow f in $N = \langle D, c, S, a, T, b \rangle$ if and only if for each $U \subseteq T$, there is a flow f_U satisfying (6.8), (6.10) and (6.11) and*

$$f_U(V, U) - f_U(U, V) \geq b(U). \tag{6.13}$$

Proof. If f is a supply-demand flow in N, then for each $U \subseteq T$ we may choose $f_U = f$ to satisfy (6.8), (6.10), (6.11) and (6.13).

Conversely, suppose that for each $U \subseteq V$ there exists an f_U satisfying (6.8), (6.10), (6.11) and (6.13). Let X be a subset of V. By Theorem 6.1.4 it suffices to show that (6.12) is satisfied. Let $S' = S \cap \overline{X}$, $W' = W \cap \overline{X}$ and $U = T \cap \overline{X}$. Since f_U satisfies (6.8), (6.10) and (6.13) we have

$$-a(S') \leq f_U(V, S') - f_U(S', V),$$

$$0 = f_U(V, W') - f_U(W', V),$$

$$b(U) \leq f_U(V, U) - f_U(U, V).$$

Adding and using (6.11), we obtain

$$b(U) - a(S') \leq f_U(V, \overline{X}) - f_U(\overline{X}, V)$$

$$= f_U(X, \overline{X}) - f_U(\overline{X}, X) \leq c(X, \overline{X}). \qquad \square$$

A corollary very similar to Corollary 6.1.5 holds if the set T is replaced by the set S.

In the next sections we shall use the flow theorems presented here in order to obtain existence theorems for matrices, graphs and digraphs.

Exercises

1. Let $N = \langle D, c, s, t \rangle$ be a capacity-constrained network and let $l : E \to Z^+$ be an integer-valued function defined on the set E of arcs of D. Prove that the maximum value of a flow f satisfying $l(x, y) \leq f(x, y) \leq c(x, y)$ for each arc (x, y) of D equals the minimum value of $c(X, \overline{X}) - l(\overline{X}, X)$ taken over all cuts separating s and t, *provided at least one such flow f exists.*
2. Use Theorem 6.1.3 to show that there is a flow f in the network N satisfying $l(x, y) \leq f(x, y) \leq c(x, y)$ for each arc (x, y) if and only if $c(X, \overline{X}) \geq l(\overline{X}, X)$ for all subsets X of vertices for which $\{s, t\} \subseteq X$ or $\{s, t\} \subseteq \overline{X}$.
3. Suppose we drop the assumptions that the capacity function and flow function are integer valued. Prove that Theorem 6.1.2 remains valid.
4. Construct an example of a capacity-constrained network N whose capacity function is integer valued for which there is a flow f of maximum value, such that $f(x, y)$ is not an integer for at least one arc (x, y). (By Theorem 6.1.2 there is also an integer-valued flow of maximum value.)

References

L.R. Ford, Jr. and D.R. Fulkerson[1962], *Flows in Networks*, Princeton University Press, Princeton.

D. Gale[1957], A theorem on flows in networks, *Pacific J. Math.*, 7, pp. 1073–1082.

A.J. Hoffman[1960], Some recent applications of the theory of linear inequalities to extremal combinatorial analysis, *Proc. Symp. in Applied Mathematics*, vol. 10, Amer. Math. Soc., pp. 113–127.

6.2   Existence Theorems for Matrices

Let $A = [a_{ij}], (i = 1, 2, \ldots, m; j = 1, 2, \ldots, n)$ be an m by n matrix whose entries are nonnegative integers. Let

$$r_i = a_{i1} + a_{i2} + \cdots + a_{in}, \quad (i = 1, 2, \ldots, m)$$

be the sum of the elements in row i of A, and let

$$s_j = a_{1j} + a_{2j} + \cdots + a_{mj}, \quad (j = 1, 2, \ldots, m)$$

be the sum of the elements in column j of A. Then

$$R = (r_1, r_2, \ldots, r_m)$$

is the *row sum vector* of A and

$$S = (s_1, s_2, \ldots, s_n)$$

is the *column sum vector* of A. The vectors R and S consist of nonnegative integers and satisfy the fundamental equation

$$r_1 + r_2 + \cdots + r_m = s_1 + s_2 + \cdots + s_n. \tag{6.14}$$

The matrix A can be regarded as the reduced adjacency matrix of a bipartite multigraph G with bipartition $\{X, Y\}$ where $X = \{x_1, x_2, \ldots, x_m\}$ and $Y = \{y_1, y_2, \ldots, y_n\}$. The multiplicity of the edge $\{x_i, y_j\}$ equals $a_{ij}, (i = 1, 2, \ldots, m; j = 1, 2, \ldots, n)$. The vector R records the degrees of the vertices in X and the vector S records the degrees of the vertices in Y. Without loss of generality we choose the ordering of the vertices in X and in Y so that

$$r_1 \geq r_2 \geq \cdots \geq r_m \quad \text{and} \quad s_1 \geq s_2 \geq \cdots \geq s_n.$$

The vectors R and S are then said to be *monotone*.

Theorem 6.2.1. *Let $R = (r_1, r_2, \ldots, r_m)$ and $S = (s_1, s_2, \ldots, s_n)$ be nonnegative integral vectors. There exists an m by n nonnegative integral matrix with row sum vector R and column sum vector S if and only if (6.14) is satisfied.*

Proof. If there exists an m by n nonnegative integral matrix with row sum vector R and column sum vector S then (6.14) holds. Conversely, suppose that (6.14) is satisfied. We inductively construct an m by n nonnegative integral matrix $A = [a_{ij}]$ with row sum vector R and column sum vector S. If $m = 1$ we let

$$A = [\; s_1 \quad s_2 \quad \cdots \quad s_n \;].$$

If $n = 1$ we let

$$A = [\; r_1 \quad r_2 \quad \cdots \quad r_m \;]^T.$$

Now we assume that $m > 1$ and $n > 1$ and proceed by induction on $m + n$. Let

$$a_{11} = \min\{r_1, s_1\}.$$

First suppose that $a_{11} = r_1$. We then let $a_{12} = \cdots = a_{1n} = 0$, and define $R' = (r_2, \ldots, r_m)$ and $S' = (s_1 - r_1, s_2, \ldots, s_n)$. We have

$$r_2 + \cdots + r_m = (s_1 - r_1) + s_2 + \cdots + s_n,$$

and by the induction assumption there exists a nonnegative integral matrix A' with row sum vector R' and column sum vector S'. The matrix

$$\begin{bmatrix} r_1 & 0 & \cdots & 0 \\ & A' & & \end{bmatrix}$$

has row sum vector R and column sum vector S. If $a_{11} = s_1$, a similar construction works. $\square$

If $m = n$, the nonnegative integral matrix A of order n with row sum vector $R = (r_1, r_2, \ldots, r_n)$ and column sum vector $S = (s_1, s_2, \ldots, s_n)$ can also be regarded as the adjacency matrix of a general digraph of order n. The set of vertices of D is $V = \{a_1, a_2, \ldots, a_n\}$ and a_{ij} equals the multiplicity of the arc (a_i, a_j), $(i, j = 1, 2, \ldots, n)$. The vector R now records the outdegrees of the vertices and is the *outdegree sequence* of D. The vector S records the indegrees of the vertices and is the *indegree sequence* of D. When dealing with digraphs we may assume without loss of generality that R or S is monotone, but we cannot in general assume that both R and S are monotone. Theorem 6.2.1 provides a necessary and sufficient condition that nonnegative integral vectors R and S of the same length be the indegree sequence and outdegree sequence, respectively, of a general digraph.

We now investigate existence questions similar to the above in which a uniform bound is placed on the elements of the matrix. Although our formulations are in terms of matrices, there are equivalent formulations in terms of bipartite multigraphs (with a uniform bound on the multiplicities of the edges) and, in the case of square matrices, in terms of general

digraphs (with a uniform bound on the mutliplicities of arcs). More general
results with nonuniform bounds can be derived in a very similar way.

Theorem 6.2.2.   *Let $R = (r_1, r_2, \ldots, r_m)$ and $S = (s_1, s_2, \ldots, s_n)$ be
nonnegative integral vectors, and let p be a positive integer. There exists an
m by n nonnegative integral matrix $A = [a_{ij}]$ such that*

$$a_{ij} \leq p, \qquad (1 \leq i \leq m, 1 \leq j \leq n)$$

$$\sum_{j=1}^{n} a_{ij} \leq r_i, \quad (1 \leq i \leq m) \qquad (6.15)$$

$$\sum_{i=1}^{m} a_{ij} \geq s_j, \quad (1 \leq j \leq n)$$

if and only if

$$p|I||J| \geq \sum_{j \in J} s_j - \sum_{i \in \bar{I}} r_i, \quad (I \subseteq \{1, 2, \ldots, m\}; J \subseteq \{1, 2, \ldots, n\}). \quad (6.16)$$

Proof. We define a capacity-constrained, supply-demand network $N =
\langle D, c, S, a, T, b \rangle$ as follows. The digraph D has order $m + n$ and its set
of vertices is $V = S \cup T$ where $S = \{x_1, x_2, \ldots, x_m\}$ is an m-set and
$T = \{y_1, y_2, \ldots, y_n\}$ is an n-set. There is an arc of D from x_i to y_j with
capacity equal to p for each $i = 1, 2, \ldots, m$ and each $j = 1, 2, \ldots, n$. There
are no other arcs in D. We define $a(x_i) = r_i, (i = 1, 2, \ldots, m)$ and $b(y_j) =
s_j, (j = 1, 2, \ldots, n)$. If f is a supply-demand flow in N, then defining

$$a_{ij} = f(x_i, y_j), \quad (i = 1, 2, \ldots, m; j = 1, 2, \ldots, n)$$

we obtain a nonnegative integral matrix A satisfying (6.15). It follows from
Theorem 6.1.4 that there is a supply-demand flow f in N if and only if
(6.16) is satisfied.   □

If in Theorem 6.2.2 both R and S are monotone, then (6.16) is equivalent to

$$pkl \geq \sum_{j=1}^{l} s_j - \sum_{i=k+1}^{m} r_i, \quad (0 \leq k \leq m, 0 \leq l \leq n). \qquad (6.17)$$

The special case of Theorem 6.2.2 obtained by choosing $p = 1$ and by
assuming (6.14) is recorded in the following corollary.

Corollary 6.2.3.   *Let $R = (r_1, r_2, \ldots, r_m)$ and $S = (s_1, s_2, \ldots, s_n)$ be
nonnegative integral vectors satisfying (6.14). There exists an m by n $(0, 1)$-
matrix with row sum vector R and column sum vector S if and only if*

$$|I||J| \geq \sum_{j \in J} s_j - \sum_{i \in \bar{I}} r_i, (I \subseteq \{1, 2, \ldots, m\}; J \subseteq \{1, 2, \ldots, n\}). \qquad (6.18)$$

The conditions given in Corollary 6.2.3 for the existence of an m by n (0,1)-matrix with row sum vector R and column sum vector S can be formulated in terms of the concepts of conjugation and majorization of vectors. Let $R = (r_1, r_2, \ldots, r_m)$ be a nonnegative integral vector of length m, and suppose that $r_k \leq n, (k = 1, 2, \ldots, m)$. The *conjugate* of R is the nonnegative integral vector $R^* = (r_1^*, r_2^*, \ldots, r_n^*)$ where

$$r_k^* = |\{i : r_i \geq k, i = 1, 2, \ldots, m\}|.$$

(There is a certain arbitrariness in the length of the conjugate R^* of R in that its length n can be any integer which is not smaller than any component of R.) The conjugate of R is monotone even if R is not. We also have the elementary relationship

$$\sum_{i=1}^{k} r_i^* = \sum_{j=1}^{m} \min\{r_j, k\}. \qquad (6.19)$$

There is a geometric way to view the conjugate vector R^*. Consider an array of m rows which has r_i 1's in the first positions of row $i, (i = 1, 2, \ldots, m)$. Then $R^* = (r_1^*, r_2^*, \ldots, r_n^*)$ is the vector of column sums of the array. For example, if $R = (5, 3, 3, 2, 1, 1)$ and $n = 5$, then using the array

$$
\begin{array}{ccccc}
1 & 1 & 1 & 1 & 1 \\
1 & 1 & 1 & & \\
1 & 1 & 1 & & \\
1 & 1 & & & \\
1 & & & & \\
1 & & & &
\end{array}
$$

we see that $R^* = (6, 4, 3, 1, 1)$.

Now let $E = (e_1, e_2, \ldots, e_n)$ and $F = (f_1, f_2, \ldots, f_n)$ be two monotone, nonnegative integral vectors. Then we write $E \preceq F$ and say that E is *majorized* by F provided the partial sums of E and F satisfy

$$e_1 + e_2 + \cdots + e_k \leq f_1 + f_2 + \cdots + f_k, \qquad (k = 1, 2, \ldots, n)$$

with equality for $k = n$.

Theorem 6.2.4.   *Let $R = (r_1, r_2, \ldots, r_m)$ and $S = (s_1, s_2, \ldots, s_n)$ be nonnegative integral vectors, and let p be a positive integer. Assume also that $r_1 + r_2 + \cdots + r_m = s_1 + s_2 + \cdots + s_n$ and that S is monotone. There exists an m by n nonnegative integral matrix $A = [a_{ij}]$ with row sum vector R and column sum vector S satisfying $a_{ij} \leq p, (1 \leq i \leq m, 1 \leq j \leq n)$ if and only if*

$$\sum_{j=1}^{k} s_j \leq \sum_{i=1}^{m} \min\{r_i, pk\}, \qquad (k = 1, 2, \ldots, n) \qquad (6.20)$$

Proof. We consider the capacity-constrained supply-demand network $N = \langle D, c, S, a, T, b \rangle$ defined in the proof of Theorem 6.2.2. It follows from Corollary 6.1.5 that the matrix A exists if and only if $r_1 + r_2 + \cdots + r_m = s_1 + s_2 + \cdots + s_n$ and for each set $U \subseteq T$ there is a flow f_U satisfying (6.8) and (6.13) each of whose values lies between 0 and p. The set of arcs (U, V) is empty and $f_U(U, V) = 0$, and a flow f_U that maximizes $f_U(V, U) = f_U(S, U)$ is obtained by defining f_U so that

$$f_U(x_i, U) = \min\{r_i, p|U|\}, \quad (i = 1, 2, \ldots, m).$$

Thus the desired matrix A exists if and only if

$$\sum_{i=1}^{m} \min\{r_i, p|U|\} \geq \sum_{\{j : y_j \in U\}} s_j \text{ for all } U \subseteq T. \qquad (6.21)$$

The left side of (6.21) is the same for all subsets U of a fixed cardinality k. Since S is monotone, for fixed k the largest value of the right side of (6.21) is $\sum_{j=1}^{k} s_j$. Hence (6.21) holds if and only if

$$\sum_{i=1}^{m} \min\{r_i, pk\} \geq \sum_{j=1}^{k} s_j, \quad (k = 1, 2, \ldots, n). \qquad (6.22)$$

$\square$

The special case of Theorem 6.2.4 obtained by taking $p = 1$ is known as the *Gale–Ryser theorem* (see Gale[1957] and Ryser[1957]).

Corollary 6.2.5.   *Let $R = (r_1, r_2, \ldots, r_m)$ and $S = (s_1, s_2, \ldots, s_n)$ be nonnegative integral vectors. Assume that S is monotone and that $r_i \leq n$, $(i = 1, 2, \ldots, m)$. There exists an m by n $(0, 1)$-matrix with row sum vector R and column sum vector S if and only if $S \preceq R^*$.*

Proof. The corollary follows from Theorem 6.2.4 by choosing $p = 1$ and using the relationship (6.19). $\square$

If $m = n$, Corollary 6.2.3 provides necessary and sufficient conditions for the existence of a digraph with prescribed indegree sequence R and outdegree sequence S. Alternative conditions are given by Corollary 6.2.5 because we may assume without loss of generality that the outdegree sequence S is monotone. We now determine conditions for existence with the added restriction that the digraph has no loops. This is equivalent to determining conditions for the existence of a square $(0,1)$-matrix with zero trace having prescribed row and column sum vectors. Following the procedure used in the proof of Theorem 6.2.4 we obtain the following theorem of Fulkerson[1960].

Theorem 6.2.6.   *Let $R = (r_1, r_2, \ldots, r_n)$ and $S = (s_1, s_2, \ldots, s_n)$ be nonnegative integral vectors and let p be a positive integer. There exists a nonnegative integral matrix $A = [a_{ij}]$ of order n such that*

$$a_{ij} \le p, \quad (i, j = 1, 2, \ldots, n),$$

$$a_{ii} = 0, \quad (i = 1, 2, \ldots, n),$$

$$\sum_{j=1}^{n} a_{ij} \le r_i, \quad (i = 1, 2, \ldots, n),$$

$$\sum_{i=1}^{n} a_{ij} \ge s_j, \quad (j = 1, 2, \ldots, n)$$

if and only if

$$\sum_{j \in J} s_j \le \sum_{j \in J} \min\{r_j, p(|J| - 1)\} + \sum_{j \in \bar{J}} \min\{r_j, p|J|\}, \quad (J \subseteq \{1, 2, \ldots, n\}).$$

$$(6.23)$$

Now assume that $p = 1$, and that both R and S are monotone. (This assumption entails some loss of generality because of the requirement that the trace is to be zero.) Then conditions (6.23) simplify considerably. Let k be an integer with $1 \le k \le n$. For $J \subseteq \{1, 2, \ldots, n\}$ with $|J| = k$, the left side of (6.23) is maximal if $J = \{1, 2, \ldots, k\}$ and the right side is minimal if $J = \{1, 2, \ldots, k\}$. Thus (6.23) holds if and only if

$$\sum_{j=1}^{k} s_j \le \sum_{j=1}^{k} \min\{r_j, k - 1\} + \sum_{j=k+1}^{n} \min\{r_j, k\}, \quad (k = 1, 2, \ldots, n). \quad (6.24)$$

Consider an array of n rows which has r_i 1's in the first positions of row i with the exception that there is a 0 in position i of row i if $r_i \ge i$, $(1 \le i \le n)$. For example, if $R = (5, 3, 3, 2, 1, 1)$ the array is

$$
\begin{array}{cccccc}
0 & 1 & 1 & 1 & 1 & 1 \\
1 & 0 & 1 & 1 & & \\
1 & 1 & 0 & 1 & & \\
1 & 1 & & & & \\
1 & & & & & \\
1 & & & & &
\end{array}
$$

Let $R^{**} = (r_1^{**}, r_2^{**}, \ldots)$ be the vector of column sums of the array. Then R^{**} is called the *diagonally restricted conjugate* of R, and it follows that

$$\sum_{i=1}^{k} r_i^{**} = \sum_{j=1}^{k} \min\{r_j, k - 1\} + \sum_{j=k+1}^{n} \min\{r_j, k\}.$$

We thus obtain from Theorem 6.2.6 the following result of Fulkerson[1960].

Theorem 6.2.7.   *Let R and S be monotone, nonnegative integral vectors of length n. There exists a $(0,1)$-matrix of order n with row sum vector R and column sum vector S and trace zero if and only if $S \preceq R^{**}$.*

Theorems 6.2.6 and 6.2.7 have been generalized in Anstee[1982] by replacing the requirement that the matrix have zero trace with the requirement that there be a prescribed zero in at most one position in each column. Existence theorems more general than Theorem 6.2.2 can be derived from more general flow theorems than those presented in section 6.1.

Algorithms for the construction of the matrices considered in this section can be found in the references. In addition they will be discussed in the book *Combinatorial Matrix Classes.*

Exercises

1. Prove that the matrix A constructed inductively in the proof of Theorem 6.2.1 contains at most $m + n - 1$ positive elements and is the reduced adjacency matrix of a bipartite graph which is a forest.
2. Let A be an m by n $(0,1)$-matrix with row sum vector R and column sum vector S. Interpret the quantity

$$|I||J| - \sum_{j \in J} s_j + \sum_{i \in \overline{I}} r_i$$

   appearing in Corollary 6.2.3 as a counting function.
3. Prove that there exists an m by n $(0,1)$-matrix with all row sums equal to the positive integer p and column sum vector equal to the nonnegative integral vector $S = (s_1, s_2, \ldots, s_n)$ if and only if $p \leq n$, $s_j \leq m$, $(j = 1, 2, \ldots, n)$ and $pm = \sum_{j=1}^{n} s_j$.
4. Generalize Theorem 6.2.2 by replacing the requirement $a_{ij} \leq p$ with $a_{ij} \leq c_{ij}$ where c_{ij}, $(1 \leq i \leq m, 1 \leq j \leq n)$ are nonnegative integers.
5. Let $0 \leq r_i' \leq r_i, 0 \leq s_j' \leq s_j, c_{ij} \geq 0, (1 \leq i \leq m, \leq j \leq n)$ be integers. Prove that there exists an m by n nonnegative integral matrix $A = [a_{ij}]$ such that

$$a_{ij} \leq c_{ij}, \quad (1 \leq i \leq m, 1 \leq j \leq n),$$

$$r_i' \leq \sum_{j=1}^{n} a_{ij} \leq r_i, \quad (1 \leq i \leq m),$$

$$s_j' \leq \sum_{i=1}^{m} a_{ij} \leq s_j, \quad (1 \leq j \leq n)$$

if and only if

$$\sum_{i \in I, j \in J} c_{ij} \geq \max \left\{ \sum_{i \in I} r_i' - \sum_{j \in \overline{J}} s_j, \sum_{j \in J} s_j' - \sum_{i \in \overline{I}} r_i \right\}$$

(Mirsky[1968]).

References

R.P. Anstee[1982], Properties of a class of (0,1)-matrices covering a given matrix, *Canad. J. Math.*, 34, pp. 438–453.

L.R. Ford, Jr. and D.R. Fulkerson[1962], *Flows in Networks*, Princeton University Press, Princeton.

D.R. Fulkerson[1960], Zero-one matrices with zero trace, *Pacific J. Math.*, 10, pp. 831–836.

D. Gale[1957], A theorem on flows in networks, *Pacific J. Math.*, 7, pp. 1073–1082.

L. Mirsky[1968], Combinatorial theorems and integral matrices, *J. Combin. Theory*, 5, pp. 30–44.

H.J. Ryser[1957], Combinatorial properties of matrices of 0's and 1's, *Canad. J. Math*, 9, pp. 371–377.

6.3   Existence Theorems for Symmetric Matrices

Let $A = [a_{ij}]$ be a symmetric matrix of order n whose entries are nonnegative integers, and let $R = (r_1, r_2, \ldots, r_n)$ be the row sum vector of A. Since A is symmetric R is also the column sum vector of A. The matrix A is the adjacency matrix of a general graph G of order n. The vertex set of G is an n-set $V = \{a_1, a_2, \ldots, a_n\}$, and a_{ij} equals the multiplicity of the edge $\{a_i, a_j\}, (i, j = 1, 2, \ldots, n)$. The vector R records the degrees of the vertices and is called the *degree sequence* of G. A reordering of the vertices of G replaces A by the symmetric matrix $P^T A P$ for some permutation matrix P of order n. Thus without loss of generality we may assume that R is monotone.

Let $R = (r_1, r_2, \ldots, r_n)$ be an arbitrary nonnegative integral vector of length n. The diagonal matrix

$$\begin{bmatrix} r_1 & 0 & \cdots & 0 \\ 0 & r_2 & \cdots & 0 \\ \vdots & \vdots & \ddots & \vdots \\ 0 & 0 & \cdots & r_n \end{bmatrix}$$

is a symmetric, nonnegative integral matrix of order n with row sum vector R. We obtain necessary and sufficient conditions for the existence of a symmetric, nonnegative integral matrix with a uniform bound on its elements and with row sum vector equal to R from Theorem 6.2.2. We first prove a lemma which allows us to dispense with the symmetry requirement.

For a real matrix $X = [x_{ij}]$ of order n we define

$$q(X) = \sum_{i=1}^{n} \sum_{j=1}^{n} |x_{ij} - x_{ji}|.$$

We note that $q(X) = 0$ if and only if X is a symmetric matrix.

Lemma 6.3.1.   *Let $R = (r_1, r_2, \ldots, r_n)$ be a nonnegative integral vector and let p be a positive integer. Assume that there is a nonnegative integral matrix $A = [a_{ij}]$ of order n whose row and column sum vectors equal R and whose elements satisfy $a_{ij} \le p, (i, j = 1, 2, \ldots, n)$. Then there exists a symmetric, nonnegative integral matrix $B = [b_{ij}]$ of order n whose row and column sum vectors equal R and whose elements satisfy $b_{ij} \le p, (i, j = 1, 2, \ldots, n)$.*

Proof. Let $B = [b_{ij}]$ be a nonnegative integral matrix of order n whose row and column sum vectors equal R and whose elements satisfy $b_{ij} \le p$, $(i, j = 1, 2, \ldots, n)$ such that $q(B)$ is minimal. Suppose that $q(B) > 0$. Let D be the digraph of order n whose vertex set is $V = \{a_1, a_2, \ldots, a_n\}$ in which there is an arc (a_i, a_j) if and only if $b_{ij} > b_{ji}, (i, j = 1, 2, \ldots, n)$. The digraph D has no loops and since $q(B) > 0$, D has at least one arc. Let a_i be a vertex whose outdegree is positive. Since r_i equals the sum of the elements in row i of B and also equals the sum of the elements in column i of B, the indegree of a_i is also positive. Conversely, if the indegree of a_i is positive, then the outdegree of a_i is also positive. It follows that there exists in D a directed cycle

$$a_{i_1} \to a_{i_2} \to \cdots \to a_{i_k} \to a_{i_1}$$

of length k for some integer k with $2 \le k \le n$. We now define a new nonnegative integral matrix $B' = [b'_{ij}]$ of order n for which $b'_{ij} \le p, (i, j = 1, 2, \ldots n)$ and for which R is both the row and column sum vector. If k is even then B' is obtained from B by decreasing the elements $b_{i_1 i_2}, b_{i_3 i_4}, \ldots,$ $b_{i_{k-1} i_k}$ by 1 and increasing the elements $b_{i_3 i_2}, b_{i_5 i_4}, \ldots, b_{i_{k-1} i_{k-2}}, b_{i_1 i_k}$ by 1. Now suppose that k is odd. If $b_{i_1 i_1} = p$ then B' is obtained from B by decreasing $b_{i_1 i_1}, b_{i_2 i_3}, b_{i_4 i_5}, \ldots, b_{i_{k-1} i_k}$ by 1 and increasing $b_{i_2 i_1}, b_{i_4 i_3}, \ldots,$ $b_{i_{k-1} i_{k-2}}, b_{i_1 i_k}$ by 1. If $b_{i_1 i_1} < p$, then B' is obtained from B by increasing $b_{i_1 i_1}, b_{i_3 i_2}, b_{i_5 i_4}, \ldots, b_{i_k i_{k-1}}$ by 1 and decreasing $b_{i_1 i_2}, b_{i_3 i_4}, \ldots, b_{i_{k-2} i_{k-1}}, b_{i_k i_1}$ by 1. The matrix B' satisfies $q(B') < q(B)$ and this contradicts our choice of B. Hence $q(B) = 0$ and the matrix B is symmetric.   □

Theorem 6.3.2.   *Let $R = (r_1, r_2, \ldots, r_n)$ be a nonnegative integral vector and let p be a positive integer. There exists a symmetric, nonnegative integral matrix $B = [b_{ij}]$ whose row sum vector equals R and whose elements satisfy $b_{ij} \le p, (i, j = 1, 2, \ldots, n)$ if and only if*

$$p|I||J| \ge \sum_{j \in J} r_j - \sum_{i \in \overline{I}} r_i, \quad (I, J \subseteq \{1, 2, \ldots, n\}). \tag{6.25}$$

Proof. The theorem is an immediate consequence of Lemma 6.3.1 and Theorem 6.2.2.   □

If R is a monotone vector then (6.25) is equivalent to

$$pkl \geq \sum_{j=1}^{l} r_j - \sum_{i=k+1}^{n} r_i, \quad (k, l = 1, 2, \ldots, n). \tag{6.26}$$

Corollary 6.3.3.   Let $R = (r_1, r_2, \ldots, r_n)$ be a monotone, nonnegative integral vector. The following are equivalent:

(i) There exists a symmetric $(0, 1)$-matrix with row sum vector equal to R.

(ii) $kl \geq \sum_{j=1}^{l} r_j - \sum_{i=k+1}^{n} r_i, \quad (k, l = 1, 2, \ldots, n).$

(iii) $R \preceq R^*$.

Proof. The equivalence of (i) and (ii) is a consequence of Theorem 6.3.2. The equivalence of (i) and (iii) is a consequence of Lemma 6.3.1 and Corollary 6.2.5 and the observation that $R \preceq R^*$ implies that $r_1 \leq n$.   □

We now consider criteria for the existence of a symmetric, nonnegative integral matrix with zero trace having a prescribed row sum vector $R = (r_1, r_2, \ldots, r_n)$ and a uniform bound on its elements. Since the sum of the entries of a symmetric integral matrix with zero trace is even, a necessary condition is that $r_1 + r_2 + \cdots + r_n$ is an even integer. The following lemma is a special case of a more general theorem of Fulkerson, Hoffman and McAndrew[1965].

Lemma 6.3.4.   Let $R = (r_1, r_2, \ldots, r_n)$ be a nonnegative integral vector such that $r_1 + r_2 + \cdots + r_n$ is an even integer, and let p be a positive integer. Assume that there exists a nonnegative integral matrix $A = [a_{ij}]$ of zero trace whose row and column sum vectors equal R and whose elements satisfy $a_{ij} \leq p, (i, j = 1, 2, \ldots, n)$. Then there exists a symmetric, nonnegative integral matrix $B = [b_{ij}]$ of zero trace whose row and column sum vectors equal R and whose elements satisfy $b_{ij} \leq p$.

Proof. Let $B = [b_{ij}]$ be a nonnegative integral matrix of zero trace such that R equals the row and column sum vectors of B, $b_{ij} \leq p, (i, j = 1, 2, \ldots, n)$ and $q(B)$ is minimal. Suppose that $q(B) > 0$. Let D be the general digraph of order n whose vertex set is $V = \{a_1, a_2, \ldots, a_n\}$ in which there is an arc (a_i, a_j) of multiplicity $a_{ij} - a_{ji}$ if $a_{ij} > a_{ji}, (i, j = 1, 2, \ldots, n)$. The digraph D has no loops and has exactly $q(B)$ arcs. Since R is both the row and column sum vector of B, the indegree of each vertex equals its outdegree. It follows that the arcs of D can be partitioned into sets each of which is the set of arcs of a directed cycle.

First suppose that there exists in D a closed directed walk $\gamma: a_{i_1} \to a_{i_2} \to \cdots \to a_{i_k} \to a_{i_1}$ of even length k in which the multiplicity of each arc on the walk does not exceed its multiplicity in D. We obtain a matrix B' from B as follows. We increase the elements $b_{i_2 i_1}, b_{i_4 i_3}, \ldots, b_{i_k i_{k-1}}$ by 1 and decrease the elements $b_{i_2i_3}, \ldots, b_{i_{k-2}i_{k-1}}, b_{i_ki_1}$ by 1. The resulting matrix B' is a nonnegative integral matrix of zero trace whose row and column sum vectors equal R and whose elements do not exceed p. Moreover, $q(B') < q(B)$ contradicting our choice of B. It follows no such directed walk γ exists in D. We now conclude that the arcs of D can be partitioned into directed cycles of odd length and no two of these directed cycles have a vertex in common.

Let $a_{i_1} \to a_{i_2} \to \cdots \to a_{i_k} \to a_{i_1}$ be a directed cycle of D of odd length k. Since $r_1 + r_2 + \cdots + r_n$ is an even integer, D has another directed cycle $a_{j_1} \to a_{j_2} \to \cdots \to a_{j_l} \to a_{j_1}$ of odd length l. Neither of the arcs (a_{i_1}, a_{j_1}) and (a_{j_1}, a_{i_1}) belongs to D and hence $b_{i_1j_1} = b_{j_1i_1}$.

First suppose that $b_{i_1j_1} = b_{j_1i_1} \geq 1$. Let B' be the matrix obtained from B by decreasing each of $b_{i_1j_1}, b_{j_1i_1}, b_{i_2i_3}, \ldots, b_{i_{k-1},i_k}, b_{j_2j_3}, \ldots, b_{j_{l-1}j_l}$ by 1 and increasing $b_{i_2i_1}, \ldots, b_{i_{k-1}i_{k-2}}, b_{i_1i_k}, b_{j_2j_1}, \ldots, b_{j_{l-1}j_{l-2}}, b_{j_1j_l}$ by 1. Then B' is a nonnegative integral matrix of zero trace whose row and column sum vectors equal R and whose elements do not exceed p, and B satisfies $q(B') < q(B)$ contradicting our choice of B. A similar construction results in a contradiction if $b_{i_1j_1} = b_{j_1i_1} = 0$. We conclude that $q(B) = 0$ and hence that B is symmetric.   □

We now obtain the conditions of Chungphaisan[1974] for the existence of a symmetric nonnegative integral matrix with a uniform bound on its elements and with a prescribed row sum vector.

Theorem 6.3.5. *Let $R = (r_1, r_2, \ldots, r_n)$ be a monotone, nonnegative integral vector such that $r_1 + r_2 + \cdots + r_n$ is an even integer, and let p be a positive integer. There exists a symmetric, nonnegative integral matrix $A = [a_{ij}]$ of order n with zero trace whose row sum vector equals R and whose elements satisfy $a_{ij} \leq p, (i, j = 1, 2, \ldots, n)$ if and only if*

$$\sum_{i=1}^{k} r_i \leq pk(k-1) + \sum_{i=k+1}^{n} \min\{r_i, pk\}, \quad (k = 1, 2, \ldots, n). \qquad (6.27)$$

Proof. It follows from Theorem 6.2.6 and Lemma 6.3.4 that a matrix satisfying the properties of the theorem exists if and only if

$$\sum_{i=1}^{k} r_i \leq \sum_{i=1}^{k} \min\{r_i, p(k-1)\} + \sum_{i=k+1}^{n} \min\{r_i, pk\}, \quad (k = 1, 2, \ldots, n).$$

$$(6.28)$$

Suppose that (6.27) holds but (6.28) does not hold for some integer k. Clearly $k > 1$. Let q be the largest integer such that $r_q \geq p(k-1)$. Then $q < k$ and hence

$$\sum_{i=1}^{k} r_i > pq(k-1) + \sum_{i=q+1}^{k} r_i + \sum_{i=k+1}^{n} r_i.$$

Hence

$$\sum_{i=1}^{q} r_i > pq(k-1) + \sum_{i=k+1}^{n} r_i.$$

On the other hand by (6.27) we have

$$\sum_{i=1}^{q} r_i \le pq(q-1) + \sum_{i=q+1}^{n} \min\{r_i, pq\}$$

$$\le pq(q-1) + \sum_{i=q+1}^{k} \min\{r_i, pq\} + \sum_{i=k+1}^{n} r_i$$

$$\le pq(q-1) + (k-q)pq + \sum_{i=k+1}^{n} r_i$$

$$\le pq(k-1) + \sum_{i=k+1}^{n} r_i.$$

This contradiction shows that (6.27) implies (6.28). The converse clearly holds and the theorem follows.   □

We now deduce the theorem of Erdös and Gallai[1960] for the existence of a symmetric (0,1)-matrix with zero trace and prescribed row sum vector (a graph with prescribed degree sequence).

Theorem 6.3.6.   *Let $R = (r_1, r_2, \ldots, r_n)$ be a monotone, nonnegative integral vector such that $r_1 + r_2 + \cdots + r_n$ is an even integer. Then the following statements are equivalent:*

(i) *There exists a symmetric $(0, 1)$-matrix with zero trace whose row sum vector equals R.*

(ii) $R \preceq R^{**}$.

(iii) $\sum_{i=1}^{k} r_i \le k(k-1) + \sum_{i=k+1}^{n} \min\{k, r_i\}, \quad (k = 1, 2, \ldots, n)$.

Proof. The theorem is a direct consequence of Theorem 6.3.5, Theorem 6.2.7 and Lemma 6.3.4.   □

Algorithms for the construction of the matrices considered in this section can be found in Havel[1955], Hakimi[1962], Fulkerson[1960], Chungphaisan[1974] and Brualdi and Michael[1989].

Exercises

1. Prove the following generalization of Lemma 6.3.4: Let C be a symmetric, nonnegative integral matrix with zero trace. Let D be the general digraph of order n whose adjacency matrix is C. Assume that for each pair of cycles

γ and γ' of D of odd length, either γ and γ' have a vertex in common or there is an edge joining a vertex of γ and a vertex of γ'. If there exists a nonnegative integral matrix B of order n with row and column sum vector R such that $B \leq C$ (entrywise), then there exists a symmetric, nonnegative integral matrix A of order n with row and column sum vector R such that $A \leq C$ (Fulkerson, Hoffman and McAndrew[1965]).

2. Prove that there exists a symmetric, nonnegative integral matrix of order n with row sum vector $R = (r_1, r_2, \ldots, r_n)$ which is the adjacency matrix of a tree of order n if and only if $r_i \geq 1, (i = 1, 2, \ldots, n)$ and $\sum_{i=1}^{n} r_i = 2(n - 1)$.

References

C. Berge[1973], *Graphs and Hypergraphs*, North-Holland, Amsterdam.

R.A. Brualdi and T.S. Michael[1989], The class of 2-multigraphs with a prescribed degree sequence, *Linear Multilin. Alg.*, 24, pp. 81–10.

V. Chungphaisan[1974], Conditions for sequences to be r-graphic, *Discrete Math.*, 7, pp. 31–39.

P. Erdös and T. Gallai[1960], *Mat. Lapok*, 11, pp. 264–274 (in Hungarian).

L.R. Ford, Jr. and D.R. Fulkerson[1962], *Flows in Networks*, Princeton University Press, Princeton.

D.R. Fulkerson[1960], Zero-one matrices with zero trace, *Pacific J. Math.*, 10, pp. 831–836.

D.R. Fulkerson, A.J. Hoffman and M.H. McAndrew[1965], Some properties of graphs with multiple edges, *Canad. J. Math.*, 17, pp. 166–177.

S.L. Hakimi[1962], On realizability of a set of integers as degrees of the vertices of a linear graph I, *J. Soc. Indust. Appl. Math.*, 10, pp. 496–506.

V. Havel[1955], A remark on the existence of finite graphs (in Hungarian), *Časopis Pěst. Mat.*, 80, pp. 477–480.

6.4 More Decomposition Theorems

Several decomposition theorems for matrices have already been established in section 4.4. In this section we obtain additional decomposition theorems by applying the network flow theorems of this chapter. We recall that Theorem 4.4.3 asserts that the m by n nonnegative integral matrix A with maximum line sum equal to k can be decomposed into k (and no fewer) subpermutation matrices $P_1, P_2, \ldots, P_k$ of size m by n. The ranks of these subpermutation matrices are unspecified.

Theorem 6.4.1. *Let A be an m by n nonnegative integral matrix, and let k and l be positive integers. Then A has a decomposition of the form*

$$A = P_1 + P_2 + \cdots + P_l$$

where $P_1, P_2, \ldots, P_l$ are subpermutation matrices of rank k if and only if each line sum of A is at most equal to l and the sum of the elements of A equals lk.

Proof. It is clear that the conditions given are necessary for there to exist a decomposition of A into l subpermutation matrices of rank k. Now suppose that no line sum of A exceeds l and that the sum of the entries of A equals lk. If necessary we augment A by including additional lines of zeros and assume that $m = n$. Let the row and column sum vectors of A be $(r_1, r_2, \ldots, r_n)$ and $(s_1, s_2, \ldots, s_n)$, respectively. By Theorem 6.2.1 there exists an n by $n - k$ nonnegative integral matrix A_1 with row sum vector $(l - r_1, l - r_2, \ldots, l - r_n)$ and column sum vector $(l, l, \ldots, l)$. There also exists an $n - k$ by n nonnegative integral matrix A_2 with row sum vector $(l, l, \ldots, l)$ and column sum vector $(l - s_1, l - s_2, \ldots, l - s_n)$. Let

$$ B = \left[\begin{array}{cc} A & A_1 \\ A_2 & O \end{array} \right] $$

where O denotes a zero matrix of order $n - k$. Then B is a nonnegative integral matrix of order $2n - k$ with all line sums equal to l. By Theorem 4.4.3 B has a decomposition

$$ B = Q_1 + Q_2 + \cdots + Q_l $$

into l permutation matrices of order $2n - k$. Each of the permutation matrices Q_i has $n - k$ 1's in positions corresponding to those of A_1 and $n - k$ 1's in positions corresponding to those of A_2. Let P_i be the leading principal submatrix of Q_i of order $n, (i = 1, 2, \ldots, l)$. Then each P_i is a subpermutation matrix of rank k and $A = P_1 + P_2 + \cdots + P_l$.   □

For each positive integer k a nonnegative integral matrix has a decomposition into subpermutation matrices of rank at most equal to k. However, not every nonnegative integral matrix can be expressed as a sum of subpermutation matrices of rank at least equal to k.

Recall that the sum of the elements of a matrix A is denoted by $\sigma(A)$.

Corollary 6.4.2.   *Let A be an m by n nonnegative integral matrix and let l be the maximum line sum of A. Let k be a positive integer. Then A has a decomposition into subpermutation matrices each of whose ranks is at least equal to k if and only if $\sigma(A) \geq lk$. Moreover, if $\sigma(A) \geq lk$, then there exists an integer $k' \geq k$ such that A has a decomposition into subpermutation matrices where the rank of each subpermutation matrix equals k' or $k' + 1$.*

Proof. If A has a decomposition of the type described in the theorem, then $\sigma(A) \geq lk$. Now suppose that $\sigma(A) \geq lk$. Let $l' = \lfloor \sigma(A)/k \rfloor$. Then $l' \geq l$ and

$$ l'k \leq \sigma(A) < (l' + 1)k, $$

and there exists an integer $k' \geq k$ such that

$$ l'k' \leq \sigma(A) < l'(k' + 1). $$

Let $\sigma(A) = l'k' + p$ where $0 \leq p < l'$ and define

$$A' = \begin{bmatrix} & & & 0 \\ & A & & \vdots \\ & & & 0 \\ 0 & \cdots & 0 & l' - p \end{bmatrix}.$$

Then A' is a nonnegative integral matrix with all line sums at most equal to l' and with sum of elements $\sigma(A') = l'(k'+1)$. By Theorem 6.4.1 A' has a decomposition into l' subpermutation matrices of rank $k' + 1$. Hence A has a decomposition into l' subpermutation matrices of rank k' or $k' + 1$.

□

Let A be an arbitrary nonnegative integral matrix. We now consider decompositions of A of the form

$$A = P_1 + P_2 + \cdots + P_l + X$$

where $P_1, P_2, \ldots, P_l$ are subpermutation matrices of a prescribed rank k and X is a nonnegative integral matrix. We define $\pi_k(A)$ to be the maximum integer l for which such a decomposition exists and we seek to determine $\pi_k(A)$. It follows from Theorem 1.2.1 that $\pi_k(A) \geq 1$ if and only if A does not have a line cover consisting of fewer than k lines. By Theorem 6.4.1 $\pi_k(A)$ equals the maximum integer l such that A has a decomposition of the form

$$A = B + X,$$

where B is a nonnegative integral matrix with sum of elements $\sigma(B) = kl$ and with line sums not exceeding l, and X is a nonnegative integral matrix. The following theorem is from Folkman and Fulkerson[1969] (see also Fulkerson[1964]).

Theorem 6.4.3. *Let $A = [a_{ij}]$ be an m by n nonnegative integral matrix and let k be a positive integer. Then*

$$\pi_k(A) = \min \left\lfloor \frac{\sigma(A')}{e + f + k - m - n} \right\rfloor$$

where the minimum is taken over all e by f submatrices A' of A with $e + f > m + n - k$.

Proof. Let l be a nonnegative integer. We define a network $N = \langle D, c, s, t \rangle$ as follows. The digraph D has order $m + n$ and its set of vertices is $V = X \cup Y$ where $X = \{x_1, x_2, \ldots, x_m\}$ is an m-set and $Y = \{y_1, y_2, \ldots, y_n\}$ is an n-set. There is an arc from x_i to y_j of capacity $a_{ij}, (i = 1, 2, \ldots, m; j = 1, 2, \ldots, n)$. There are also arcs from s to $x_i, (i = 1, 2, \ldots, m)$ and from y_j to $t, (j = 1, 2, \ldots, n)$ each of capacity l. By Theorem 6.4.1 $\pi_k(A) \geq l$ if and only if there exists a flow in N with value lk. Applying Theorem 6.1.2 we see that a flow with value lk exists if and only if

$$\sigma(A') + l(m - e) + l(n - f) \geq lk$$

for every e by f submatrix A' of A, $(e = 0, 1, \ldots, m; f = 0, 1, \ldots, n)$. The theorem now follows.   □

In Theorem 6.4.1 we determined when a nonnegative integral matrix has a decomposition into subpermutation matrices of a specified rank k. A more general question was considered by Folkman and Fulkerson[1969]. Let $K = (k_1, k_2, \ldots, k_l)$ be a monotone vector of positive integers, and let A be an m by n nonnegative integral matrix. A K-decomposition of A is a decomposition

$$A = P_1 + P_2 + \cdots + P_l$$

where P_i is an m by n subpermutation matrix of rank k_i, $(i = 1, 2, \ldots, l)$. Theorem 6.4.1 determines when the matrix A has a $(k, k, \ldots, k)$-decomposition. Since every nonnegative integral matrix is the sum of subpermutation matrices of rank 1, we can obtain from Theorem 6.4.3 necessary and sufficient conditions for A to have a K-decomposition if the vector K has the form $(k, k, \ldots, k, 1, 1, \ldots, 1)$, that is, if K has at most two different components one of which equals 1. The general case in which K has only two different components is settled in the following theorem of Folkman and Fulkerson[1969], which we state without proof.

Theorem 6.4.4.   *Let $K = (k_1, k_2, \ldots, k_l)$ be a monotone vector of positive integers, and let $K^* = (k_1^*, k_2^*, \ldots)$ be the conjugate of K. Let A be an m by n nonnegative integral matrix. If A has a K-decomposition, then*

$$\sigma(A') \geq \sum_{j \geq m-e+n-f+1} k_j^* \tag{6.29}$$

for all e by f submatrices A' of A, $(e = 0, 1, \ldots, m; f = 0, 1, \ldots, n)$ with equality for $A' = A$. If the components of K take on at most two different values and (6.29) holds, with equality for $A' = A$, then A has a K-decomposition.

We conclude this section by stating without proof another decomposition theorem of Folkman and Fulkerson[1969] which is used in their proof of Theorem 6.4.4. It can be proved using the network flow theorems of Section 6.1.

Theorem 6.4.5.   *Let A be an m by n nonnegative integral matrix with row sum vector $R = (r_1, r_2, \ldots, r_m)$ and column sum vector $S = (s_1, s_2, \ldots, s_n)$. Let $R' = (r_1', r_2', \ldots, r_m')$, $R'' = (r_1'', r_2'', \ldots, r_m'')$, $S' = (s_1', s_2', \ldots, s_n')$, and $S'' = (s_1'', s_2'', \ldots, s_n'')$ be nonnegative integral vectors such that $r_i \leq r_i' + r_i''$, $(i = 1, 2, \ldots, m)$ and $s_j \leq s_j' + s_j''$, $(j = 1, 2, \ldots, n)$. Let σ' and*

σ'' *be nonnegative integers such that* $\sigma(A) = \sigma' + \sigma''$. *Then there exist* m *by* n *nonnegative integral matrices* A' *and* A'' *such that* $A = A' + A''$ *where* $\sigma(A') = \sigma'$, $\sigma(A'') = \sigma''$, *the components of the row and column sum vectors of* A' *do not exceed the corresponding components of* R' *and* S', *respectively, and the components of the row and column sum vectors of* A'' *do not exceed the corresponding components of* R'' *and* S'', *respectively, if and only if*

$$\sigma' - \sum_{i \in \overline{I}} r_i' - \sum_{j \in \overline{J}} s_j' \le \sigma(A[I, J]) \le \sum_{i \in I} r_i' + \sum_{j \in J} s_j''.$$

and

$$\sigma'' - \sum_{i \in \overline{I}} r_i'' - \sum_{j \in \overline{J}} s_j'' \le \sigma(A[I, J]) \le \sum_{i \in I} r_i'' + \sum_{j \in J} s_j'$$

for all $I \subseteq \{1, 2, \ldots, m\}$ *and all subsets* $J \subseteq \{1, 2, \ldots, n\}$. *Here* $A[I, J]$ *denotes the submatrix of* A *with row indices in* I *and column indices in* J.

The proof of the Theorem 6.4.5 proceeds by defining a digraph with certain lower and upper bounds on arc flows and then applying the circulation theorem, Theorem 6.1.3.

Exercises

1. Let A be an m by n nonnegative integral matrix and let p be the maximal line sum of A. Let k be a nonnegative integer. Prove that A has a decomposition into subpermutation matrices of ranks k or $k + 1$ if and only if

$$pk \le \sigma(A) \le \lfloor \sigma(A)/k \rfloor (k + 1).$$

2. Prove Theorem 6.4.5.

References

J. Folkman and D.R. Fulkerson[1969], Edge colorings in bipartite graphs, *Combinatorial Mathematics and Their Applications* (R.C. Bose and T. Dowling, eds.), University of North Carolina Press, Chapel Hill, pp. 561–577.

D.R. Fulkerson[1964], The maximum number of disjoint permutations contained in a matrix of zeros and ones, *Canad. J. Math.*, 10, pp. 729–735.

6.5 A Combinatorial Duality Theorem

Let $A = [a_{ij}]$, $(1 \le i \le m; 1 \le j \le n)$ be a nonnegative integral matrix and let k be a positive integer. We define

$$\sigma_k(A) = \max\{\sigma(P_1 + P_2 + \cdots + P_k)\}$$

where the maximum is taken over all m by n subpermutation matrices $P_1, P_2, \ldots, P_k$ such that

$$P_1 + P_2 + \cdots + P_k \leq A \text{ (entrywise)}.$$

Thus $\sigma_k(A)$ equals the maximum sum of the ranks of k subpermutation matrices whose sum does not exceed A. If A is a $(0,1)$-matrix, then $\sigma_k(A)$ equals the maximum sum of the ranks of k "disjoint" subpermutation matrices contained in A. By Theorem 4.4.3

$$\sigma_k(A) = \max\{\sigma(X)\}$$

where the maximum is taken over all m by n nonnegative integral matrices X such that $X \leq A$ and each line sum of X is at most equal to k. In terms of the bipartite graph G whose reduced adjacency matrix is A, $\sigma_k(A)$ equals the maximum number of edges of G which can be covered by k matchings. The integer $\sigma_1(A)$ equals the term rank $\rho(A)$ of A. If k equals the maximum line sum of A, then $\sigma_k(A)$ equals $\sigma(A)$, the sum of the entries of A. We define $\sigma_0(A)$ to be equal to 0, and let

$$\underline{\sigma}(A) = (\sigma_0(A), \sigma_1(A), \sigma_2(A), \ldots).$$

The sequence $\underline{\sigma}(A)$ is an infinite nondecreasing sequence with maximal element equal to $\sigma(A)$ and is called the *matching sequence* of A. In this section we discuss some elegant results of Saks[1986] concerning the matching sequence of a nonnegative integral matrix.

The following theorem of Vogel[1963] contains as a special case an evaluation of the terms of the matching sequence.

Theorem 6.5.1. *Let $A = [a_{ij}]$ be an m by n nonnegative integral matrix, and let $R = (r_1, r_2, \ldots, r_m)$ and $S = (s_1, s_2, \ldots, s_n)$ be nonnegative integral vectors. Then the maximum value of $\sigma(X)$ taken over all m by n nonnegative integral matrices $X = [x_{ij}]$ such that*

$$X \leq A; \quad \sum_{j=1}^{n} x_{ij} \leq r_i, \ (i = 1, 2, \ldots, m); \quad \sum_{i=1}^{m} x_{ij} \leq s_j, \ (j = 1, 2, \ldots, n)$$

equals

$$\min_{I \subseteq \{1,2,\ldots,m\}, J \subseteq \{1,2,\ldots,n\}} \left\{ \sum_{i \in I} r_i + \sum_{j \in J} s_j + \sum_{i \in \overline{I}, j \in \overline{J}} a_{ij} \right\}.$$

Proof. The proof is a straightforward consequence of the maxflow-mincut theorem, Theorem 6.1.2. We define a capacity-constrained network $N = \langle D, c, s, t \rangle$ as follows. The digraph D has order $m + n + 2$ and its set of vertices is $V = \{s, t\} \cup X \cup Y$ where $X = \{x_1, x_2, \ldots, x_m\}$ and

$Y = \{y_1, y_2, \ldots, y_n\}$. There is an arc from s to x_i of capacity $c(s, x_i) = r_i, (i = 1, 2, \ldots, m)$, an arc from y_j to t of capacity $c(y_j, t) = s_j$, $(j = 1, 2, \ldots, n)$ and an arc from x_i to y_j of capacity $c(x_i, y_j) = a_{ij}$, $(i = 1, 2, \ldots, m; j = 1, 2, \ldots, n)$. There are no other arcs in D. If f is a flow in N of value v, then defining

$$x_{ij} = f(x_i, y_j), \quad (i = 1, 2, \ldots, m; j = 1, 2, \ldots, n)$$

we obtain an m by n nonnegative integral matrix $X = [x_{ij}]$ such that $x_{ij} \leq a_{ij}$ for all i and j, the row sum vector R' and column sum vector S' of X satisfy $R' \leq R$ and $S' \leq S$, and $\sigma(X) = v$. Conversely, given such an X there is a flow with value $\sigma(X)$. Let $(Z, \overline{Z})$ be a cut separating s and t where

$$\overline{Z} \cap X = \{x_i : i \in I\} \quad \text{and} \quad Z \cap Y = \{y_j : j \in J\}.$$

Then the capacity of $(Z, \overline{Z})$ is

$$c(Z, \overline{Z}) = \sum_{i \in I} r_i + \sum_{j \in J} s_j + \sum_{i \in \overline{I}, j \in \overline{J}} a_{ij}.$$

The theorem now follows.   □

Let $I \subseteq \{1, 2, \ldots, m\}$ and $J \subseteq \{1, 2, \ldots, n\}$. Then we define

$$\sigma_{I,J}(A) = \sigma(A) - \sigma(A[\overline{I}, \overline{J}]),$$

the sum of the elements of A which belong to the union of the rows indexed by I and the columns indexed by J. Recall that $A[\overline{I}, \overline{J}]$ denotes the submatrix of A with row indices in $\overline{I}$ and column indices in $\overline{J}$. For $p \geq 0$, we let

$$\tau_p(A) = \max\{\sigma_{I,J}(A) : I \subseteq \{1, 2, \ldots, m\}, J \subseteq \{1, 2, \ldots, n\}, |I| + |J| = p\},$$

$$(6.30)$$

the maximum sum of the elements of A lying in the union of p lines. We note that $\tau_p(A) = \sigma(A)$ for all $p \geq m + n$. The infinite nondecreasing sequence

$$\underline{\tau}(A) = (\tau_0(A), \tau_1(A), \tau_2(A), \ldots)$$

is called the *covering sequence* of A. The matching sequence of a nonnegative integral matrix can be obtained from its covering sequence.

Corollary 6.5.2.   *Let A be an m by n nonnegative integral matrix and let k be a nonnegative integer. Then*

$$\sigma_k(A) = \min\{\sigma(A) + kp - \tau_p(A) : p \geq 0\}, \quad (k \geq 0).$$

Proof. By Theorem 6.5.1 with $r_i = k, (i = 1, 2, \ldots, m)$, and $s_j = k, (j = 1, 2, \ldots, n)$, $\sigma_k(A)$ equals the minimum value of the quantities

$$k(|I| + |J|) + \sum_{i \in \overline{I}, j \in \overline{J}} a_{ij} = \sigma(A) + k(|I| + |J|) - \sigma_{I,J}(A),$$

over all subsets I of $\{1, 2, \ldots, m\}$ and J of $\{1, 2, \ldots, n\}$. Setting $p = |I| + |J|$ we see that $\sigma_k(A)$ equals

$$\min_{p \geq 0} \min\{\sigma(A) + kp - \sigma_{I,J}(A) : I \subseteq \{1, 2, \ldots, m\};$$
$$J \subseteq \{1, 2, \ldots, n\}, |I| + |J| = p\},$$

and the result follows from (6.30). □

We now digress and study some properties of integer sequences. Let n be a nonnegative integer. We denote by T_n the set of all infinite integer sequences

$$\underline{t} = (t_0, t_1, t_2, \ldots.)$$

satisfying

(i) $0 = t_0 \leq t_1 \leq t_2 \leq \cdots \leq n$, and
(ii) $t_l = n$ for some integer l (and hence for all integers greater than l).

We partially order the set T_n by defining $\underline{s} \leq_T \underline{t}$ provided $s_k \leq t_k$ for all $k \geq 0$. This partial order determines a lattice with meet and join given as follows:

$$s \wedge_T t = (\min\{s_0, t_0\}, \min\{s_1, t_1\}, \min\{s_2, t_2\}, \ldots.),$$

$$s \vee_T t = (\max\{s_0, t_0\}, \max\{s_1, t_1\}, \max\{s_2, t_2\}, \ldots.).$$

We note that $(0, n, n, n, \ldots.)$ is the unique maximal element of T_n.

A *composition* of a positive integer n is an infinite sequence

$$\underline{\lambda} = (\lambda_1, \lambda_2, \lambda_3, \ldots.)$$

of nonnegative integers with sum $\sum_{k \geq 1} \lambda_k = n$. Notice that there is no monotonicity assumption on the terms of a composition. The set of all compositions of the integer n is denoted by C_n. We partially order the set C_n by defining $\underline{\mu} \leq_C \underline{\lambda}$ provided

$$\sum_{i=1}^{k} \mu_i \leq \sum_{i=1}^{k} \lambda_i, \quad (k \geq 1).$$

Let $\underline{t} = (t_0, t_1, t_2, \ldots.)$ be a sequence of integers. The *difference sequence*

$$\delta \underline{t} = ((\delta \underline{t})_1, (\delta \underline{t})_2, (\delta \underline{t})_3, \ldots.)$$

is defined by $(\delta \underline{t})_k = t_k - t_{k-1}, (k \geq 1)$. If $\underline{t}$ is in T_n then $\delta \underline{t}$ is in C_n. Conversely, if $\underline{\lambda}$ is in C_n and we define $\underline{t} = (0, \lambda_1, \lambda_1 + \lambda_2, \dots)$, then $\delta t = \underline{\lambda}$. Moreover it follows that $\underline{s} \leq_T \underline{t}$ if and only if $\delta \underline{s} \leq_C \delta \underline{t}$. Hence $\delta : T_n \to C_n$ defines an isomorphism of the partially ordered sets T_n and C_n. Since T_n is a lattice, C_n is also a lattice and we denote its meet and join by $\wedge_C$ and $\vee_C$, respectively.

A *partition* of the integer n is a composition $\underline{\lambda}$ of n which satisfies the monotonicity assumption $\lambda_1 \geq \lambda_2 \geq \lambda_3 \geq \cdots$. The partitions of n form a finite subset P_n of the set C_n of compositions of n. A sequence $\underline{t}$ in T_n is called *convex* provided that

$$2t_k \geq t_{k+1} + t_{k-1}, (k = 1, 2, \dots).$$

The set of convex sequences in T_n is denoted by X_n. A sequence $\underline{t}$ in T_n belongs to X_n if and only if the difference sequence $\delta \underline{t}$ belongs to P_n. Thus $\delta : X_n \to P_n$ defines an isomorphism of the partially ordered sets X_n and P_n. In particular, X_n is a finite set.

The join $\vee_T$ of two convex sequences in X_n need not be a convex sequence. For instance, the join of the convex sequences

$$\underline{s} = (0, 4, 8, 12, 12, 12, \dots), \qquad \underline{t} = (0, 5, 7, 9, 11, 12, 12, \dots)$$

in X_{12} is

$$\underline{s} \vee_T \underline{t} = (0, 5, 8, 12, 12, 12, \dots)$$

and is not convex. Thus X_n is not in general a sublattice of the lattice T_n, and P_n is not in general a sublattice of C_n. However, X_n is closed under the meet operation $\wedge_T$ of T_n, and hence P_n is closed under the meet operation $\wedge_C$ of C_n.

Lemma 6.5.3. *If $\underline{s}$ and $\underline{t}$ are in X_n then their meet $\underline{s} \wedge_T \underline{t}$ is also in X_n. If $\underline{\lambda}$ and $\underline{\mu}$ are in P_n then their meet $\underline{\lambda} \wedge_C \underline{\mu}$ is also in P_n.*

Proof. We have

$$\begin{aligned}
2(\underline{s} \wedge_T \underline{t})_k &= 2 \min\{s_k, t_k\} \\
&= \min\{2s_k, 2t_k\} \geq \min\{s_{k+1} + s_{k-1}, t_{k+1} + t_{k-1}\} \\
&\geq \min\{s_{k+1}, t_{k+1}\} + \min\{s_{k-1}, t_{k-1}\} \\
&= (\underline{s} \wedge_T \underline{t})_{k+1} + (\underline{s} \wedge_T \underline{t})_{k-1}
\end{aligned}$$

and the result follows. $\qquad \square$

Let $\underline{t}$ belong to T_n and consider the nonempty set

$$\{\underline{s} : \underline{s} \in X_n, \underline{t} \leq_T \underline{s}\} \tag{6.31}$$

of convex sequences above $\underline{t}$. It follows from Lemma 6.5.3 that the set (6.31) has a unique minimal element, namely

$$\underline{t}^X = \wedge_T\{\underline{s} : \underline{s} \in X_n, \underline{t} \leq_T \underline{s}\},$$

and this minimal element is convex. The convex sequence $\underline{t}^X$ is called the *convex closure* of $\underline{t}$. Similarly it follows that for a composition $\underline{\lambda}$ in C_n there is a unique minimal partition $\underline{\lambda}^P \geq_C \underline{\lambda}$, called the *partition closure* of $\underline{\lambda}$. We have

$$(\delta\underline{t})^P = \delta\underline{t}^X, \quad (\underline{t} \in T_n). \tag{6.32}$$

We next determine the relationship between two convex sequences whose difference sequences are conjugate partitions.

Lemma 6.5.4.   *Let $\underline{s}$ and $\underline{t}$ be convex sequences in X_n. The following are equivalent:*

(i) $\delta\underline{s}$ and $\delta\underline{t}$ are conjugate partitions of n;
(ii) $s_k = \min\{n + kp - t_p : p \geq 0\}$, $(k = 0, 1, 2, \ldots)$;
(iii) $t_k = \min\{n + kp - s_p : p \geq 0\}$, $(k = 0, 1, 2, \ldots)$.

Proof. For $p \geq 0$ we have $t_p = \sum_{i=0}^{p}(\delta\underline{t})_i$ and hence

$$n + kp - t_p = \sum_{i=1}^{p} k + \sum_{i \geq p+1}(\delta\underline{t})_i.$$

It follows from this equation that

$$\min\{n + kp - t_p : p \geq 0\} = \sum_{l \geq 1}\min\{(\delta\underline{t})_l, k\} = \sum_{i=1}^{k}\max\{j : (\delta\underline{t})_j \geq i\}.$$

Thus *(ii)* is equivalent to

$$(\delta\underline{s})_k = \max\{j : (\delta\underline{t})_j \geq k\} = ((\delta\underline{t})^*)_k, \quad (k = 0, 1, 2, \ldots).$$

This proves the equivalence of *(i)* and *(ii)* and hence also the equivalence of *(i)* and *(iii)*.   □

By Lemma 6.5.4 the function $\Phi : X_n \to X_n$ defined by

$$(\Phi\underline{t})_k = \min\{n + kp - t_p : p \geq 0\}, \quad (k \geq 0); \quad \underline{t} \in X_n$$

is the bijection on X_n corresponding to the bijection $\Psi : P_n \to P_n$ determined by conjugation on P_n, that is,

$$\Phi = \delta\Psi\delta^{-1},$$

where $\Psi\underline{\lambda} = \underline{\lambda}^*, (\underline{\lambda} \in P_n).$

The function Φ can be extended to all of T_n by defining

$$(\Phi\underline{t})_k = \min\{n + kp - t_p : p \geq 0\}, \quad (k \geq 0),$$

for each $\underline{t}$ in T_n. We now show that $\Phi\underline{t}$ is always convex.

Lemma 6.5.5.   *For all $\underline{t}$ in T_n, $\Phi\underline{t}$ is in X_n.*

Proof. We have

$$(\Phi\underline{t})_0 = \min\{n - t_p : p \geq 0\} = 0,$$

since $t_p = n$ for some $p \geq 0$;

$$(\Phi\underline{t})_k = \min\{n + kp - t_p\} \leq n + k0 - t_0 = n,$$

with equality holding for $k = n$;

$$\begin{aligned}
(\Phi\underline{t})_{k+1} &= \min\{n + (k+1)p - t_p : p \geq 0\} \\
&\geq \min\{n + kp - t_p : p \geq 0\} = (\Phi\underline{t})_R, \quad (k \geq 0).
\end{aligned}$$

Hence $\Phi\underline{t}$ is in T_n. We also have

$$\begin{aligned}
2(\Phi\underline{t})_k &= 2\min\{n + kp - t_p : p \geq 0\} \\
&= \min\{n + (k+1)p - t_p + n + (k-1)p + t_p : p \geq 0\} \\
&\geq \min\{n + (k+1)p - t_p; p \geq 0\} + \min\{n + (k-1)p - t_p : p \geq 0\} \\
&= (\Phi\underline{t})_{k+1} + (\Phi\underline{t})_{k-1}, (k \geq 1).
\end{aligned}$$

Hence $\Phi\underline{t}$ is in X_n.   □

In summary, $\Phi : T_n \to X_n$ is a function which is a bijection on X_n. Moreover, it is a consequence of the definition of Φ that Φ is order-reversing, that is, $\underline{s} \leq_T \underline{t}$ implies that $\Phi\underline{t} \leq_T \Phi\underline{s}$.

Lemma 6.5.6.   *For all $\underline{t}$ in T_n,*

$$\underline{t} \leq_T \Phi^2\underline{t}.$$

Proof. We have

$$\begin{aligned}
(\Phi^2\underline{t})_k &= \min\{n + kp - (\Phi\underline{t})_p : p \geq 0\} \\
&= \min\{n + kp - \min\{n + pq - t_q : q \geq 0\} : p \geq 0\} \\
&= \min\{\max\{p(k - q) + t_q : q \geq 0\}; p \geq 0\} \\
&\geq \min\{t_k : p \geq 0\} = t_k, \quad (k \geq 0).
\end{aligned}$$
                                                                        □

We now show that $\Phi^2 \underline{t}$ is the convex closure of a sequence $\underline{t}$ in T_n.

Theorem 6.5.7.   *For all $\underline{t}$ in T_n,*

$$\Phi^2 \underline{t} = \underline{t}^X.$$

Proof. Using Lemma 6.5.5 and Lemma 6.5.6 we see that $\Phi^2 \underline{t}$ is a convex sequence and

$$\underline{t} \leq_T \underline{t}^X \leq_T \Phi^2 \underline{t}.$$

Since Φ is order reversing we have

$$\Phi \underline{t} \leq_T \Phi^3 \underline{t} \leq_T \Phi \underline{t}^X \leq_T \Phi \underline{t}.$$

Thus $\Phi^3 \underline{t} = \Phi \underline{t}^X$. Since Φ is a bijection on X_n, we conclude that $\Phi^2 \underline{t} = \underline{t}^X$.   □

The isomorphism δ of T_n and C_n now yields the following.

Theorem 6.5.8.   *Let $\underline{t}$ be a sequence in T_n. Then*

$$\delta \Phi \underline{t} = ((\delta \underline{t})^P)^*.$$

Proof. By Theorem 6.5.7 $\Phi^2 \underline{t} = \underline{t}^X$ and hence $\delta \Phi^2 \underline{t} = \delta \underline{t}^X$. By Lemmas 6.5.4 and 6.5.5, $\delta \Phi^2 \underline{t} = (\delta \Phi \underline{t})^*$ and hence

$$(\delta \Phi \underline{t})^* = \delta \underline{t}^X.$$

By (6.32) $\delta \underline{t}^X = (\delta \underline{t})^P$ and thus $(\delta \Phi \underline{t})^* = (\delta \underline{t})^P$.   □

We now return to the m by n nonnegative integral matrix A and its matching sequence $\underline{\sigma}(A) = (\sigma_0(A), \sigma_1(A), \sigma_2(A), \ldots.)$ and covering sequence $\underline{\tau}(A) = (\tau_0(A), \tau_1(A), \tau_2(A), \ldots.)$. The sequences $\underline{\sigma}(A)$ and $\underline{\tau}(A)$ belong to $T_{\sigma(A)}$.

Theorem 6.5.9.   *Let A be an m by n nonnegative integral matrix. Then*

$$\delta \underline{\sigma}(A) = ((\delta \underline{\tau}(A))^P)^*.$$

In particular, the sequence of differences of the matching sequence $\underline{\sigma}(A)$ is convex.

Proof. By Corollary 6.5.2, $\underline{\sigma}(A) = \Phi \underline{\tau}(A)$, and hence by Lemma 6.5.5, $\underline{\sigma}(A)$ is a convex sequence. The theorem now follows by applying Theorem 6.5.8.   □

In words, Theorem 6.5.9 says that the sequence of differences of the matching sequence is the conjugate of the partition closure of the sequence of differences of the covering sequence. The sequence of differences of the covering sequences is not in general convex. For instance, let

$$A = \begin{bmatrix} 1 & 1 & 1 & 1 \\ 1 & 1 & 1 & 1 \\ 1 & 0 & 0 & 0 \\ 0 & 1 & 0 & 0 \\ 0 & 0 & 1 & 0 \\ 0 & 0 & 0 & 1 \end{bmatrix}.$$

Then

$$\underline{\tau}(A) = (0, 4, 8, 9, 12, 12, \ldots.)$$

and

$$\delta\underline{\tau}(A) = (4, 4, 1, 3, 0, 0, \ldots.).$$

The partition closure of $\delta\underline{\tau}(A)$ is

$$(4, 4, 2, 2, 0, 0, \ldots.),$$

which is self-conjugate and equals the sequence of differences of the matching sequence $\underline{\sigma}(A)$.

More general results can be found in Saks[1986].

Exercises

1. Show how König's theorem (Theorem 1.2.1) follows from Theorem 6.5.1.
2. Let $\underline{\lambda}$ be a composition of n. Prove that the following procedure always results in the partition closure $\underline{\lambda}^P$ of $\underline{\lambda}$: Choose any j such that $\lambda_j < \lambda_{j+1}$. Replace λ_j by $\lambda_j + 1$ and λ_{j+1} by $\lambda_{j+1} - 1$. Repeat until a partition of n is obtained (Saks[1986]).
3. Determine the partition closure of the composition $(2, 3, 6, 2, 4, 4, 5, 0, 0, \ldots)$ of 26.
4. Determine the matching sequence and covering sequence of the m by n matrix of all 1's. Verify Theorem 6.5.9 in this case.
5. Determine the matching sequence and covering sequence of the matrix

$$A = \begin{bmatrix} 1 & 1 & 0 & 0 & 1 \\ 0 & 0 & 1 & 1 & 0 \\ 1 & 0 & 0 & 1 & 1 \\ 1 & 1 & 0 & 0 & 0 \\ 0 & 0 & 1 & 1 & 0 \\ 1 & 1 & 0 & 0 & 1 \end{bmatrix}$$

and verify Theorem 6.5.9 in this case.

References

M. Saks[1986], Some sequences associated with combinatorial structures, *Discrete Math.*, 59, pp. 135–166.

W. Vogel[1963], Bemerkungen zur Theorie der Matrizen aus Nullen und Einsen, *Archiv der Math.*, 14, pp. 139–144.

7

The Permanent

7.1 Basic Properties

Let $A = [a_{ij}]$, $(i = 1, 2, \ldots, m; j = 1, 2, \ldots, n)$ be a matrix of size m by n with elements in a field F and assume that $m \leq n$. As defined in Chapter 1 the *permanent* of A is

$$\operatorname{per}(A) = \sum a_{1i_1} a_{2i_2} \cdots a_{mi_m}, \qquad (7.1)$$

where the summation extends over all the m-permutations $(i_1, i_2, \ldots, i_m)$ of the integers $1, 2, \ldots, n$. There is a nonzero product in the summation (7.1) if and only if there exist m nonzero elements of A no two of which are on the same line. By Theorem 1.2.1 all of the products in the summation (7.1) are zero if and only if A has a p by q zero submatrix with $p + q = n + 1$.

The permanent of A is unchanged under arbitrary permutations of the lines of A and also under transposition of A if $m = n$. The permanent of the matrix obtained from A by multiplying the elements of some row by the scalar c equals $c\operatorname{per}(A)$. In addition, if the kth row α of A is the sum of two row vectors α' and α'', then $\operatorname{per}(A) = \operatorname{per}(A') + \operatorname{per}(A'')$ where A' is obtained from A by replacing α with α' and A'' is obtained from A by replacing α with α''. Thus the permanent is a linear function of each of the rows of the matrix A.

If A has the block form

$$\begin{bmatrix} A_1 & O \\ A_3 & A_2 \end{bmatrix}$$

where A_1 is a square matrix, then

$$\operatorname{per}(A) = \operatorname{per}(A_1)\operatorname{per}(A_2).$$

For a square matrix the definition of the permanent is the same as that of the determinant apart from a factor of ± 1 preceding each of the products in the summation (7.1). However, this similarity of the permanent to the determinant does not extend to an effective computational procedure for the permanent which is analogous to that for the determinant. This is because the permanent is in general greatly altered by the addition of a multiple of one row to another.

The Laplace expansion for the determinant has a simple counterpart for the permanent. Let $A(i,j)$ denote the matrix of size $m-1$ by $n-1$ obtained from A by deleting row i and column j, $(i=1,2,\ldots,m; j=1,2,\ldots,n)$. It follows as a direct consequence of the definition of the permanent that

$$\operatorname{per}(A) = \sum_{j=1}^{n} a_{ij}\operatorname{per}(A(i,j)), \quad (i=1,2,\ldots,m). \tag{7.2}$$

The expansion of the permanent in (7.2) is called the *Laplace expansion by row* i. If $m=n$ there is also a Laplace expansion by column i for the permanent.

We now describe a procedure of Ryser[1963] for the evaluation of the permanent. It is the best known general computational procedure for the permanent (see Nijenhuis and Wilf[1978]).

Theorem 7.1.1. *Let* $A = [a_{ij}]$ *be a matrix of size* m *by* n *with* $m \le n$. *For each integer* r *between* $n-m$ *and* $n-1$ *let* A_r *denote an* m *by* $n-r$ *submatrix of* A *obtained by deleting* r *columns of* A. *Let* $\prod(A_r)$ *denote the product of the row sums of* A_r. *Let* $\sum \prod(A_r)$ *denote the sum of the products* $\prod(A_r)$ *taken over all choices for* A_r. *Then*

$$\operatorname{per}(A) = \sum \prod (A_{n-m}) - \binom{n-m+1}{1} \sum \prod (A_{n-m+1}) +$$

$$\binom{n-m+2}{2} \sum \prod (A_{n-m+2}) - \cdots + (-1)^{m-1} \binom{n-1}{m-1} \sum \prod (A_{n-1}). \tag{7.3}$$

Proof. Let U denote the set of all sequences

$$j_1, j_2, \ldots, j_m \text{ where } 1 \le j_i \le n, \quad (i=1,2,\ldots,m). \tag{7.4}$$

For each sequence (7.4) let

$$f(j_1, j_2, \ldots, j_m) = a_{1j_1} a_{2j_2} \cdots a_{mj_m}.$$

Let P_i denote the set of all sequences (7.4) that do *not* contain the integer i, $(i=1,2,\ldots,n)$. Then

$$\operatorname{per}(A) = \sum f(j_1, j_2, \ldots, j_m)$$

where the summation extends over all those sequences (7.4) which belong to exactly $n - m$ of the sets $P_1, P_2, \ldots, P_n$. For each choice of $i_1, i_2, \ldots, i_r$ with

$$1 \leq i_1 < i_2 < \cdots < i_r \leq n, \tag{7.5}$$

let

$$f(P_{i_1}, P_{i_2}, \ldots, P_{i_r}) = \sum f(j_1, j_2, \ldots, j_m)$$

where the summation extends over all those sequences (7.4) which are in the intersection $P_{i_1} \cap P_{i_2} \cap \cdots \cap P_{i_r}$. Let A_r denote the submatrix of A obtained by deleting columns $i_1, i_2, \ldots, i_r$. Then

$$\prod(A_r) = f(P_{i_1}, P_{i_2}, \ldots, P_{i_r})$$

and the sum

$$\sum f(P_{i_1}, P_{i_2}, \ldots, P_{i_r})$$

taken over all $i_1, i_2, \ldots, i_r$ satisfying (7.5) equals

$$\sum \prod(A_r).$$

We now apply the inclusion-exclusion principle (Ryser[1963]) and use the fact that

$$\operatorname{per}(A) = \sum_{1 \leq i_1 < i_2 < \cdots < i_{n-m} \leq n} f(P_{i_1}, P_{i_2}, \ldots, P_{i_{n-m}})$$

to obtain the equation (7.3).   □

Corollary 7.1.2.   *Let A be a square matrix of order n. Then the permanent of A equals*

$$\prod(A) - \sum\prod(A_1) + \sum\prod(A_2) - \cdots + (-1)^{n-1}\sum\prod(A_{n-1}). \tag{7.6}$$

If A is the 2 by 2 matrix

$$\begin{bmatrix} a_{11} & a_{12} \\ a_{21} & a_{22} \end{bmatrix},$$

then (7.6) gives the following formula for the permanent:

$$\operatorname{per}(A) = (a_{11} + a_{12})(a_{21} + a_{22}) - (a_{11}a_{21} + a_{12}a_{22}).$$

Exercises

1. Use Corollary 7.1.2 in order to evaluate the permanent of the matrix J_n of all 1's thereby obtaining an identity for $n!$.
2. Use Corollary 7.1.2 in order to evaluate the permanent of the (0,1)-matrix of order n all of whose elements equal 1 except for p elements in the first row.
3. Compare the number of multiplications required to evaluate the permanent of a matrix using the definition (7.1) of the permanent and the formula (7.3).
4. Let A be a (0,1)-matrix of order n and let α and β denote, respectively, the first row and second row of A. Let A_1 denote the (0,1)-matrix obtained from A by replacing α by the row which has 1's in those positions in which at least one of α and β has a 1 and replacing β by the row which has 1's in those positions in which both α and β have 1's. Let A_2 be the (0,1)-matrix obtained from A by replacing α by the row which has 1's in those positions in which α has a 1 and β has a zero and replacing β by the row which has 1's in those positions in which β has a 1 and α has a 0. Prove that

$$\operatorname{per}(A) = \operatorname{per}(A_1) + \operatorname{per}(A_2)$$

(Kallman[1982]).

References

R. Kallman[1982], A method for finding permanents of 0,1 matrices, *Maths. of Computation*, 38, pp. 167–170.

H. Minc[1978], *Permanents*, Addison-Wesley, Reading, Mass.

A. Nijenhuis and H.S. Wilf[1978], *Combinatorial Algorithms*, 2d ed., Academic Press, New York.

H.J. Ryser[1963], *Combinatorial Mathematics*, Carus Mathematical Monograph No. 14, Math. Assoc. of America, Washington, D.C.

7.2   Permutations with Restricted Positions

We now let A be the incidence matrix for m subsets $X_1, X_2, \ldots, X_m$ of the n-set $X = \{1, 2, \ldots, n\}$. As noted in Theorem 1.2.3, $\operatorname{per}(A)$ equals the number of distinct SDR's for this configuration of subsets, that is, $\operatorname{per}(A)$ equals the number of m-permutations $i_1, i_2, \ldots, i_m$ of the n-set X in which i_j is restricted by the condition

$$i_j \in X_j, \quad (j = 1, 2, \ldots, m).$$

This simple observation accounts for the importance of the permanent as a counting function in combinatorics.

A classical combinatorial enumeration problem known as "le problème des recontres" asks for the number of permutations $i_1, i_2, \ldots, i_n$ of the n-set $\{1, 2, \ldots, n\}$ such that $i_j \neq j, (j = 1, 2, \ldots, n)$. Such permutations have no element in their natural position and are called *derangements* of order

n. Let D_n denote the number of derangements of order $n, (n \geq 1)$. If we choose

$$X_i = \{1, 2, \ldots, n\} - \{i\}, \quad (i = 1, 2, \ldots, n),$$

then D_n equals the number of SDR's of this configuration of sets and hence

$$D_n = \text{per}(J_n - I_n),$$

the permanent of the n by n matrix

$$J_n - I_n = \begin{bmatrix} 0 & 1 & \cdots & 1 \\ 1 & 0 & \cdots & 1 \\ \vdots & \vdots & \ddots & \vdots \\ 1 & 1 & \cdots & 0 \end{bmatrix}.$$

Applying (7.6) we obtain the formula

$$D_n = \sum_{r=0}^{n-1} (-1)^r \binom{n}{r} (n-r)^r (n-r-1)^{n-r}, \quad (n \geq 1). \qquad (7.7)$$

A different formula for D_n can be obtained by applying the inclusion-exclusion principle to the set of all permutations of $\{1, 2, \ldots, n\}$ and the sets $P_1, P_2, \ldots, P_n$ where P_i consists of those permutations which have i in position i, $(i = 1, 2, \ldots, n)$. This yields

$$D_n = n! \sum_{r=0}^{n} (-1)^r \frac{1}{r!}. \qquad (7.8)$$

Another classical combinatorial enumeration problem that can be formulated as a permanent of a (0,1)-matrix is the "problème des ménages." This problem asks for the number M_n of ways to seat n married couples at a round table with men and women in alternate positions and with no husband next to his wife. The wives may be seated first and this may be done in $2(n!)$ ways. Each husband is then excluded from the two seats next to his wife. The number of ways the husbands may be seated is the same for each seating arrangement of the wives. Let

$$A_i = \{1, 2, \ldots, n\} - \{i, i+1\}, \quad (i = 1, 2, \ldots, n)$$

where $n + 1$ is interpreted as 1. Then

$$M_n = 2(n!)U_n \qquad (7.9)$$

where U_n is the number of permutations $i_1, i_2, \ldots, i_n$ of $\{1, 2, \ldots, n\}$ satisfying $i_j \in A_i, (i = 1, 2, \ldots, n)$. The numbers U_n are called the *ménage numbers.* Let C_n denote the permutation matrix of order n with 1's in positions $(1,2), (2,3), \ldots, (n-1,n), (n,1)$. We have

$$U_n = \operatorname{per}(J_n - I_n - C_n).$$

Before obtaining a formula for the ménage numbers U_n by evaluating the permanent of the matrix $J_n - I_n - C_n$, we derive a general formula for the evaluation of permanents of matrices of 0's and 1's.

Let k be a nonnegative integer and let n be a positive integer. The collection of all k-subsets of the n-set $\{1, 2, \ldots, n\}$ is denoted by $\mathcal{P}_{k,n}$. Let A be an m by n matrix over a field F. As usual for α in $\mathcal{P}_{k,m}$ and β in $\mathcal{P}_{k,n}$, $A[\alpha, \beta]$ is the k by k submatrix of A determined by rows i with $i \in \alpha$ and columns j with $j \in \beta$. The permanent $\operatorname{per}(A[\alpha, \beta])$ is called a *permanental k-minor* of A, or sometimes a *permanental minor* of A. The sum of the permanental k-minors of A is

$$p_k(A) = \sum_{\beta \in \mathcal{P}_{k,n}} \sum_{\alpha \in \mathcal{P}_{k,m}} \operatorname{per}(A[\alpha, \beta]). \tag{7.10}$$

We define $p_0(A) = 0$, and note that if $m \le n$ then $p_m(A) = \operatorname{per}(A)$ and $p_k(A) = 0$ for $k > m$. If the matrix A is a $(0,1)$-matrix, then $p_k(A)$ counts the number of different selections of k 1's of A with no two of the 1's on the same line.

The next theorem allows us to evaluate the permanent of an m by n $(0,1)$-matrix A in terms of the permanental minors of the *complementary matrix* $J_{m,n} - A$.

Theorem 7.2.1. *Let A be an m by n $(0,1)$-matrix satisfying $m \le n$. Then*

$$\operatorname{per}(A) = \sum_{k=0}^{m} (-1)^k p_k(J_{m,n} - A) \frac{(n-k)!}{(n-m)!}. \tag{7.11}$$

Proof. Let A be the incidence matrix of the configuration $X_1, X_2, \ldots, X_m$ of subsets of the n-set $X = \{1, 2, \ldots, n\}$. We shall apply the inclusion-exclusion principle to the set $S_{m,n}$ of all m-permutations of $\{1, 2, \ldots, n\}$. Let P_j denote the subset of $S_{m,n}$ consisting of all those m-permutations $i_1, i_2, \ldots, i_m$ such that

$$i_j \in \overline{X}_j = \{1, 2, \ldots, n\} - X_j, \quad (j = 1, 2, \ldots, m).$$

Then $\operatorname{per}(A)$ equals the number of m-permutations in $S_{m,n}$ which are in none of the sets $P_1, P_2, \ldots, P_m$. Let α be a k-subset of $\{1, 2, \ldots, m\}$ and let $g(\alpha)$ denote the number of m-permutations in $\cap_{i \in \alpha} P_i$. Then

$$g(\alpha) = \frac{(n-k)!}{(n-m)!} \text{per}((J_{m,n} - A)[\alpha, \{1, 2, \ldots, n\}])$$

$$= \frac{(n-k)!}{(n-m)!} \sum_{\beta \in \mathcal{P}_{k,n}} \text{per}((J_{m,n} - A)[\alpha, \beta]).$$

Hence

$$\sum_{\alpha \in \mathcal{P}_{k,m}} \cdot g(\alpha) =$$

$$\frac{(n-k)!}{(n-m)!} \sum_{\alpha \in \mathcal{P}_{k,m}} \sum_{\beta \in \mathcal{P}_{k,n}} \text{per}((J_{m,n} - A)[\alpha, \beta]) = \frac{(n-k)!}{(n-m)!} p_k(J_{m,n} - A).$$

The formula (7.11) now follows from the inclusion-exclusion principle. $\square$

The formula (7.8) for $D_n = \text{per}(J_n - I_n)$ follows from (7.11) by taking $m = n$ and $A = J_n - I_n$. The matrix $J_{m,n} - A$ in the right-hand side of (7.11) is then I_n and $p_k(I_n) = \binom{n}{k}$.

We now return to the ménage numbers U_n. In obtaining a formula for the numbers U_n we shall make use of the results of Kaplansky[1943] given in the next lemma.

Lemma 7.2.2. *Let $f(n, k)$ denote the number of ways to select k objects, no two adjacent, from n objects arranged in a line and let $g(n, k)$ denote the number of ways when the objects are arranged in a circle. Then*

$$f(n, k) = \binom{n - k + 1}{k}, \quad (k = 1, 2, \ldots, n) \tag{7.12}$$

and

$$g(n, k) = \frac{n}{n - k} \binom{n - k}{k}, \quad (k = 1, 2, \ldots, n). \tag{7.13}$$

Proof. First suppose that the n objects are arranged in a line. We have $f(n, 1) = n = \binom{n}{1}, (n \geq 1)$, and $f(n, n) = 0 = \binom{1}{n}, (n \geq 2)$. Now assume that $1 \leq k \leq n$. The selections may be partitioned into those that contain the first object [there are $f(n - 2, k - 1)$ of these] and those that do not [there are $f(n - 1, k)$]. Hence

$$f(n, k) = f(n - 1, k) + f(n - 2, k - 1).$$

Using this recurrence relation and arguing inductively we obtain

$$f(n,k) = \binom{n-k}{k} + \binom{n-k}{k-1} = \binom{n-k+1}{k}.$$

Now suppose that the objects are arranged in a circle. Partitioning the selections as above, we obtain

$$\begin{aligned} g(n,k) &= f(n-3,k-1) + f(n-1,k) \\ &= \binom{n-k-1}{k-1} + \binom{n-k}{k} \\ &= \frac{n}{n-k}\binom{n-k}{k}. \end{aligned}$$

□

The following formula is due to Touchard[1934].

Theorem 7.2.3. *The ménage numbers are given by the formula*

$$U_n = \sum_{k=0}^{n}(-1)^k \frac{2n}{2n-k}\binom{2n-k}{k}(n-k)!. \tag{7.14}$$

Proof. We apply Theorem 7.2.1 and obtain

$$U_n = \mathrm{per}(J_n - I_n - C_n) = \sum_{k=0}^{n}(-1)^k p_k(I_n + C_n)(n-k)!.$$

But $p_k(I_n + C_n) = g(2n,k)$ where the function g was defined in Lemma 7.2.2. By (7.13)

$$g(2n,k) = \frac{2n}{2n-k}\binom{2n-k}{k},$$

and (7.14) follows.

□

We return to a general m by n (0,1)-matrix A satisfying $m \le n$. We may identify A with an m by n chessboard in which the square in row i and column j has been removed from play for all those i and j satisfying $a_{ij} = 0$. The number $r_k = p_k(A)$ counts the number of ways to place k identical rooks on this board so that no rook can attack another (that is, no two rooks lie on the same line), and is called the kth *rook number* of A, ($k = 0, 1, \ldots, m$). The $(m+1)$-tuple $(r_0, r_1, \ldots, r_m)$ is the *rook vector* of A and

$$r(x) = r_0 + r_1 x + \cdots + r_m x^m$$

is the *rook polynomial* of A. The rook polynomial of $A = I_n$ is

$$\binom{n}{0} + \binom{n}{1} x + \cdots + \binom{n}{n} x^n$$

and that of $A = I_n + C_n$ is

$$g(2n, 0) + g(2n, 1)x + \cdots + g(2n, n)x^n.$$

Theorem 7.2.1 shows how to compute the mth rook number of A from the rook vector of the complementary matrix $J_{mn} - A$. The entire rook vector of the matrix A may be obtained in this way.

Theorem 7.2.4. *Let A be an m by n $(0,1)$-matrix with $m \le n$. Then*

$$p_t(A) = \sum_{k=0}^{t} (-1)^k p_k(J_{m,n} - A) \binom{m-k}{t-k} \frac{(n-k)!}{(n-t)!}, \quad (t = 0, 1, \ldots, m).$$
(7.15)

Proof. Using Theorem 7.2.1 we calculate that

$$
\begin{aligned}
p_t(A) &= \sum_{\alpha \in \mathcal{P}_{t,m}} \mathrm{per}(A[\alpha, \{1, 2, \ldots, n\}]) \\
&= \sum_{\alpha \in \mathcal{P}_{t,m}} \sum_{k=0}^{t} (-1)^k p_k(J_{t,n} - A[\alpha, \{1, 2, \ldots, n\}]) \frac{(n-k)!}{(n-t)!} \\
&= \sum_{k=0}^{t} \sum_{\alpha \in \mathcal{P}_{t,m}} (-1)^k p_k(J_{t,n} - A[\alpha, \{1, 2, \ldots, n\}]) \frac{(n-k)!}{(n-t)!} \\
&= \sum_{k=0}^{t} (-1)^k p_k(J_{mn} - A) \binom{m-k}{t-k} \frac{(n-k)!}{(n-t)!}.
\end{aligned}
$$
$\square$

The computation of the rook vector of an m by n $(0,1)$-matrix is in general a difficult problem. There is, however, a special class of $(0,1)$-matrices whose rook polynomials have an exceedingly simple form.

Let $b_1, b_2, \ldots, b_m$ be integers with $0 \le b_1 \le b_2 \le \cdots \le b_m$. The m by b_m $(0,1)$-matrix $A = [a_{ij}]$ defined by

$$a_{ij} = 1 \text{ if and only if } 1 \le j \le b_i, \quad (i = 1, 2, \ldots, m)$$

is a *Ferrers matrix* and is denoted by $F(b_1, b_2, \ldots, b_m)$. In what follows the number of columns of the Ferrers matrix can be taken to be any integer $n \ge b_m$ with no change in the conclusions. We define

$$[x]_k = x(x-1) \cdots (x-k+1), \quad (k \ge 1) \quad \text{and} \quad [x]_0 = 1.$$

Theorem 7.2.5. *Let $F(b_1, b_2, \ldots, b_m)$ be a Ferrers matrix with rook vector $(r_0, r_1, \ldots, r_m)$. Then*

$$\sum_{k=0}^{m} r_k [x]_{m-k} = \prod_{i=1}^{m} (x + q_i) \qquad (7.16)$$

where $q_i = b_i - i + 1, (i = 1, 2, \ldots, m)$.

Proof. Let x be a nonnegative integer and consider the Ferrers matrix

$$F(x + b_1, x + b_2, \ldots, x + b_m) = \begin{bmatrix} J_{m,x} & A \end{bmatrix}.$$

We evaluate the mth rook number r'_m of $F(x + b_1, x + b_2, \ldots, x + b_m)$ in two different ways. Counting the number of ways to select m 1's of $F(x + b_1, x + b_2, \ldots, x + b_m)$, no two from the same line, according to the number of 1's selected from $J_{m,x}$, we obtain

$$r'_m = \sum_{k=0}^{m} r_k [x]_{m-k}.$$

We may also evaluate r'_m by first choosing a 1 in row 1, then choosing a 1 in row 2, ..., then a 1 in row m in such a way that no two 1's chosen belong to the same column. Because

$$x + b_1 \leq x + b_2 \leq \cdots \leq x + b_m$$

we obtain

$$\begin{aligned} r'_m &= (x + b_1)(x + b_2 - 1) \cdots (x + b_m - (m-1)) \\ &= (x + q_1)(x + q_2) \cdots (x + q_m). \end{aligned}$$

Equating these two counts we obtain (7.16).   □

The following two corollaries are direct consequences of Theorem 7.2.5.

Corollary 7.2.6.

$$\operatorname{per}(F(b_1, b_2, \ldots, b_m)) = \prod_{i=1}^{m} (b_i - i + 1).$$

Corollary 7.2.7. *Let $F(b_1, b_2, \ldots, b_m)$ and $F(c_1, c_2, \ldots, c_m)$ be two Ferrers matrices, and let $q_i = b_i - i + 1$ and $q'_i = c_i - i + 1$ for $i = 1, 2, \ldots, m$. Then $F(b_1, b_2, \ldots, b_m)$ and $F(c_1, c_2, \ldots, c_m)$ have the same rook polynomial if and only if the numbers $q_1, q_2 \ldots, q_m$ are identical with the numbers $q'_1, q'_2, \ldots, q'_m$ including multiplicities.*

One may also define the *rook polynomial* of an m by n matrix A over a field F by

$$r(x) = \sum_{k=0}^{m} p_k(A) x^k.$$

We conclude this section with the statement of the following theorem of Nijenhuis[1976].

Theorem 7.2.8.   *If A is a nonnegative matrix then all the roots of its rook polynomial are real numbers.*

Exercises

1. Obtain formula (7.8) directly from the inclusion-exclusion principle.
2. Determine the rook vector of the matrix

$$A = \begin{bmatrix} 1 & 1 & 0 & 0 & 0 & 0 & 0 & 0 \\ 0 & 1 & 1 & 0 & 0 & 0 & 0 & 0 \\ 0 & 0 & 1 & 0 & 0 & 0 & 0 & 0 \\ 0 & 0 & 0 & 1 & 1 & 0 & 0 & 0 \\ 0 & 0 & 0 & 0 & 1 & 0 & 0 & 0 \\ 0 & 0 & 0 & 0 & 0 & 1 & 0 & 0 \\ 0 & 0 & 0 & 0 & 0 & 0 & 1 & 1 \\ 0 & 0 & 0 & 0 & 0 & 0 & 1 & 1 \end{bmatrix}.$$

3. Let A be the (0,1)-matrix of order $n = 2m$ given by

$$A = J_n - (J_2 \oplus J_2 \oplus \cdots \oplus J_2).$$

Use Theorem 7.2.1 to obtain a formula for $\mathrm{per}(A)$.
4. Let $a_i, (i = 1, 2, \ldots, r)$ and $b_i, (i = 1, 2, \ldots, r)$ be positive integers such that $a_1 + a_2 + \cdots + a_r = n$ and $b_1 + b_2 + \cdots + b_r \leq n$. Let $A = A_{(a_1, a_2, \ldots, a_r; b_1, b_2, \ldots, b_r)}$ denote the (0,1)-matrix of order n obtained from the matrix J_n of all 1's of order n by choosing for each $i = 1, 2, \ldots, r$ "line disjoint" submatrices of sizes a_i by b_i and replacing their 1's by 0's. Prove that

$$\mathrm{per}(A) = \sum (-1)^{i_1 + i_2 + \cdots + i_r} (n - i_1 - i_2 - \cdots - i_r) \prod_{k=1}^{r} \binom{a_k}{i_k} \binom{b_k}{i_k} i_k!$$

where the summation extends over all $i_1, i_2, \ldots, i_r$ satisfying $i_k \leq a_k, (k = 1, 2, \ldots, r)$ (Kaplansky[1939,1944] and Chung, Diaconis, Graham and Mallows[1981]).
5. Prove that the rook vector of the Ferrers matrix $F(0, 1, 2, \ldots, m - 1)$ of order m equals $(S(m, m), S(m, m - 1), \ldots, S(m, 1), S(m, 0))$ where $S(m, k)$ denotes a Stirling number of the second kind (see Stanley[1986] for the definition of the Stirling numbers of the second kind).

References

F.R.K. Chung, P. Diaconis, R.L. Graham and C.L. Mallows[1981], On the permanents of complements of the direct sum of identity matrices, *Advances in Applied Math.*, 2, pp. 121–137.

I. Kaplansky[1939], A generalization of the "Problème des recontres," *Amer. Math. Monthly*, 46, pp. 159–161.

[1943], Solution of the "Problème des ménages," *Bull. Amer. Math. Soc.*, 49, pp. 784–785.

[1944], Symbolic solution of certain problems in permutations, *Bull. Amer. Math. Soc.*, 50, pp. 906–914.

A. Nijenhuis[1976], On permanents and the zeros of rook polynomials, *J. Combin. Theory, Ser. A*, 21, pp. 240–244.

H. Minc[1978], *Permanents*, Addison-Wesley, Reading, Mass.

H.J. Ryser[1963], *Combinatorial Mathematics*, Carus Mathematical Monograph No. 14, Math. Assoc. of America, Washington, D.C.

R.P. Stanley[1986], *Enumerative Combinatorics*, Wadsworth and Brooks/Cole, Monterey, Calif.

J. Touchard[1934], Sur un problème des permutations, *C. R. Acad. Sci. Paris*, 198, pp. 631–633.

7.3 Matrix Factorization of the Permanent and the Determinant

Let

$$A = \begin{bmatrix} a_{11} & a_{12} & a_{13} \\ a_{21} & a_{22} & a_{23} \\ a_{31} & a_{32} & a_{33} \end{bmatrix}$$

be a matrix of order 3. We observe that the result of the following matrix multiplication

$$\begin{bmatrix} a_{11} & a_{12} & a_{13} \end{bmatrix} \begin{bmatrix} a_{22} & a_{23} & 0 \\ a_{21} & 0 & a_{23} \\ 0 & a_{21} & a_{22} \end{bmatrix} \begin{bmatrix} a_{33} \\ a_{32} \\ a_{31} \end{bmatrix} \tag{7.17}$$

is a matrix of order 1 whose unique element equals the permanent of A. Each of the three matrices in this factorization of $\mathrm{per}(A)$ depends only on one row of the matrix A. A matrix factorization for the determinant of A is given by

$$\begin{bmatrix} a_{11} & a_{12} & a_{13} \end{bmatrix} \begin{bmatrix} a_{22} & a_{23} & 0 \\ -a_{21} & 0 & a_{23} \\ 0 & -a_{21} & -a_{22} \end{bmatrix} \begin{bmatrix} a_{33} \\ -a_{32} \\ a_{31} \end{bmatrix}. \tag{7.18}$$

The factorization (7.18) is obtained from the factorization (7.17) by affixing minus signs to some of the elements of the matrix factors. These two factorizations for the permanent and the determinant of a matrix of order 3 are instances of factorizations of the permanent and determinant that exist for all square matrices. These factorizations were discovered by Jurkat and Ryser[1966], but we shall follow the simplified account of Brualdi and Shader[1990] (see also Brualdi[1990]).

Let n be a positive integer. We partially order the collection $\mathcal{P}_n = \mathcal{P}_{n,n}$ of all subsets of $\{1, 2, \ldots, n\}$ by inclusion and thereby obtain a partially ordered set. A *chain* in this partially ordered set is a sequence of distinct subsets

$$X_0 \subset X_1 \subset \cdots \subset X_k \tag{7.19}$$

which are related by inclusion as shown. The *length of the chain* (7.19) is k. The maximal length of a chain is n. Let

$$Y_0 \subset Y_1 \subset \cdots \subset Y_n \tag{7.20}$$

be a chain of length n. Then Y_j is in $\mathcal{P}_{j,n}, (j = 0, 1, \ldots, n)$, and there is a unique element i_j such that

$$Y_j = Y_{j-1} \cup \{i_j\}, \quad (j = 1, 2, \ldots, n). \tag{7.21}$$

Moreover,

$$i_1, i_2, \ldots, i_n \tag{7.22}$$

is a permutation of $\{1, 2, \ldots, n\}$. Conversely, given a permutation (7.22) of $\{1, 2, \ldots, n\}$, if we define $Y_0 = \emptyset$ and recursively define Y_j using (7.21) then we obtain a chain (7.20) of length n. Thus the chains (7.20) of length n are in one-to-one correspondence with the permutations (7.22) of $\{1, 2, \ldots, n\}$.

Now assume that the collection $\mathcal{P}_{j,n}$ of subsets of $\{1, 2, \ldots, n\}$ of size j have been linearly ordered in some way for each $j = 0, 1, \ldots, n$. To fix the notation, let us choose for each j the lexicographic order and let the elements of $\mathcal{P}_{j,n}$ in the lexicographic order be

$$Z_1^{(j)}, Z_2^{(j)}, \ldots, Z_{q_{j,n}}^j, \quad (j = 0, 1, \ldots, n)$$

where $q_{j,n}$ denotes $\binom{n}{j}$.

Let $A = [a_{ij}], (i, j = 1, 2, \ldots, n)$ be a matrix of order n over a field F (a ring will do). We use the elements of the matrix A in order to define a *weighted incidence matrix*

$$W^{(j)} = \left[w_{k,l}^{(j)} \right], \quad (k = 1, 2, \ldots, q_{j-1,n}; l = 1, 2, \ldots, q_{j,n})$$

of size $q_{j-1,n}$ by $q_{j,n}$ for the inclusions that hold between the sets in $\mathcal{P}_{j-1,n}$ and the sets in $\mathcal{P}_{j,n}, (j = 1, 2, \ldots, n)$. We define

$$w_{kl}^{(j)} = \begin{cases} 0 & \text{if } Z_k^{(j-1)} \not\subset Z_l^{(j)} \\ a_{jt} & \text{if } Z_k^{(j-1)} \cup \{t\} = Z_l^{(j)}. \end{cases} \tag{7.23}$$

We remark that only the elements of row j of A are used to define the matrix $W^{(j)}$.

We now define the *weight of the chain* (7.20) corresponding to the permutation (7.22) to be

$$a_{1i_1} a_{2i_2} \cdots a_{ni_n}.$$

Thus the sum of the weights of all the chains (7.20) of length n in $\mathcal{P}_n$ equals the permanent of the matrix A.

We now compute the matrix product

$$W^{(1)} W^{(2)} \ldots W^{(n)}. \tag{7.24}$$

The matrix (7.24) is a matrix of size 1 by 1 whose unique element equals

$$\sum_{i_1=1}^{q_{1,n}} \sum_{i_2=1}^{q_{2,n}} \cdots \sum_{i_{n-1}=1}^{q_{n-1,n}} w_{1i_1}^{(1)} w_{i_1 i_2}^{(2)} \cdots w_{i_{n-1}1}^{(n)}. \tag{7.25}$$

The product

$$w_{1i_1}^{(1)} w_{i_1 i_2}^{(2)} \cdots w_{i_{n-1}1}^{(n)} \tag{7.26}$$

equals zero if

$$Z_1^{(0)} = \emptyset, Z_{i_1}^{(1)}, Z_{i_2}^{(2)}, \ldots, Z_1^{(n)} = \{1, 2, \ldots, n\} \tag{7.27}$$

is not a chain. If, however, (7.27) is a chain, then the product (7.26) equals its weight. Hence the matrix (7.24) of order 1 has unique element equal to $\text{per}(A)$ and

$$\text{per}(A) = W^{(1)} W^{(2)} \ldots W^{(n)} \tag{7.28}$$

is a matrix factorization of the permanent of A.

We now alter the definition of the weighted incidence matrices $W^{(j)}$ as given in (7.23) by defining

$$w_{kl}^{(j)} = \begin{cases} 0 & \text{if } Z_k^{(j-1)} \not\subset Z_l^{(j)} \\ (-1)^{c_j} a_{jt} & \text{if } Z_k^{(j-1)} \cup \{t\} = Z_l^{(j)} \end{cases} \tag{7.29}$$

where c_j elements of $Z_k^{(j-1)}$ are greater than t. The *weight of the chain* (7.20) corresponding to the permutation (7.22) is now

$$(-1)^{\mathrm{inv}(i_1,i_2,\ldots,i_n)}a_{1i_1}a_{2i_2}\cdots a_{ni_n}$$

where $\mathrm{inv}(i_1,i_2,\ldots,i_n)$ equals the number of inversions in the permutation $(i_1,i_2,\ldots,i_n)$. Thus the sum of all the weights of the chains (7.20) of length n in $\mathcal{P}_n$ equals the determinant of A. Arguing as above we now obtain the matrix factorization

$$\det(A) = W^{(1)}W^{(2)}\ldots W^{(n)} \tag{7.30}$$

of the determinant of A.

The matrices $W^{(j)}$ which occur in the factorizations (7.28) and (7.30) of the permanent and determinant have an inductive structure which facilitates their computation. We first treat the $W^{(j)}$'s that arise in the factorization of the permanent. The matrix $W^{(j)}$ depends only on row j of A, and we now denote row j of A by

$$x = (x_1, x_2, \ldots, x_n)$$

and write

$$W^{(j)} = W^{(j)}(x).$$

We have

$$W^{(1)}(x) = \begin{bmatrix} x_1 & x_2 & \cdots & x_n \end{bmatrix}$$

and

$$W^{(n)}(x) = \begin{bmatrix} x_n \\ x_{n-1} \\ \vdots \\ x_1 \end{bmatrix}.$$

Let j be an integer with $1 < j < n$. Let

$$U_1^{(j-1)}, U_2^{(j-1)}, \ldots, U_{q_{j-2,n-1}}^{(j-1)}$$

be the sets of $\mathcal{P}_{j-1,n}$ containing 1, arranged in lexicographic order, and let

$$V_1^{(j-1)}, V_2^{(j-1)}, \ldots, V_{q_{j-1,n-1}}^{(j-1)}$$

be the sets in $\mathcal{P}_{j-1,n}$ not containing 1, arranged in lexicographic order. Then

$$U_1^{(j-1)}, U_2^{(j-1)}, \ldots, U_{q_{j-2,n-1}}^{(j-1)}, V_1^{(j-1)}, V_2^{(j-1)}, \ldots, V_{q_{j-1,n-1}}^{(j-1)}$$

are the sets in $\mathcal{P}_{j-1,n}$ arranged in lexicographic order. In a similar way we obtain the sets

$$U_1^{(j)}, U_2^{(j)}, \ldots, U_{q_{j-1,n-1}}^{(j)}, V_1^{(j)}, V_2^{(j)}, \ldots, V_{q_{j,n-1}}^{(j)}$$

of $\mathcal{P}_{j,n}$ arranged in lexicographic order. Let

$$x^* = (x_2, x_3, \ldots, x_n).$$

It now follows from (7.23) that

$$W^{(j)}(x) = \begin{bmatrix} W^{(j-1)}(x^*) & O \\ x_1 I_{q_{j-1,n-1}} & W^{(j)}(x^*) \end{bmatrix}, \quad (j = 1, 2, \ldots, n-1) \quad (7.31)$$

where O denotes a zero matrix of size $q_{j-2,n-1}$ by $q_{j,n-1}$. The inductive structure of $W^{(j)}(x)$ given in (7.31) implies that exactly $n - (j - 1)$ components of x occur in each row of $W^{(j)}(x)$ and they occur in the same order as their occurrence in x. Also exactly j components of x occur in each column of $W^{(j)}(x)$ and they occur in the reverse order to their occurrence in x.

There is a similar inductive structure for the matrices $W^{(j)}(x)$ that occur in the factorization of the determinant. In this case we have

$$W^{(1)}(x) = \begin{bmatrix} x_1 & x_2 & \cdots & x_n \end{bmatrix}$$

and

$$W^{(n)}(x) = \begin{bmatrix} x_n \\ -x_{n-1} \\ \cdots \\ (-1)^{n-1} x_1 \end{bmatrix}.$$

The j components of x that occur in each column of $W^{(j)}(x)$ occur in the reverse order to their occurrence in x. It follows from (7.29) that the signs in front of these j components of x alternate and that the sign of the topmost component is $+1$ whereas the sign of the bottommost component is $(-1)^{j-1}$. These observations allow us to obtain the inductive structure

$$W^{(j)}(x) = \begin{bmatrix} W^{(j-1)}(x^*) & O \\ (-1)^{j-1} x_1 I_{q_{j-1,n-1}} & W^{(j)}(x^*) \end{bmatrix}, \quad (j = 1, 2, \ldots, n-1)$$

in this case.

Let C be a rectangular complex matrix. The l_2-norm of C is the square root of the largest eigenvalue of the positive semidefinite matrix CC^* and is denoted by $||C||$. Here C^* denotes the conjugate transpose of the matrix C. The l_2-norm is a matrix norm, that is, $||C_1 C_2|| \le ||C_1|| ||C_2||$ whenever the product $C_1 C_2$ is defined. Jurkat and Ryser[1966] used the factorization (7.28) of the permanent and an evaluation of $||W^{(j)}(x)||$ in the case that x is a (0,1)-vector to obtain an upper bound for the permanent of a

(0,1)-matrix of order n in terms of its row sum vector $(r_1, r_2, \ldots, r_n)$. This bound is not as good as the one given in Theorem 7.4.5 in the next section. They also used the factorization (7.30) of the determinant and the evaluation

$$\|W^{(j)}(x)\| = \sqrt{x_1^2 + x_2^2 + \cdots + x_n^2}, \quad (j = 1, 2, \ldots, n)$$

for x a complex vector in order to obtain Hadamard's determinant inequality

$$|\det(A)| \leq \prod_{i=1}^{n} \sqrt{a_{i1}^2 + a_{i2}^2 + \cdots + a_{in}^2}.$$

Exercises

1. Compute the matrix factorizations of the permanent and determinant of a matrix of order 4.
2. In the matrix factorizations $W^{(1)} W^{(2)} \cdots W^{(n)}$ of the permanent and the determinant, determine the elements of the matrix $W^{(r)} \cdots W^{(s)}$ where r and s are integers with $0 \leq r < s \leq n$.

References

R.A. Brualdi[1990], The many facets of combinatorial matrix theory, in *Matrix Theory and Applications*, C.R. Johnson ed., *Proc. Symposia Pure and Applied Math.*, vol. 40, Amer. Math. Soc., Providence.

R.A. Brualdi and B.L. Shader[1990], Matrix factorizations of determinants and permanents, *J. Combin. Theory, Ser. A*, 54, pp. 132–134.

W.B. Jurkat and H.J. Ryser[1966], Matrix factorizations of determinants and permanents, *J. Algebra*, 3, pp. 1–27.

7.4   Inequalities

Let A be a (0,1)-matrix of size m by n with $m \leq n$. The permanent of A satisfies the inequality

$$0 \leq \mathrm{per}\,(A) \leq n(n-1) \cdots (n - m + 1) \tag{7.32}$$

Equality holds on the right in (7.32) if and only if A is the matrix $J_{m,n}$ of all 1's. Equality holds on the left if and only if the term rank of A is strictly less than m. In this section we shall improve the inequalities (7.32) by taking into account the number of 1's in each row of A, the number of 1's in each column of A, or the total number of 1's of A.

The following theorem of Ostrand[1970] is a strengthening of earlier results of Hall[1948], Jurkat and Ryser[1966] and Rado[1967].

Theorem 7.4.1. *Let $A = [a_{ij}]$ be a $(0,1)$-matrix of size m by n with* per $(A) > 0$. *Let the row sums of A be $r_1, r_2, \ldots, r_m$ and assume that the rows of A have been arranged so that $r_1 \leq r_2 \leq \cdots \leq r_m$. Then*

$$\operatorname{per}(A) \geq \prod_{i=1}^{m} \max\{1, r_i - i + 1\}. \tag{7.33}$$

Proof. We prove (7.33) by induction on m. If $m = 1$ then (7.33) clearly holds. Now assume that $m > 1$. Because per(A) > 0, if P and Q are permutation matrices such that

$$PAQ = \begin{bmatrix} A_1 & O \\ A_2 & A_3 \end{bmatrix}, \tag{7.34}$$

the number k of rows and the number l of columns of A_1 satisfies $l \geq k$. We consider two cases.

Case 1. Whenever (7.34) holds with $k < m$ we have $l > k$.

Without loss of generality we assume that the r_1 1's in the first row of A occur in the first r_1 columns. The Laplace expansion by row 1 yields

$$\operatorname{per}(A) = \sum_{j=1}^{r_1} a_{1j}\operatorname{per}(A(1,j)) \tag{7.35}$$

where $A(1, j)$ is the submatrix of A obtained by deleting row 1 and column j. The assumption in this case and Theorem 1.2.3 imply that

$$\operatorname{per}(A(1,j)) > 0, \quad (j = 1, 2, \ldots, r_1).$$

The number $r_i(j)$ of 1's in row i of $A(1, j)$ satisfies

$$r_i(j) \geq r_{i+1} - 1, \quad (i = 1, 2, \ldots, m - 1; j = 1, 2, \ldots, r_1). \tag{7.36}$$

Let

$$r_1'(j), r_2'(j), \ldots, r_{m-1}'(j)$$

denote a rearrangement of $r_1(j), r_2(j), \ldots, r_{m-1}(j)$ satisfying

$$r_1'(j) \leq r_2'(j) \leq \cdots \leq r_{m-1}'(j), \quad (j = 1, 2, \ldots, r_1). \tag{7.37}$$

Applying the inductive hypothesis to each of the matrices $A(1, j), (j = 1, 2, \ldots, r_1)$, we obtain

$$\operatorname{per}(A(1,j)) \geq \prod_{i=1}^{m-1} \max\{1, r_i'(j) - i + 1\}$$

$$\geq \prod_{i=1}^{m-1} \{1, r_{i+1} - 1 - i + 1\} = \prod_{i=2}^{m} \max\{1, r_i - i + 1\}. \tag{7.38}$$

The second inequality in (7.38) follows from (7.36) and the fact that the rearrangement of the rows that achieves (7.37) can be accomplished by rearrangements only among those rows of $A(1,j)$ that correspond to rows of A with the same number of 1's. Combining (7.35) and (7.38) we obtain

$$\text{per}(A) \geq r_1 \prod_{i=2}^{m} \max\{1, r_i - i + 1\}$$

and (7.33) holds in this case.

Case 2. There exist permutation matrices P and Q and an integer k with $1 \leq k \leq n-1$ such that (7.34) holds where A_1 has order k.

There exists a k-subset α of the rows of A and a k-subset β of the columns such that the submatrix $A[\alpha, \overline{\beta}]$ of A determined by the rows in α and the columns not in β is a zero matrix. We have

$$\text{per}(A) = \text{per}(A[\alpha, \beta])\text{per}(A[\overline{\alpha}, \overline{\beta}]). \qquad (7.39)$$

Let $\alpha = \{i_1, i_2, \ldots, i_k\}$ and let $\overline{\alpha} = \{i_{k+1}, i_{k+2}, \ldots, i_m\}$ where $i_1 \leq i_2 \leq \cdots \leq i_k$ and $i_{k+1} \leq i_{k+2} \leq \cdots \leq i_m$. Because $A[\alpha, \overline{\beta}] = O$ and $r_1 \leq r_2 \leq \cdots \leq r_m$ we have $r_i \leq k, (i = 1, 2, \ldots, i_k)$ and hence

$$\max\{1, r_i - i + 1\} = 1, \quad (i = k, k+1, \ldots, i_k). \qquad (7.40)$$

Applying the inductive hypothesis and using (7.40) we obtain

$$\text{per}(A[\alpha, \beta]) \geq \prod_{j=1}^{k} \max\{1, r_{i_j} - j + 1\}$$

$$\geq \prod_{j=1}^{k} \max\{1, r_j - j + 1\} = \prod_{j=1}^{i_k} \max\{1, r_j - j + 1\}. \qquad (7.41)$$

Let the row sums of $A[\overline{\alpha}, \overline{\beta}]$ be $r'_{k+1}, r'_{k+2}, \ldots, r'_m$ arranged so that $r'_{k+1} \leq r'_{k+2} \leq \cdots \leq r'_m$. Applying the inductive hypothesis again and using (7.40) we obtain

$$\text{per}(A[\overline{\alpha}, \overline{\beta}]) \geq \prod_{i=1}^{m-k} \max\{1, r'_{k+i} - i + 1\}$$

$$\geq \prod_{j=i_k+1}^{m} \max\{1, r'_j - (j-k) + 1\} \geq \prod_{j=i_k+1}^{m} \max\{1, r_j - j + 1\}. \qquad (7.42)$$

Using (7.39), (7.41) and (7.42) we obtain (7.33). Hence the theorem holds by induction. □

Let A be a Ferrers matrix $F(b_1, b_2, \ldots, b_m)$ as defined in section 7.2. The row sums of A satisfy $b_1 \leq b_2 \leq \cdots \leq b_m$ and $\mathrm{per}(A) > 0$ if and only if $b_i \geq i, (i = 1, 2, \ldots, m)$. By Corollary (7.2.6), $\mathrm{per}(A) = \prod_{i=1}^{m}(b_i - i + 1)$ and hence equality holds in (7.33). Equality also holds in (7.33) for all permutation matrices.

The following result is due to Hall[1948].

Corollary 7.4.2. *Let A be a $(0,1)$-matrix of size m by n with $m \leq n$, and let t be a positive integer such that each row of A contains at least t 1's. If $t < m$ and $\mathrm{per}(A) > 0$, then $\mathrm{per}(A) \geq t!$. If $t \geq m$, then $\mathrm{per}(A) \geq t!/(t - m)!$.*

Proof. If $t < m$ and $\mathrm{per}(A) > 0$, then by Theorem 7.4.1

$$\mathrm{per}(A) \geq \prod_{i=1}^{m} \max\{1, t - i + 1\} = t!.$$

Now suppose that $t \geq m$. It follows from Theorem 1.2.3 that $\mathrm{per}(A) > 0$. By Theorem 7.4.1

$$\mathrm{per}(A) \geq \prod_{i=1}^{m} \max\{1, t - i + 1\} = t!/(t - m)!. \qquad \square$$

Inequality (7.33) gives the best known general lower bound for the permanent, if not zero, of a $(0,1)$-matrix in terms of the numbers r_i of 1's in its rows. We now obtain the best known general upper bound $\prod_{i=1}^{n}(r_i!)^{1/r_i}$ for the permanent of a $(0,1)$-matrix A of order n in terms of the r_i's. This bound was conjectured by Minc[1963] as a generalization of a conjecture of Ryser[1960]. Minc[1974] proved the conjecture under the assumption that no row of A has more than eight 1's. Weaker upper bounds were obtained by Jurkat and Ryser[1966], Minc[1963,1967], Wilf[1968] and Nijenhuis and Wilf[1970]. The conjecture was first proved by Brégman [1973]. A simpler proof was obtained by Schrijver[1978] and it is this proof that we present. We first prove two lemmas. We adopt the convention that $0^0 = 1$.

Lemma 7.4.3. *Let $t_1, t_2, \ldots, t_m$ be nonnegative numbers. Then*

$$\left(\frac{1}{m} \sum_{i=1}^{m} t_i\right)^{\sum_{i=1}^{m} t_i} \leq \prod_{i=1}^{m} t_i^{t_i}. \tag{7.43}$$

Proof. The function $f(x) = x \log x$ is a convex function on $(0, \infty)$ and hence satisfies

$$f\left(\frac{1}{m} \sum_{i=1}^{m} t_i\right) \leq \frac{1}{m} \sum_{i=1}^{m} f(t_i).$$

This yields

$$\frac{1}{m}\left(\sum_{i=1}^{m} t_i\right) \log\left(\frac{1}{m}\sum_{i=1}^{m} t_i\right) \le \frac{1}{m}\sum_{i=1}^{m} t_i \log t_i,$$

from which (7.43) for positive t_i's follows. By taking limits we see that (7.43) holds for nonnegative t_i's. □

Lemma 7.4.4. *Let $A = [a_{ij}]$ be a $(0,1)$-matrix of order n. Let T be the set of all permutations σ of $\{1, 2, \ldots, n\}$ satisfying $\prod_{i=1}^{n} a_{i\sigma(i)} = 1$. Then*

$$\prod_{\{(i,k):a_{ik}=1\}} \text{per}(A(i,k))^{\text{per}(A(i,k))} = \prod_{\sigma \in T}\prod_{i=1}^{n} \text{per}(A(i,\sigma(i))). \qquad (7.44)$$

Proof. The only factors $\text{per}(A(i,k))$ that occur on either side of (7.44) are those for which $a_{ik} = 1$. If $a_{ik} = 1$, then the number of times that $\text{per}(A(i,k))$ occurs as a factor on the right of (7.44) equals the number of permutations σ of $\{1, 2, \ldots, n\}$ satisfying $\sigma(i) = k$ and $\prod_{i=1}^{n} a_{i\sigma(i)} = 1$; this number equals $\text{per}(A(i,k))$. □

Theorem 7.4.5. *Let $A = [a_{ij}]$ be a $(0,1)$-matrix with row sums $r_1, r_2, \ldots, r_n$. Then*

$$\text{per}(A) \le \prod_{i=1}^{n} (r_i!)^{\frac{1}{r_i}}. \qquad (7.45)$$

Proof. The inequality holds if $n = 1$. We suppose that $n > 1$ and proceed by induction on n. Let T be defined as in Lemma 7.4.4. The Laplace expansion of the permanent by rows and the inequality (7.43) and the identity (7.44) yield

$$(\text{per}(A))^{n\,\text{per}(A)} = \prod_{i=1}^{n} (\text{per}(A))^{\text{per}(A)}$$

$$= \prod_{i=1}^{n}\left(\sum_{k=1}^{n} a_{ik}\text{per}(A(i,k))\right)^{\sum_{k=1}^{n} a_{ik}\text{per}(A(i,k))}$$

$$\le \prod_{i=1}^{n}\left(r_i^{\text{per}(A)}\prod_{\{k:a_{ik}=1\}} \text{per}(A(i,k))^{\text{per}(A(i,k))}\right)$$

$$= \prod_{i=1}^{n} r_i^{\text{per}(A)}\prod_{\{(i,k):a_{ik}=1\}} \text{per}(A(i,k))^{\text{per}(A(i,k))}$$

$$= \prod_{i=1}^{n} r_i^{\text{per}(A)}\prod_{\sigma \in T}\prod_{i=1}^{n} \text{per}(A(i,\sigma(i))). \qquad (7.46)$$

Let σ be a permutation in T. Applying the inductive hypothesis to $\text{per}(A(i, \sigma(i))), (i = 1, 2, \ldots, n)$ we obtain

$$\prod_{i=1}^{n} \text{per}(A(i, \sigma(i)))$$

$$\leq \prod_{i=1}^{n} \left(\prod_{\{j: j \neq i, a_{j\sigma(i)} = 0\}} (r_j!)^{1/r_j} \right) \left(\prod_{\{j: j \neq i, a_{j\sigma(i)} = 1\}} ((r_j - 1)!)^{1/(r_j - 1)} \right)$$

$$= \prod_{j=1}^{n} \left(\prod_{\{i: j \neq i, a_{j\sigma(i)} = 0\}} (r_j!)^{1/r_j} \right) \left(\prod_{\{i: j \neq i, a_{j\sigma(i)} = 1\}} ((r_j - 1)!)^{1/(r_j - 1)} \right)$$

$$= \prod_{j=1}^{n} (r_j!)^{(n-r_j)/r_j} ((r_j - 1)!)^{(r_j - 1)/(r_j - 1)}$$

$$= \prod_{j=1}^{n} (r_j!)^{(n-r_j)/r_j} (r_j - 1)!. \tag{7.47}$$

Using (7.47) in (7.46) we obtain

$$(\text{per}(A)^n \text{per}(A) \leq \prod_{i=1}^{n} r_i^{\text{per}(A)} \left(\prod_{j=1}^{n} (r_j!)^{(n-r_j)/r_j} (r_j - 1)! \right)^{\text{per}(A)}$$

$$= \left(\prod_{j=1}^{n} (r_j!)^{(n-r_j)/r_j} r_j! \right)^{\text{per}(A)} = \left(\prod_{j=1}^{n} (r_j!)^{1/r_j} \right)^{n\text{per}(A)}$$

Hence $\text{per}(A) \leq \prod_{j=1}^{n} (r_j!)^{1/r_j}$, and the theorem follows by induction. $\quad\square$

The inequality (7.45) is an improvement for (0,1)-matrices of the upper bound

$$\text{per}(A) \leq r_1 r_2 \cdots r_n$$

which is valid more generally for nonnegative matrices of order n. We remark that the inequalities obtained from these inequalities by replacing the row sums $r_1, r_2, \ldots, r_n$ of A with its column sums $s_1, s_2, \ldots, s_n$ also hold. Thus for A a (0,1)-matrix we have

$$\text{per}(A) \leq \min \left\{ \prod_{i=1}^{n} (r_i!)^{1/r_i}, \prod_{i=1}^{n} (s_i!)^{1/s_i} \right\}.$$

If A is a nonnegative matrix, then

$$\text{per}(A) \leq \min\left\{\prod_{i=1}^{n} r_i, \prod_{i=1}^{n} s_j\right\}.$$

The following theorem of Jurkat and Ryser[1967] improves this last inequality.

Theorem 7.4.6. *Let $A = [a_{ij}]$ be a nonnegative matrix of order n with row sums $r_1, r_2, \ldots, r_n$ and column sums $s_1, s_2, \ldots, s_n$. Assume that the rows and columns of A have been arranged so that $r_1 \leq r_2 \leq \cdots \leq r_n$ and $s_1 \leq s_2 \leq \cdots \leq s_n$. Then*

$$\text{per}(A) \leq \prod_{i=1}^{n} \min\{r_i, s_i\}.$$

Proof. We first observe that an easy induction shows that

$$\prod_{i=1}^{n} \min\{r_i, s_{j_i}\} \leq \prod_{i=1}^{n} \min\{r_i, s_i\}$$

holds for each permutation $j_1, j_2, \ldots, j_n$ of $\{1, 2, \ldots, n\}$.

The theorem holds if $n = 1$. We assume that $n > 1$ and proceed by induction on n. Without loss of generality we assume that $r_1 \leq s_1$. From the Laplace expansion by row 1 we obtain

$$\text{per}(A) = \sum_{j=1}^{n} a_{1j}\text{per}(A(1,j)).$$

Let the row sums and the column sums of $A(1,j)$ be, respectively,

$$r_2(j), r_3(j), \ldots, r_n(j), \quad (j = 1, 2, \ldots, n)$$

and

$$s_1(j), \ldots, s_{j-1}(j), s_{j+1}(j), \ldots, s_n(j), \quad (j = 1, 2, \ldots, n).$$

Let $i_1, i_2, \ldots, i_{n-1}$ be a permutation of $\{2, 3, \ldots, n\}$ such that

$$r_{i_1}(j) \leq r_{i_2}(j) \leq \cdots \leq r_{i_{n-1}}(j).$$

Let $k_1, k_2, \ldots, k_{n-1}$ be a permutation of $\{1, \ldots, j-1, j+1, \ldots, n\}$ such that

$$s_{k_1}(j) \leq s_{k_2}(j) \leq \cdots \leq s_{k_{n-1}}(j).$$

We have $r_{i_t}(j) \leq r_{i_t}$ and $s_{k_t}(j) \leq s_{k_t}$ for $t = 1, 2, \ldots, n-1$ and $j = 1, 2, \ldots, n$. Hence by the inductive hypothesis

$$\text{per}(A(1,j)) \leq \prod_{t=1}^{n-1} \min\{r_{i_t}, s_{k_t}\}, \quad (j = 1, 2, \ldots, n).$$

Therefore

$$\mathrm{per}(A) = \sum_{j=1}^{n} a_{1j}\mathrm{per}(A(1,j)) \le \left(\sum_{j=1}^{n} a_{1j}\right) \prod_{t=1}^{n-1} \min\{r_{i_t}, s_{k_t}\}$$

$$= r_1 \prod_{t=1}^{n-1} \min\{r_{i_t}, s_{k_t}\}$$

$$\le r_1 \prod_{i=2}^{j} \min\{r_i, s_{i-1}\} \prod_{i=j+1}^{n} \min\{r_i, s_i\}$$

$$\le r_1 \prod_{i=2}^{j} \min\{r_i, s_i\} \prod_{i=j+1}^{n} \min\{r_i, s_i\}$$

$$= r_1 \prod_{i=2}^{n} \min\{r_i, s_i\}.$$

Because $r_1 \le s_1$ the theorem now follows. $\quad\square$

Theorem 7.4.5 implies the validity of a conjecture of Ryser[1960]. Let n and k be integers with $1 \le k \le n$, and let $\mathcal{A}_{n,k}$ denote the set of all $(0,1)$-matrices of order n with exactly k 1's in each line. By Theorem 7.4.2 the permanent of each matrix A in $\mathcal{A}_{n,k}$ satisfies $\mathrm{per}(A) \ge k!$.

Theorem 7.4.7. *If k is a divisor of n, then the maximum permanent of a matrix in $\mathcal{A}_{n,k}$ equals $(k!)^{n/k}$.*

Proof. Suppose that k is a divisor of n and that A is a matrix in $\mathcal{A}_{n,k}$. By (7.45), $\mathrm{per}(A) \le (k!)^{n/k}$. The matrix $A = J_k \oplus \cdots \oplus J_k$, which is the direct sum of n/k matrices each of which equals the all 1's matrix J_k of order k, satisfies $\mathrm{per}(A) = (k!)^{n/k}$. $\quad\square$

Brualdi, Goldwasser and Michael[1988] generalized Theorem 7.4.7 by showing that $\mathrm{per}(A) \le (k!)^{n/k}$ provided that k is a divisor of n and A is a $(0,1)$-matrix of order n and the *average* number of 1's in each row equals k.

Now let

$$\beta(n, k) = \max\{\mathrm{per}(A) : A \in \mathcal{A}_{n,k}\}$$

denote the largest permanent achieved by a matrix in $\mathcal{A}_{n,k}$. If k is a divisor of n, then by Theorem 7.4.7, $\beta(n, k) = (k!)^{n/k}$. If $k = 2$ then Merriell[1980] proved that

$$\beta(n, 2) = 2^{\lfloor n/2 \rfloor}$$

holds for all n. Brualdi, Goldwasser and Michael[1988] showed that if $A \in \mathcal{A}_{n,2}$ satisfies $per(A) = 2^{\lfloor n/2 \rfloor}$, then there are permutation matrices P and Q of order n such that either

$$PAQ = J_2 \oplus J_2 \oplus \cdots \oplus J_2, \quad (n \text{ even})$$

or

$$PAQ = (J_3 - I_3) \oplus J_2 \oplus \cdots \oplus J_2, \quad (n \text{ odd}).$$

If $k = 3$ and 3 is not a divisor of n, then Merriell determined $\beta(n,3)$ as follows:

$$\beta(3t + 1, 3) = 6^{t-1}9, \quad (t \geq 1),$$

$$\beta(5, 3) = 13, \beta(3t + 2) = 6^{t-2}9^2, \quad (t \geq 2).$$

The matrix

$$(J_4 - I_4) \oplus J_3 \oplus \cdots \oplus J_3$$

in $\mathcal{A}_{3t+1,3}$ has permanent equal to $6^{t-1}9, (t \geq 1)$. The matrix

$$\begin{bmatrix} 0 & 0 & 1 & 1 & 1 \\ 1 & 0 & 0 & 1 & 1 \\ 1 & 1 & 0 & 0 & 1 \\ 1 & 1 & 1 & 0 & 0 \\ 0 & 1 & 1 & 1 & 0 \end{bmatrix}$$

in $\mathcal{A}_{5,3}$ has permanent equal to 13. The matrix

$$(J_4 - I_4) \oplus (J_4 - I_4) \oplus J_3 \oplus \cdots \oplus J_3$$

in $\mathcal{A}_{3t+2,3}$ has permanent equal to $6^{t-2}9^2, (t \geq 2)$.

If $k = 4$ then Bol'shakov[1986] determined that

$$\beta(4t + 1, 4) = 24^{t-1}44, \quad (t \geq 1).$$

The matrix

$$(J_5 - I_5) \oplus J_4 \oplus \cdots \oplus J_4$$

in $\mathcal{A}_{4t+1}$ has permanent equal to $24^{t-1}44$.

If $k = n - 1$ then $\beta(n, n-1) = D_n$, the nth derangement number, and every matrix in $\mathcal{A}_{n.k}$ has permanent equal to D_n. Now let $k = n - 2$. If $n \geq 8$, then Brualdi, Goldwasser and Michael[1988] showed that if A is a matrix in $\mathcal{A}_{n,n-2}$ satisfying $per(A) = \beta(n, n-2)$, then there exist permutation matrices P and Q of order n such that

$$PAQ = J_n - (J_2 \oplus J_2 \oplus \cdots \oplus J_2), \quad (n \text{ even}), \tag{7.48}$$

or

$$PAQ = J_n - ((J_3 - I_3) \oplus J_2 \oplus \cdots \oplus J_2), \quad (n \text{ odd}). \tag{7.49}$$

A simple expression for $\beta(n, n-2)$ is not known, but recurrence relations can be obtained from (7.48) and (7.49). If $n < 8$ the matrices A in $\mathcal{A}_{n,n-2}$ with $\text{per}(A) = \beta(n, n-2)$ are given in Brualdi, Goldwasser and Michael[1988].

Now let

$$\lambda(n, k) = \min\{\text{per}(A) : A \in \mathcal{A}_{n,k}\}, \quad (k = 1, 2, \ldots, n)$$

denote the smallest permanent achieved by a matrix in $\mathcal{A}_{n,k}$. We have $\lambda(n, 1) = 1$ and because the matrix $I_n + C_n{}^1$ in $\mathcal{A}_{n,2}$ has permanent equal to 2, $\lambda(n, 2) = 2$. We also have $\lambda(n, n) = n!$ and $\lambda(n, n-1) = D_n$, the nth derangement number. Henderson[1975] has shown that $\lambda(n, n-2) = U_n$, the nth ménage number, if n is even and $\lambda(n, n-2) = -1 + U_n$ if n is odd. The matrix $J_n - I_n - C_n$ in $\mathcal{A}_{n,n-2}$ has permanent equal to U_n. If n is odd and $n = 2k+1$, the matrix $J_n - ((I_k + C_k) \oplus (I_{k+1} + C_{k+1}))$ has permanent equal to $-1 + U_n$.

The exact value of $\lambda(n, 3)$ is not known in general. However the following bound of Voorhoeve[1979] gives the exponential lower bound

$$\lambda(n, 3) \geq 6 \left(\frac{4}{3}\right)^{n-3}, \quad (n \geq 3).$$

This bound holds for a wider class of matrices whose introduction facilitates its proof.

Let $\mathcal{B}_{n,k}$ denote the class of all nonnegative integral matrices of order n each of whose line sums equals k. The class $\mathcal{A}_{n,k}$ consists of all (0,1)-matrices in $\mathcal{B}_{n,k}$. Let

$$\lambda^*(n, k) = \min\{\text{per}(A) : A \in \mathcal{B}_{n,k}\}, \quad (k = 1, 2, \ldots, n)$$

denote the smallest permanent achieved by a matrix in $\mathcal{B}_{n,k}$. We have $\lambda(n, k) \geq \lambda^*(n, k)$.

Theorem 7.4.8.   *For all $n \geq 3$,*

$$\lambda^*(n, 3) \geq 6 \left(\frac{4}{3}\right)^{n-3}. \tag{7.50}$$

Proof. Let $\mathcal{B}'_{n,3}$ denote the class of matrices obtained by subtracting 1 from a positive entry of matrices in $\mathcal{B}_{n,3}$. Thus a nonnegative integral matrix of order n belongs to $\mathcal{B}'_{n,3}$ if and only if its row sums and its

---

[1] Recall that C_n is the permutation matrix of order n with 1's in positions $(1, 2), (2, 3), \ldots, (n-1, n), (n, 1)$.

column sums are $3, \ldots, 3, 2$ in some order. We denote the smallest permanent achieved by a matrix in $\mathcal{B}'_{n,3}$ by $\lambda'(n, 3)$. We prove (7.50) by showing that

$$\lambda^*(n, 3) \geq \frac{3}{2}\lambda'(n, 3), \quad (n \geq 3), \tag{7.51}$$

and

$$\lambda'(n, 3) \geq \frac{4}{3}\lambda'(n - 1, 3), \quad (n \geq 4). \tag{7.52}$$

Let A be a matrix in $\mathcal{B}_{n,3}$. Without loss of generality we assume that the first row of A equals

$$a_1, a_2, a_3, 0, \ldots, 0$$

where a_1, a_2 and a_3 are nonnegative integers with $a_1 + a_2 + a_3 = 3$. We have

$$(2a_1, 2a_2, 2a_3)$$

$$= a_1(a_1 - 1, a_2, a_3) + a_2(a_1, a_2 - 1, a_3) + a_3(a_1, a_2, a_3 - 1). \tag{7.53}$$

Because the permanent is a linear function of each of the rows of a matrix, it follows from (7.53) that

$$2\mathrm{per}(A) = a_1\mathrm{per}(A_1) + a_2\mathrm{per}(A_2) + a_3\mathrm{per}(A_3) \tag{7.54}$$

where A_1, A_2 and A_3 are obtained from A by subtracting one, respectively, from the elements of A in positions (1,1), (1,2) and (1,3). We note that one or two of the a_i's may be zero, but then the term $a_i\mathrm{per}(A_i) = 0$. If $a_i \neq 0$ then A_i belongs to the class $\mathcal{B}'_{n,3}$, and it follows from (7.54) that

$$2\mathrm{per}(A) \geq (a_1 + a_2 + a_3)\lambda'(n, 3) = 3\lambda'(n, 3).$$

Hence (7.51) holds.

Now assume that $n \geq 4$, and let A denote a matrix in $\mathcal{B}'_{n,3}$ satisfying $\mathrm{per}(A) = \lambda'(n, 3)$. Without loss of generality we assume that the first row sum of A equals 2 and that the first row of A is either $1, 1, 0, \ldots, 0$ or $2, 0, \ldots, 0$.

First assume that the first row of A is $1, 1, 0, \ldots, 0$. We write

$$A = \begin{bmatrix} 1 & 1 & 0 & \cdots & 0 \\ u & v & & B & \end{bmatrix}.$$

Because the permanent is a linear function of each of the columns of a square matrix, we have $\text{per}(A) = \text{per}(C)$ where

$$C = \begin{bmatrix} u+v & B \end{bmatrix},$$

a matrix of order $n-1$. The sum s of the entries of $u+v$ equals 3 or 4. If $s = 3$, then C is in the class $\mathcal{B}_{n-1,3}$ and by (7.51)

$$\lambda'(n,3) = \text{per}(A) = \text{per}(C) \geq \lambda^*(n-1,3)$$

$$\geq \frac{3}{2}\lambda'(n-1,3) \geq \frac{4}{3}\lambda'(n-1,3).$$

Now suppose that $s = 4$. Without loss of generality we assume that

$$u+v = \begin{bmatrix} d_1 & d_2 & d_3 & d_4 & 0 & \cdots & 0 \end{bmatrix}^{\mathrm{T}},$$

where d_1, d_2, d_3, d_4 are nonnegative integers satisfying $d_1 + d_2 + d_3 + d_4 = 4$. We have

$$(3d_1, 3d_2, 3_3, 3d_4)$$

$$= d_1(d_1 - 1, d_2, d_3, d_4) + d_2(d_1, d_2 - 1, d_3, d_4) + d_3(d_1, d_2, d_3 - 1, d_4)$$

$$+ d_4(d_1, d_2, d_3, d_4 - 1).$$

Using the linearity of the permanent again we obtain

$$3\text{per}(C) = \sum_{i=1}^{4} d_i \text{per}(C_i)$$

where C_i is the matrix obtained by subtracting one from the element d_i in the first column of $C, (i = 1, 2, 3, 4)$. If $d_i \neq 0$ then C_i is a matrix in the class $\mathcal{B}'_{n-1,3}$ and hence

$$3\lambda'(n,3) = 3\text{per}(A) = 3\text{per}(C) \geq \sum_{i=1}^{4} d_i \lambda'(n-1,3) = 4\lambda'(n-1,3).$$

Therefore $\lambda'(n,3) \geq (4/3)\lambda'(n-1,3)$ if $s = 4$.

Now assume that the first row of A is $(2, 0, \ldots, 0)$. We then write

$$A = \begin{bmatrix} 2 & 0 & \cdots & 0 \\ u & & B & \end{bmatrix}.$$

We have $\text{per}(A) = 2\text{per}(B)$ where either B is in the class $\mathcal{B}_{n-1,3}$ or B is in the class $\mathcal{B}'_{n-1,3}$. Hence

$$\lambda'(n,3) = \text{per}(A) \geq 2\min\{\lambda(n-1,3), \lambda'(n-1,3)\} = 2\lambda'(n-1,3)$$

and hence

$$\lambda'(n,3) \geq \frac{4}{3}\lambda'(n-1,3).$$

Thus (7.52) holds for all $n \geq 4$. The proof of the theorem is completed by noting that $\lambda'(3,3) = 4$.   □

Let A be a matrix in the class $\mathcal{B}_{n,k}$ where $n \geq 3$ and $k \geq 3$. Then there exist matrices A_1 in $\mathcal{B}_{n,3}$ and A_2 in $\mathcal{B}_{n,k-3}$ such that $A = A_1 + A_2$. Hence

$$\text{per}(A) \geq \text{per}(A_1) \geq \lambda^*(n,3) \geq 6\left(\frac{4}{3}\right)^{n-3},$$

and it follows that

$$\lambda^*(n,k) \geq 6\left(\frac{4}{3}\right)^{n-3} \geq \left(\frac{4}{3}\right)^n, \quad (n \geq 3, k \geq 3). \tag{7.55}$$

Let

$$\theta_k = \liminf_{n\to\infty} \lambda^*(n,k)^{1/n}, \quad (k \geq 3). \tag{7.56}$$

The number θ_k gives the best exponential lower bound

$$\text{per}(A) \geq \theta_k^n$$

for matrices A in $\mathcal{B}_{n,k}$ with n suffcently large. It follows from (7.55) that

$$\theta_k \geq \frac{4}{3}, \quad (k \geq 3). \tag{7.57}$$

The following upper bound for θ_k is due to Schrijver and Valiant[1980].

Theorem 7.4.9.   *For all $k \geq 3$,*

$$\theta_k \leq \frac{(k-1)^{k-1}}{k^{k-2}}.$$

Proof. Let $X = \{1, 2, 3, \ldots, nk\}$ and let $\mathcal{X}_{n,k}$ denote the collection of all ordered partitions $\mathcal{U} = (U_1, U_2, \ldots, U_n)$ of X into n sets of size k. An elementary combinatorial count shows that the number $c(n,k)$ of ordered partitions in $\mathcal{X}_{n,k}$ satisfies

$$c(n,k) = \frac{(nk)!}{(k!)^n}, \tag{7.58}$$

Let $\mathcal{V} = (V_1, V_2, \ldots, V_n)$ denote another ordered partition in $\mathcal{X}_{n,k}$. A set $R = \{x_1, x_2, \ldots, x_n\}$ of size n which consists of one element from each of the sets $U_1, U_2, \ldots, U_n$ is called a *transversal* of $\mathcal{U}$. The collection of transversals of $\mathcal{U}$ is denoted by $\mathcal{T}_{\mathcal{U}}$. A transversal of $\mathcal{U}$ which is also a transversal of $\mathcal{V}$ is called a *common transversal* of $\mathcal{U}$ and $\mathcal{V}$. (Thus the elements of a common transversal can be ordered to give an SDR of $\mathcal{U}$, and they can also be ordered to give an SDR of $\mathcal{V}$.) We denote the collection of all common transversals of $\mathcal{U}$ and $\mathcal{V}$ by $\mathcal{T}_{\mathcal{U},\mathcal{V}}$, and denote their number by $t_{\mathcal{U},\mathcal{V}}$. The number $t_{\mathcal{U},\mathcal{V}}$ can be computed as a permanent. Let $E = [e_{ij}]$ be the nonnegative integral matrix of order n defined by

$$e_{ij} = |U_i \cap V_j|, \quad (i, j, = 1, 2, \ldots, n).$$

The matrix E belongs to the class $\mathcal{B}_{n,k}$. If $j_1, j_2, \ldots, j_n$ is a permutation of $\{1, 2, \ldots, n\}$, then $e_{1j_1} e_{2j_2} \cdots e_{nj_n}$ counts the number of common transversals $\{x_1, x_2, \ldots, x_n\}$ of $\mathcal{U}$ and $\mathcal{V}$ in which $x_i \in U_i \cap V_{j_i}, (i = 1, 2, \ldots, n)$. Hence

$$t_{\mathcal{U},\mathcal{V}} = \operatorname{per}(E) \geq \theta_k^n. \tag{7.59}$$

We now fix the ordered partition $\mathcal{U}$ and allow the ordered partition $\mathcal{V}$ to vary over the set $\mathcal{X}_{n,k}$. It follows from (7.58) and (7.59) that

$$\sum_{\mathcal{V} \in \mathcal{X}_{n,k}} t_{\mathcal{U},\mathcal{V}} \geq \frac{(nk)!}{(k!)^n} \theta_k^n. \tag{7.60}$$

We also have

$$\sum_{\mathcal{V} \in \mathcal{X}_{n,k}} t_{\mathcal{U},\mathcal{V}} = \sum_{\mathcal{V} \in \mathcal{X}_{n,k}} \sum_{R \in \mathcal{T}_{\mathcal{U},\mathcal{V}}} 1 = \sum_{R \in \mathcal{T}_{\mathcal{U}}} \sum_{\{\mathcal{V} : \mathcal{V} \in \mathcal{X}_{n,k}, R \in \mathcal{T}_{\mathcal{V}}\}} 1$$

$$= \sum_{R \in \mathcal{T}_{\mathcal{U}}} n!c(n, k-1) = k^n n!c(n, k-1). \tag{7.61}$$

In the previous calculation we have used the facts that a transversal of $\mathcal{U}$ is a common transversal of $\mathcal{U}$ and $\mathcal{V}$ for exactly $n!c(n, k-1)$ ordered partitions $\mathcal{V}$ in $\mathcal{X}_{n,k}$ and that $\mathcal{U}$ has exactly k^n transversals. We now use (7.58) (with k replaced with $k-1$), (7.60) and (7.61) and obtain

$$k^n n! \frac{(nk-n)!}{(k-1)!^n} \geq \frac{(nk)!}{k!^n} \theta_k^n$$

and hence

$$\theta_k^n \leq \frac{k^{2n}}{\binom{nk}{n}}. \tag{7.62}$$

Applying Stirling's formula in (7.62) we obtain $\theta_k \leq (k-1)^{k-1}/k^{k-2}$. $\quad \square$

Corollary 7.4.10.

$$\theta_3 = \frac{4}{3}.$$

Proof. By (7.57), $\theta_3 \geq 4/3$. By Theorem 7.4.9, $\theta_3 \leq 4/3$. $\square$

It has been *conjectured* by Schrijver and Valiant[1980] that

$$\theta_k = (k-1)^{k-1}/k^{k-2}, k \geq 3.$$

The solution of the van der Waerden conjecture for the minimum permanent of a doubly stochastic matrix of order n (Egoryčev[1981] and Falikman[1981]) yields the bound $\theta_k \geq k/e$ which is better than the bound in (7.57) for all $k \geq 4$. In addition, Schrijver[1983] has shown that $\theta_4 \geq 3/2$ and $\theta_6 \geq 20/9$.

We now consider lower and upper bounds for the permanent of a (0,1)-matrix which depend on the total number of 1's in the matrix (and not on how these 1's are distributed in the rows and columns of the matrix).

We recall from Chapter 4 the definitions of a fully indecomposable and nearly decomposable matrix. A (0,1)-matrix A of order n is fully indecomposable provided A does not have an r by $n - r$ zero submatrix for any integer r with $1 \leq r \leq n - 1$. The fully indecomposable matrix A is nearly decomposable provided the replacement of a 1 of A with a 0 always results in a matrix which is not fully indecomposable. By Theorem 1.2.1 (cf. Theorem 4.2.2) the (0,1)-matrix A of order n is fully indecomposable if and only if

$$\text{per}(A(i,j)) > 0, \text{ for all } i, j = 1, 2, \dots, n. \tag{7.63}$$

A fully indecomposable (0,1)-matrix has a nonzero permanent. Moreover, it follows from (7.63) that if the fully indecomposable matrix A has row sums $r_1, r_2, \dots, r_n$ and column sums $s_1, s_2, \dots, s_n$, then

$$\text{per}(A) \geq \max\{r_1, r_2, \dots, r_n, s_1, s_2, \dots, s_n\}. \tag{7.64}$$

Now let A be a (0,1)-matrix of order n with a nonzero permanent, and let $A_1, A_2, \dots, A_t, (t \geq 1)$, be the fully indecomposable components of A (cf. Theorem 4.2.6). Then

$$\text{per}(A) = \prod_{i=1}^{t} \text{per}(A_i). \tag{7.65}$$

Hence bounds for the permanent of a fully indecomposable matrix will give bounds for the permanent of any matrix with a nonzero permanent. Minc[1969] obtained the following lower bound for the permanent of a fully indecomposable (0,1)-matrix. We follow the proof of Hartfiel[1970].

Theorem 7.4.11. *Let A be a fully indecomposable $(0,1)$-matrix of order n with exactly $\sigma(A)$ 1's. Then*

$$\operatorname{per}(A) \geq \sigma(A) - 2n + 2. \tag{7.66}$$

Proof. We first establish (7.66) for nearly decomposable matrices A by induction on n. If $n = 1$ then (7.66) clearly holds. Now assume that $n \geq 2$. It follows from Theorem 4.3.4 that we may assume that

$$A = \begin{bmatrix} 1 & 0 & 0 & \cdots & 0 & 0 & \\ 1 & 1 & 0 & \cdots & 0 & 0 & \\ 0 & 1 & 1 & \cdots & 0 & 0 & \\ \vdots & \vdots & \vdots & \ddots & \vdots & \vdots & F_1 \\ 0 & 0 & 0 & \cdots & 1 & 0 & \\ 0 & 0 & 0 & \cdots & 1 & 1 & \\ & & F_2 & & & & B \end{bmatrix}$$

where B is a nearly decomposable matrix of order m with $1 \leq m \leq n - 1$, and F_1 and F_2 each contains exactly one 1. Applying the inductive hypothesis and (7.63) to B we obtain

$$\operatorname{per}(A) \geq \operatorname{per}(B) + 1 \geq \sigma(B) - 2m + 2 + 1 = \sigma(A) - 2n + 2.$$

Hence (7.66) holds if A is nearly decomposable.

Now assume that A is not nearly decomposable. Because A is fully indecomposable there exists a $(0,1)$-matrix C of order n such that $A - C$ is a nearly decomposable $(0,1)$-matrix. Applying (7.63) to A and (7.66) to $A - C$, we obtain

$$\operatorname{per}(A) \geq \operatorname{per}(A-C) + \sigma(C) \geq \sigma(A-C) - 2n + 2 + \sigma(C) = \sigma(A) - 2n + 2.$$

$\square$

Corollary 7.4.12. *Let A be a fully indecomposable, nonnegative integral matrix of order n the sum of whose elements equals $\sigma(A)$. Then*

$$\operatorname{per}(A) \geq \sigma(A) - 2n + 2. \tag{7.67}$$

Proof. If A is a $(0,1)$-matrix we apply Theorem 7.4.11. Suppose that some entry a_{rs} of A is greater than 1. Let B be the matrix obtained from A by subtracting 1 from a_{rs}. Then $\sigma(B) < \sigma(A)$ and B is a fully indecomposable, nonnegative integral matrix. Arguing by induction on the sum of elements we obtain

$$\operatorname{per}(A) = \operatorname{per}(B) + \operatorname{per}(A(r,s)) \geq \sigma(B) - 2n + 2 + 1 = \sigma(A) - 2n + 2.$$

$\square$

Brualdi and Gibson[1977] characterized the fully indecomposable, non-negative integral matrices A of order n for which equality holds in (7.67) as follows. Assume that $n \geq 2$. Then $\text{per}(A) = \sigma(A) - 2n + 2$ if and only if there exists an integer p with $0 \leq p \leq n - 1$ and permutation matrices P and Q of order n such that

$$PAQ = \begin{bmatrix} A_3 & A_1 \\ A_2 & O \end{bmatrix} \tag{7.68}$$

where A_3 is a nonnegative integral matrix of size $n - p$ by $p + 1$, and A_1^T and A_2 are (0,1)-matrices with exactly two 1's in each row. The full indecomposability assumption on A implies that the matrices A_1^T and A_2 are incidence matrices of graphs which are trees.

We now turn to an upper bound of Foregger[1975] for the permanent of a fully indecomposable, nonnegative integral matrix. The following lemma is a step in its proof.

Lemma 7.4.13. *Let $A = [a_{ij}]$ be a fully indecomposable, nonnegative integral matrix of order $n \geq 2$. Then there exists an integer $j \geq 0$ and a fully indecomposable $(0,1)$-matrix B of order n with $B \leq A$ such that*

$$\text{per}(A) \leq 2^j \text{per}(B) - (2^j - 1) \text{ where } \sigma(A) - \sigma(B) = j. \tag{7.69}$$

Proof. If A is a (0,1)-matrix, then $B = A$ and $j = 0$ satisfy (7.69). Now suppose that there exist an element a_{rs} of A with $a_{rs} \geq 2$. Let A' be the matrix obtained from A by subtracting 1 from a_{rs}. Because A is fully indecomposable and $n \geq 2$, there exists an integer $t \neq s$ such that $a_{rt} \geq 1$. From the Laplace expansion by row r we obtain

$$\text{per}(A) = \sum_{k=1}^{n} a_{rk} \text{per}(A(r,k))$$

$$\geq a_{rs} \text{per}(A(r,s)) + a_{rt} \text{per}(A(r,t)) \geq 2\text{per}(A(r,s)) + 1.$$

Hence

$$\text{per}(A(r,s)) \leq \frac{(\text{per}(A) - 1)}{2}. \tag{7.70}$$

We also have

$$\text{per}(A) = \text{per}(A') + \text{per}(A(r,s)). \tag{7.71}$$

From (7.70) and (7.71) we get

$$\text{per}(A) \leq 2\text{per}(A') - 1. \tag{7.72}$$

The proof of the lemma is now completed by induction on the sum

$$\sum (a_{ij} - 1 : a_{ij} \geq 2, i = 1, 2, \ldots, n, j = 1, 2, \ldots, n). \qquad \square$$

Theorem 7.4.14.   *Let A be a fully indecomposable, nonnegative integral matrix of order n, the sum of whose elements equals $\sigma(A)$. Then*

$$\mathrm{per}\,(A) \le 2^{\sigma(A)-2n} + 1. \tag{7.73}$$

Proof. If $n = 1$ the inequality holds. We now assume that $n \ge 2$, and proceed by induction on n. If A is a (0,1)-matrix and each row sum of A is at least equal to three, then (7.45) implies that we have strict inequality in (7.73).

First suppose that A is a (0,1)-matrix with at least one row sum equal to 2. Without loss of generality we assume that row 1 of A equals $1, 1, 0, \ldots, 0$. The linearity of the permanent implies that

$$\mathrm{per}(A) = \mathrm{per}(A(1,1)) + \mathrm{per}(A(1,2)) = \mathrm{per}(A')$$

where A' is the matrix of order $n - 1$ obtained from A by adding column 1 to column 2 and then deleting row 1 and column 1. The matrix A' is fully indecomposable and $\sigma(A') = \sigma(A) - 2$. Using the inductive assumption we obtain

$$\mathrm{per}(A) = \mathrm{per}(A') \le 2^{\sigma(A')-2(n-1)} + 1 = 2^{\sigma(A)-2n} + 1.$$

Now suppose that A has at least one element which is strictly larger than 1. Let B be a (0,1)-matrix and j an integer satisfying the conclusions of Lemma 7.4.13. Applying what we have just shown to the fully indecomposable (0,1)-matrix B, we obtain

$$\mathrm{per}(A) \le 2^j \mathrm{per}(B) - (2^j - 1)$$

$$\le 2^j(2^{\sigma(B)-2n} + 1) - (2^j - 1) = 2^{\sigma(A)-2n} + 1.$$

Hence the theorem follows by induction.   □

The fully indecomposable, nonnegative integral matrices A for which equality holds in (7.73) have been characterized by Foregger[1975]. Up to row and column permutations such matrices are equal to a matrix of the form

$$
\begin{bmatrix}
I_{k_1} + C_{k_1} & O & \cdots & O & E_1 \\
E_2 & I_{k_2} + C_{k_2} & \cdots & O & O \\
\vdots & \vdots & \ddots & \vdots & \vdots \\
O & O & \cdots & I_{k_{p-1}} + C_{k_{p-1}} & O \\
O & O & \cdots & E_p & I_{k_p} + C_{k_p}
\end{bmatrix}
$$

where p is a positive integer, E_i is a (0,1)-matrix containing exactly one 1, and C_{k_i} as usual denotes the permutation matrix of order k_i with 1's in positions $(1,2),(2,3),\ldots,(k_i-1,k_i),(k_i,1)$. (If $k_i = 1$, then $C_{k_i} = [1]$ and $I_{k_i} + C_{k_i} = [2]$.)

As noted in the proof of Theorem 7.4.14, if A is a fully indecomposable $(0,1)$-matrix with all row sums at least equal to three, then (7.73) is implied by (7.45). If some row sum equals two, then (7.73) may be better than (7.45). For example, let

$$A = \begin{bmatrix} 1 & 0 & 1 & 0 \\ 1 & 1 & 0 & 1 \\ 0 & 1 & 1 & 0 \\ 0 & 0 & 1 & 1 \end{bmatrix}.$$

Then $\mathrm{per}(A) = 3$ and (7.73) gives $\mathrm{per}(A) \le 3$. However, (7.45) gives $\mathrm{per}(A) \le 2^{3/2}3^{1/3} = 4.079\ldots$. The inequality (7.73) holds for integral matrices A, but (7.45) need not hold if A is not a $(0,1)$-matrix.

Donald et al.[1984] have improved (7.7.3) by showing that a fully inde-composable, nonnegative integral matrix A with row sums $r_1,\ldots,r_n$ and column sums $s_1,\ldots,s_n$ satisfies

$$\mathrm{per}(A) \le 1 + \min\left\{\prod_i(r_i-1), \prod_i(s_i-1)\right\}.$$

To conclude this section we derive the following theorem of Brualdi and Gibson[1977] in which the full indecomposability assumption in Theorem 7.4.14 is replaced by the assumption of total support. We recall from section 4.2 that a matrix of total support is, up to permutation of rows and columns, a direct sum of $t \ge 1$ fully indecomposable matrices, and these t matrices are the fully indecomposable components of A.

Theorem 7.4.15.   *Let A be a nonnegative integral matrix of order n with total support, and let t be the number of fully indecomposable components of A. Then*

$$\mathrm{per}\,(A) \le 2^{\sigma(A)-2n+t}. \tag{7.74}$$

Proof. Without loss of generality we assume that $A = A_1 \oplus A_2 \oplus \cdots \oplus A_t$ where A_i is a fully indecomposable matrix of order $n_i \ge 1, (i = 1,2,\ldots,t)$. We have

$$\mathrm{per}(A) = \prod_{i=1}^{t} \mathrm{per}(A_i).$$

If for some i we have $n_i = 1$ and $A_i = [1]$, then $2^{\sigma(A_i)-2n_i+1} = 1$ and $\mathrm{per}(A_i) = 1$. It follows that we may assume that $\sigma(A_i) - 2n_i \ge 0, (i = 1,2,\ldots,t)$. If for some i we have $n_i = 1$ and $A_i = [2]$, then $2^{\sigma(A_i)-2n_i+1} = 2$

and $\text{per}(A_i) = 2$. Hence we may now assume that $\sigma(A_i) - 2n_i \geq 1$, $(i = 1, 2, \ldots, t)$. Using Theorem 7.4.14 we obtain

$$\text{per}(A) = \prod_{i=1}^{t} \text{per}(A_i) \leq \prod_{i=1}^{t} (2^{\sigma(A_i) - 2n_i} + 1)$$

$$\leq 2^{(\sum_{i=1}^{t} \sigma(A_i) - 2n_i) + t - 1} + 1 = 2^{\sigma(A) - 2n + t - 1} + 1 < 2^{\sigma(A) - 2n + t}. \qquad \square$$

It follows from the proof of Theorem 7.4.15 that equality holds in (7.74) if and only if each fully indecomposable component of A is either a matrix of order 1 whose unique element is 1 or 2, or $I_k + C_k$ for some $k \geq 2$.

Exercises

1. Let A be a (0,1)-matrix of size m by n with $m \leq n$. Prove that

$$\text{per}(A) \leq n(n-1) \cdots (n - m + 1)$$

with equality if and only if $A = J_{m,n}$.

2. Let A be a fully indecomposable (0,1)-matrix of order n and let r be the maximum row sum of A. Prove that $\text{per}(A) \geq r$ with equality if and only if at least $n - 1$ of the row sums of A equal 2 (Minc[1973]).

3. Let A be a fully indecomposable (0,1)-matrix of order n and let r be the minimal row sum of A. Prove that

$$\text{per}(A) \geq \sigma(A) - 2n + 2 + \sum_{k=1}^{r-1} (k! - 1)$$

(Gibson[1972]).

4. Use Theorem 7.4.5 to show that if A is a (0,1)-matrix of order n with row sums $r_1, r_2, \ldots, r_n$, then

$$\text{per}(A) \leq \prod_{i=1}^{n} \frac{r_i + 1}{2}$$

(Minc[1963]).

5. Determine all fully indecomposable (0,1)-matrices with permanent equal to 3.

6. Determine all fully indecomposable (0,1)-matrices with permanent equal to 4.

7. Let k be an integer with $0 \leq k \leq 2^{n-1}$. Show that there is a (0,1)-matrix of order n with permanent equal to k (Brualdi and Newman[1965]).

8. Let p be a prime number and let A be a circulant, nonnegative integral matrix of order p. Let r be the common value of the row sums of A. Prove that

$$\text{per}(A) \equiv r \pmod{p}$$

(Brualdi and Newman[1965]).

9. Let A be a (0,1)-matrix of order n with exactly n 0's. Prove that $\text{per}(A) \leq D_n$, the nth derangement number, with equality if and only if there are permutation matrices P and Q such that $PAQ = J - I$ (Brualdi, Goldwasser and Michael[1988]).

10. Let A be a $(0,1)$-matrix of order n of the form

$$\begin{bmatrix} A_3 & A_1 \\ A_2 & O \end{bmatrix}$$

where O denotes a zero matrix of size p by $n - p - 1$, and A_1^T and A_2 are incidence matrices of trees. Prove that $\operatorname{per}(A) = \sigma(A_3)$, the number of 1's of A_3 (Brualdi and Gibson[1977]).

11. Let $n = 2k + 1$. Show that the permanent of the matrix $J_n - ((I_k + C_k) \oplus (I_{k+1} + C_{k+1}))$ of order n with exactly two 0's in each row and column equals $-1 + U_n$, where U_n denotes the nth ménage number (Henderson[1975]).

12. Let B be a nonnegative matrix of order n each of whose line sums is at most 1. Use Corollary 7.1.2 to prove that $\operatorname{per}(I - B) \geq 0$ (Gibson[1966]).

References

V.I. Bol'shakov[1986], Upper values of a permanent in Λ_n^k, *Combinatorial analysis, No. 7 (Russian)*, Moskov. Gos. Univ., Moscow, pp. 92–118 and 164–165.
[1986], The spectrum of the permanent on Λ_n^k, *Proceedings of the All-Union seminar on discrete mathematics and its applications, Moscow (1984) (Russian)*, Moskov. Gos. Univ. Mekh.-Mat. Fak., Moscow, pp. 65–73.

L.M. Brégman[1973], Certain properties of nonnegative matrices and their permanents, *Dokl. Akad. Nauk SSSR*, 211, pp. 27–30 (*Soviet Math. Dokl.*, 14, pp. 945–949).

R.A. Brualdi and P.M. Gibson[1977], Convex polyhedra of doubly stochastic matrices I. Applications of the permanent function, *J. Combin. Theory, Ser. A*, 22, pp. 194–230.

R.A. Brualdi and M. Newman[1965], Some theorems on the permanent, *J. Res. National Bur. Stands.*, 69B, pp. 159–163.

R.A. Brualdi, J.L. Goldwasser and T.S. Michael[1988], Maximum permanents of matrices of zeros and ones, *J. Combin. Theory, Ser. A*, 47, pp. 207–245.

R.A. Brualdi and B.L. Shader[1990], Matrix factorizations of determinants and permanents, *J. Combin. Theory, Ser. A*, 54, pp. 132–134.

J. Donald, J. Elwin, R. Hager and P. Salomon[1984], A graph theoretic upper bound on the permanent of a nonnegative integer matrix. I, *Linear Alg. Applics.*, 61, pp. 187–198.

G.P. Egoryčev[1981], A solution of van der Waerden's permanent problem, *Dokl. Akad. Nauk SSSR*, 258, pp. 1041–1044 (*Soviet Math. Dokl.*, 23, pp. 619–622).

D.I. Falikman[1981], A proof of van der Waerden's conjecture on the permanent of a doubly stochastic matrix, *Mat. Zametki*, 29, pp. 931–938 (*Math. Notes*, 29, pp. 475–479).

T.H. Foregger[1975], An upper bound for the permanent of a fully indecomposable matrix, *Proc. Amer. Math. Soc.*, 49, pp. 319–324.

P.M. Gibson[1966], A short proof of an inequality for the permanent function, *Proc. Amer. Math. Soc.*, 17, pp. 535–536.
[1972], A lower bound for the permanent of a $(0,1)$-matrix, *Proc. Amer. Math. Soc.*, 33, pp. 245–246.

M. Hall, Jr.[1948], Distinct representatives of subsets, *Bull. Amer. Math. Soc.*, 54, pp. 922–926.

D.J. Hartfiel[1970], A simplified form for nearly reducible and nearly decomposable matrices, *Proc. Amer. Math. Soc.*, 24, pp. 388–393.

J.R. Henderson[1975], Permanents of (0,1)-matrices having at most two 0's per line, *Canad. Math. Bull.*, 18, pp. 353–358.

W.B. Jurkat and H.J. Ryser[1966], Matrix factorizations of permanents and determinants, *J. Algebra*, 3, pp. 1–27.

  [1967], Term ranks and permanents of nonnegative matrices, *J. Algebra*, 5, pp. 342–357.

D. Merriell[1980], The maximum permanent in Λ_n^k, *Linear Multilin. Alg.*, 9, pp. 81–91.

H. Minc[1963], Upper bound for permanents of (0,1)-matrices, *Bull. Amer. Math. Soc.*, 69, pp. 789–791.

  [1967], A lower bound for permanents of (0,1)-matrices, *Proc. Amer. Math. Soc.*, 18, pp. 1128–1132.

  [1969], On lower bounds for permanents of (0,1)-matrices, *Proc. Amer. Math. Soc.*, 22, pp. 117–123.

  [1973], (0,1)-matrices with minimal permanents, *Israel J. Math.*, 77, pp. 27–30.

  [1974], An unresolved conjecture on permanents of (0,1)-matrices, *Linear Multilin. Alg.*, 2, pp. 57–64.

  [1978], *Permanents*, Addison-Wesley, Reading, Mass.

A. Nijenhuis and H. Wilf[1970], On a conjecture of Ryser and Minc, *Nederl. Akad. Wetensch. Proc., Ser. A*, 73 (=*Indag. Math.*, 32), pp. 151–158.

P.A. Ostrand[1970], Systems of distinct representatives II, *J. Math. Anal. Applics.*, 32, pp. 1–4.

R. Rado[1967], On the number of systems of distinct representatives, *J. London Math. Soc.*, 42, pp. 107–109.

H.J. Ryser[1960], Matrices of zeros and ones, *Bull. Amer. Mat. Soc.*, 66, pp. 442–464.

A. Schrijver[1978], A short proof of Minc's conjecture, *J. Combin. Theory, Ser. A*, 25, pp. 80–83.

  [1983], Bounds on permanents, and the number of 1-factors and 1-factorizations of bipartite graphs, *Surveys in Combinatorics*, Cambridge University Press, Cambridge, pp. 107–134.

A. Schrijver and W.G. Valiant[1980], On lower bounds for permanents, *Indag. Math.*, 42, pp. 425–427.

M. Voorhoeve[1979], A lower bound for the permanents of certain (0,1)-matrices, *Indag. Math.*, 41, pp. 83–86.

H.S. Wilf[1968], A mechanical counting method and combinatorial applications, *J. Combin. Theory*, 4, pp. 246–258.

7.5   Evaluation of Permanents

Let $X = [x_{ij}]$, $(i, j = 1, 2, \ldots, n)$ be a real matrix of order n. The definition

$$\mathrm{per}(X) = \sum x_{1\pi(1)} x_{2\pi(2)} \cdots x_{n\pi(n)}$$

of the permanent of X is very much like the definition

$$\det(X) = \sum (\mathrm{sign}\pi) x_{1\pi(1)} x_{2\pi(2)} \cdots x_{n\pi(n)}$$

of the determinant. Both summations are over all permutations π of $\{1, 2, \ldots, n\}$. In the case of the determinant a negative sign is affixed to those summands which correspond to even permutations. The similarity of the two definitions and the existence of efficient computational procedures for determinants suggest the possibility of directly affixing negative signs to some of the elements of the matrix X in order to obtain a matrix $X' = [x'_{ij}]$ with $x'_{ij} = \pm x_{ij}, (i, j = 1, 2, \ldots, n)$ which satisfies

$$\det(X') = \text{per}(X).$$

Let $E = [e_{ij}]$ be a matrix of order n each of whose elements equals 1 or -1, and let

$$E * X = [e_{ij}x_{ij}], \quad (i, j = 1, 2, \ldots, n),$$

be the *elementwise* product of E and X. We say that the matrix E *converts the permanent of n by n matrices into the determinant* provided

$$\text{per}(X) = \det(E * X) \tag{7.75}$$

for all matrices X of order n. If (7.75) holds then we have an effective procedure to calculate the permanent of any real matrix X of order n. Since (7.75) is to hold for all matrices X of order n, (7.75) is a polynomial identity in the n^2 elements of the matrix X. Hence the $(1, -1)$-matrix E converts the permanent of n by n matrices into the determinant if and only if

$$e_{1\pi(1)}e_{2\pi(2)} \cdots e_{n\pi(n)} = \text{sign}(\pi(1), \pi(2), \ldots, \pi(n)) \tag{7.76}$$

for each permutation $(\pi(1), \pi(2), \ldots, \pi(n))$ of $\{1, 2, \ldots, n\}$. Equivalently, the $(1, -1)$-matrix E converts the permanent of n by n matrices into the determinant if and only if

$$\det(E) = n!.$$

Let $n = 2$ and let

$$E = \begin{bmatrix} 1 & -1 \\ 1 & 1 \end{bmatrix}.$$

Then $\det(E) = 2$ and it follows that (7.75) holds. Pólya[1913] (see also Szegö[1913]) observed that for $n \geq 3$ there is no matrix E which converts the permanent of n by n matrices into the determinant.

Theorem 7.5.1. *Let $n \geq 3$. Then there does not exist a $(1, -1)$-matrix E which converts the permanent of n by n matrices into the determinant.*

Proof. First suppose that $n = 3$ and that $E = [e_{ij}]$ is a $(1, -1)$-matrix of order 3. There are three even permutations of $\{1, 2, 3\}$ and also three odd permutations, and both the even permutations and the odd permutations partition the elements of E. Since the even permutations partition the elements of E we have

$$\prod_{i,j=1}^{3} e_{ij} = (e_{11}e_{22}e_{33})(e_{12}e_{23}e_{31})(e_{13}e_{21}e_{31}) = (1)(1)(1) = 1.$$

Since the odd permutations partition the elements of E we also have

$$\prod_{i,j=1}^{3} e_{ij} = (e_{11}e_{23}e_{32})(e_{13}e_{22}e_{31})(e_{12}e_{21}e_{33}) = (-1)(-1)(-1) = -1.$$

Hence E does not convert the permanent of 3 by 3 matrices into the determinant. If $n > 3$, then by considering matrices of the form $X = I_{n-3} \oplus X'$, where X' is a matrix of order 3, we also conclude that there is no $(1, -1)$-matrix E which converts the permanent of n by n matrices into the determinant. □

Let $\mathcal{M}_n$ denote the linear space of n by n matrices over the real field. Let E be a $(1, -1)$-matrix of order n. The mapping $T : \mathcal{M}_n \to \mathcal{M}_n$ defined by $T(X) = E * X$ for all X in $\mathcal{M}_n$ is an instance of a linear transformation. Marcus and Minc[1961] generalized Theorem 7.5.1 by showing that for $n \geq 3$ no linear transformation $T : \mathcal{M}_n \to \mathcal{M}_n$ satisfies

$$\text{per}(X) = \det(T(X)), X \in \mathcal{M}_n. \tag{7.77}$$

Theorem 7.5.1 was further generalized by von zur Gathen[1987a] who showed that there is no affine transformation T on $\mathcal{M}_n$ for which (7.77) holds. These results all hold for matrices over an arbitrary infinite field of characteristic different from 2.

We now consider the possibility of converting the permanent into the determinant on "coordinate" subspaces of the linear space $\mathcal{M}_n$. Let $A = [a_{ij}]$ be a $(0,1)$-matrix of order n. Then

$$\mathcal{M}_n(A) = \{A * X : X \in \mathcal{M}_n\}$$

is the linear subspace of $\mathcal{M}_n$ consisting of the matrices which have 0's in those positions in which A has 0's (and possibly other 0's as well). If $\text{per}(A) = 0$, then $\text{per}(A * X) = \det(A * X) = 0$ for all X in $\mathcal{M}_n$. As a result *we henceforth assume that* $\text{per}(A) \neq 0$. Let E be an n by n $(0, 1, -1)$-matrix, and assume that an element of E equals 0 if and only if the corresponding element of A equals 0. Thus if $|E|$ denotes the matrix obtained from E by replacing each element with its absolute value, then $|E| = A$. We say that the matrix E *converts the permanent of matrices in* $\mathcal{M}_n(A)$ *into the determinant* provided

$$\text{per}(Y) = \det(E * Y) \tag{7.78}$$

for all matrices Y in $\mathcal{M}_n(A)$. Since A and E have 0's in exactly the same positions, (7.78) is equivalent to

$$\mathrm{per}(A * X) = \det(E * X) \qquad (7.79)$$

for all matrices X in $\mathcal{M}_n$. Equation (7.79) is a polynomial identity in the elements of X corresponding to the 1's of A. Hence (7.79) holds if and only if

$$\mathrm{per}(A) = \det(E).$$

Since $|E| = A$, we conclude that E converts the permanent of matrices in $\mathcal{M}_n(A)$ into the determinant if and only if

$$\mathrm{per}(|E|) = \det(E). \qquad (7.80)$$

Thus the problem of finding coordinate subspaces of $\mathcal{M}_n$ on which the permanent can be evaluated as a determinant is equivalent to the problem of finding $(0, 1, -1)$-matrices E of order n satisfying (7.80).

For $n = 3$ the matrix

$$E = \begin{bmatrix} 1 & 1 & 0 \\ -1 & 1 & 1 \\ 1 & -1 & 1 \end{bmatrix}$$

satisfies (7.80) and hence converts the permanent of 3 by 3 matrices with a 0 in position (1,3) into the determinant.

For a real number x, the *sign* of x is defined by

$$\mathrm{sign}(x) = \begin{cases} 1 & \text{if } x > 0 \\ 0 & \text{if } x = 0 \\ -1 & \text{if } x < 0. \end{cases}$$

The *sign pattern* of the real matrix $X = [x_{ij}]$ of order n is the $(0, 1, -1)$-matrix

$$\mathrm{sign}(X) = [\mathrm{sign}(x_{ij})], (i, j = 1, 2, \ldots, n)$$

obtained by replacing each of the elements of X with its sign. The matrix X is *sign-nonsingular* provided that each matrix with the same sign pattern as X is nonsingular. A matrix obtained from a sign-nonsingular matrix by arbitrary line permutations or by transposition is also sign-nonsingular. In discussing sign-nonsingular matrices there is no loss in generality in restricting our attention to $(0, 1, -1)$-matrices X, that is matrices X for which $X = \mathrm{sign}(X)$. We now show the equivalence of sign-nonsingular matrices and matrices which convert the permanent into the determinant (Brualdi[1988] and Brualdi and Shader[1991]).

Theorem 7.5.2. *The $(0, 1, -1)$-matrix X of order n is sign-nonsingular* *if and only if X converts the permanent of matrices in $\mathcal{M}_n(|X|)$ into the determinant or the negative of the determinant, that is if and only if*

$$\operatorname{per}(|X|) = \pm \det(X) \neq 0. \tag{7.81}$$

Proof. If (7.81) holds then there is a nonzero term in the determinant expansion of X and all nonzero terms have the same sign, and it follows that X is sign-nonsingular. Now suppose that $X = [x_{ij}]$ is sign-nonsingular. Let $(i_1, i_2, \ldots, i_n)$ be a permutation of $\{1, 2, \ldots, n\}$ for which $x_{1i_1} x_{2i_2} \cdots x_{ni_n} \neq 0$. Let $Y = [y_{ij}]$ be a nonnegative matrix for which y_{ij_i} equals 1 for $i = 1, 2, \ldots, n$ and y_{rs} equals a positive number ϵ otherwise. Then $Z = X * Y$ has the same sign pattern as X and for ϵ sufficiently small

$$\operatorname{sign}(\det(Z)) = \operatorname{sign}(i_1, i_2, \ldots, i_n) x_{1i_1} x_{2i_2} \cdots x_{ni_n}.$$

It now follows by continuity that each nonzero term in the determinant expansion of X has the same sign, and hence (7.81) holds.   □

It follows from Theorem 7.5.2 that a sign-nonsingular matrix gives both a coordinate subspace $\mathcal{M}_n(A)$ of $\mathcal{M}_n$ on which the permanent can be evaluated as a determinant and a prescription for doing so. We now turn to demonstrating that under the assumption of full indecomposability this prescription is unique up to diagonal equivalence. First we prove the following lemma which contains a result first proved by Sinkhorn and Knopp[1969] (see also Ryser[1973] and Engel and Schneider[1973]).

Lemma 7.5.3.   *Let $A = [a_{ij}]$ be a fully indecomposable matrix for which there exists a nonzero number d such that for each permutation $(i_1, i_2, \ldots, i_n)$ of $\{1, 2, \ldots, n\}$*

$$a_{1i_1} a_{2i_2} \cdots a_{ni_n} = 0 \text{ or } d.$$

Then there exist nonsingular diagonal matrices D_1 and D_2 such that $D_1 A D_2$ is a $(0,1)$-matrix. Moreover, the matrices D_1 and D_2 are unique up to multiplication of D_1 by a nonzero scalar θ and multiplication of D_2 by θ^{-1}.

Proof. We prove the lemma by induction on n. If $n = 1$ the lemma clearly holds. Now assume that $n > 1$. By Theorem 4.2.8 we may assume that

$$A = \begin{bmatrix} A_1 & O & \cdots & O & E_1 \\ E_2 & A_2 & \cdots & O & O \\ \vdots & \vdots & \ddots & \vdots & \vdots \\ O & O & \cdots & A_{m-1} & O \\ O & O & \cdots & E_m & A_m \end{bmatrix},$$

where $m \geq 2$, A_i is fully indecomposable of order $n_i, (i = 1, 2, \ldots, m)$, and E_i contains at least one nonzero element $e_i, (i = 1, 2, \ldots, m)$. By the induction hypothesis we may assume that each of the matrices A_i is a $(0,1)$-matrix and that $d = 1$. Since each A_i is fully indecomposable it follows from Theorem 4.2.2 that the nonzero elements of E_i have a common value e_i and that $e_1 e_2 \cdots e_m = 1$. Letting

$$D_2 = e_1 I_{n_1} \oplus (e_1 e_2) I_{n_2} \oplus \cdots \oplus (e_1 e_2 \cdots e_m) I_{n_m} \quad \text{and} \quad D_1 = D_2^{-1},$$

we see that $D_1 A D_2$ is a $(0,1)$-matrix. The uniqueness statement in the lemma follows easily from the inductive hypothesis.   □

The following three theorems are from Brualdi and Shader[1991].

Theorem 7.5.4.   *Let A be a $(0,1)$-matrix with total support and let X and Y be sign-nonsingular matrices with $|X| = |Y| = A$. Then there exist diagonal matrices D_1 and D_2 of order n whose diagonal elements equal ± 1 such that $Y = D_1 X D_2$. If A is a fully indecomposable matrix then D_1 and D_2 are unique up to a scalar factor of -1.*

Proof. Assume that A is fully indecomposable. It follows from Theorem 7.5.2 that both X and Y satisfy the hypotheses of Lemma 7.5.3 with $d = 1$. Hence there exist nonsingular diagonal matrices D_1 and D_2, unique up to multiplication by θ and θ^{-1}, respectively, such that $X = D_1 Y D_2$. Let

$$D_1 = \text{diag}(d_1, d_2, \ldots, d_n) \quad \text{and} \quad D_2 = \text{diag}(d_1', d_2', \ldots, d_n')$$

be diagonal matrices, and let $K = \{i : |d_i| = |d_1|, 1 \le i \le n\}$ and $L = \{j : |d_j'| = |d_1|^{-1}, 1 \le j \le n\}$. Since X and Y are $(0, 1, -1)$-matrices, it follows that the submatrices $A[K, \overline{L}]$ and $A[\overline{K}, L]$ of A in rows K and columns $\overline{L}$ and rows $\overline{K}$ and columns L, respectively, are zero matrices. Because A is fully indecomposable, we have $K = L = \{1, 2, \ldots, n\}$. Let $D_1^* = d_1^{-1} D_1$ and $D_2^* = d_1 D_2$. Then D_1^* and D_2^* are diagonal matrices whose main diagonal elements equal ± 1 such that $Y = D_1^* X D_2^*$. The theorem now follows.   □

Theorem 7.5.5.   *Let A be a fully indecomposable $(0,1)$-matrix of order n, and let $X = [x_{ij}]$ be a sign-nonsingular matrix with $|X| = A$. Let B be a $(0,1)$-matrix obtained from A by replacing a 0 in position (u, v) with a 1. Then the following are equivalent:*

(i)   *There exists a sign-nonsingular matrix $Z = [z_{ij}]$ with $|Z| = B$.*

(ii) *There exists a sign-nonsingular matrix $\hat{Z} = [\hat{z}_{ij}]$ which can be obtained from X by changing x_{uv} to 1 or -1.*

(iii) *The matrix obtained from X by deleting row u and column v is a sign-nonsingular matrix.*

Proof. Statements (ii) and (iii) are clearly equivalent and (ii) implies (i). Now suppose that (i) holds. Let $\tilde{Z}$ be the matrix obtained from Z by replacing z_{uv} with 0. Then $|\tilde{Z}| = A$ and $\tilde{Z}$ is a sign-nonsingular matrix.

By Theorem 7.5.4 there exist diagonal matrices D_1 and D_2 with diagonal elements equal to ± 1 such that $D_1 \tilde{Z} D_2 = X$. The matrix $D_1 Z D_2$ now satisfies (ii).   □

A similar proof allows one to extend Theorem 7.5.5 to matrices with total support.

Theorem 7.5.6.   *Let $A = [a_{ij}]$ be a $(0,1)$-matrix with total support and assume that $A = A_1 \oplus \cdots \oplus A_k \oplus A_{k+1}$ where the matrices $E_1, \ldots, E_k$ are fully indecomposable. Let $X = [x_{ij}] = X_1 \oplus \cdots \oplus X_k \oplus X_{k+1}$ be a sign-nonsingular matrix with $|X| = A$. Let*

$$
B = \begin{bmatrix}
A_1 & O & O & \cdots & O & F_k \\
F_1 & A_2 & O & \cdots & O & O \\
O & F_2 & A_3 & \cdots & O & O \\
\vdots & \vdots & \vdots & \ddots & \vdots & \vdots \\
O & O & O & \cdots & A_{k-1} & O \\
O & O & O & \cdots & F_{k-1} & A_k
\end{bmatrix} \oplus A_{k+1}
$$

where the matrix F_i is a $(0,1)$-matrix with exactly one 1 and this 1 is in position (u_i, v_i) of $F_i, (i = 1, 2, \ldots, k)$. Then the following are equivalent:

   (i) *There exists a sign-nonsingular matrix Z with $|Z| = B$.*
   (ii) *There exists a sign-nonsingular matrix $\hat{Z} = [\hat{z}_{ij}]$ with $|\hat{Z}| = B$ such that $\hat{z}_{ij} = x_{ij}$ for all (i, j) for which $a_{ij} \neq 0$.*
   (iii) *For $i = 1, 2, \ldots, k$ the matrix obtained from X_i by deleting row u_{i-1} and column v_i is a sign-nonsingular matrix (here we interpret u_0 as u_k).*

Let A be a $(0,1)$-matrix of order n with total support. It follows from Theorem 4.3.4 that there exist permutation matrices P and Q of order n such that starting with the identity matrix I_n, we can obtain PAQ by applying the constructions for the matrices B in Theorems 7.5.5 and 7.5.6. Let $B_0 = I_n, B_1, \ldots, B_l = PAQ$ be one such way. Theorems 7.5.5 and 7.5.6 imply that given any conversion $\hat{B}_i$ of B_i, there is a conversion of B_{i+1} if and only if there is a conversion $\hat{B}_{i+1}$ of B_{i+1} which extends the conversion $\hat{B}_i$. Starting with the conversion $\hat{B}_0 = I_n$ of A_0, we attempt to extend a conversion of A_i to a conversion of A_{i+1}. If at some point we are unable to do so, then Theorems 7.5.5 and 7.5.6 imply that no conversion of A exists. This provides us with an algorithm which avoids backtracking to determine whether or not there exists a sign-nonsingular matrix X with $|X| = A$, equivalently to determine whether or not the permanent of the matrices in $\mathcal{M}_n(A)$ can be converted into the determinant and a prescription for doing so.

Little[1975] characterized the coordinate subspaces $\mathcal{M}_n(A)$ of $\mathcal{M}_n$ on which the permanent can be converted into the determinant, and thus the

(0,1)-matrices $|E|$ which result from sign-nonsingular $(0, 1, -1)$-matrices E. We now describe without proof his characterization.

Let $Y = [y_{ij}]$ be a matrix of order n, and suppose that there exist integers p, q, r, and s with $1 \le p, q, r, s \le n$ and $p \ne q$ and $r \ne s$ such that $y_{pj} = 0, (j \ne r, s)$, $y_{ir} = 0, (i \ne p, q)$, $y_{pr} = 1$ and $y_{qs} = 0$. Let Y' be the matrix of order $n - 1$ obtained from Y by replacing y_{qs} by $y_{ps}y_{qr}$ and deleting both row p and column q. The matrix Y' is said to be obtained from Y by *contraction*. It follows from the Laplace expansion for the permanent that

$$\mathrm{per}(Y) = \mathrm{per}(Y').$$

If Y is a (0,1)-matrix then Y' is also a (0,1)-matrix.

Theorem 7.5.7. *Let A be a $(0, 1)$-matrix of order n. There exists a $(0, 1, -1)$-matrix E of order n such that $|E| = A$ and $\mathrm{per}(A) = \det(E)$ if and only if there do not exist permutation matrices P and Q of order n and a $(0, 1)$-matrix B with $B \le A$ such that for some integer $k \le n - 3$,*

$$PBQ = \begin{bmatrix} I_k & O \\ O & B' \end{bmatrix} \tag{7.82}$$

where the all 1's matrix J_3 of order 3 can be obtained by a sequence of contractions starting with the matrix B'.

For example, let

$$A = \begin{bmatrix} 1 & 1 & 0 & 0 & 0 \\ 1 & 0 & 1 & 0 & 0 \\ 0 & 1 & 0 & 1 & 1 \\ 0 & 0 & 1 & 1 & 1 \\ 0 & 0 & 1 & 1 & 1 \end{bmatrix}. \tag{7.83}$$

Then with $B = A$ in Theorem 7.5.7, $P = Q = I_5$ and $k = 0$ [the identity matrix I_k in (7.82) is vacuous], we obtain

$$\begin{bmatrix} 1 & 1 & 0 & 0 \\ 1 & 0 & 1 & 1 \\ 0 & 1 & 1 & 1 \\ 0 & 1 & 1 & 1 \end{bmatrix}$$

after one contraction and J_3 after two contractions. Thus there is no matrix which converts the permanent into the determinant on the coordinate subspace $\mathcal{M}_5(A)$.

It follows from Theorem 7.5.7 that the permanent can be converted into the determinant on a coordinate subspace $\mathcal{M}_n(A)$ if and only if it can be converted on each coordinate subspace $\mathcal{M}_n(B)$ where B is a $(0,1)$-matrix with $B \leq A$ and $\operatorname{per}(B) = 6$.

We now discuss some connections between sign-nonsingular matrices and digraphs. Let $E = [e_{ij}]$ be a $(0, 1, -1)$-matrix of order n. A necessary condition for E to be sign-nonsingular is that $\operatorname{per}(|E|) \neq 0$. Since neither line permutations nor the multiplication of certain lines by -1 affect the sign-nonsingularity of E, we henceforth assume that the elements on the main diagonal of E are all equal to -1. Let $D(E)$ be the digraph of E with vertices $1, 2, \ldots, n$. We use the elements of E in order to assign weights to the arcs of $D(E)$. The resulting *weighted digraph* $D_w(E)$ has an arc $i \rightarrow j$ from vertex i to vertex j if and only if $e_{ij} \neq 0$ and the *weight* of this arc equals $e_{ij},(i, j = 1, 2, \ldots, n)$. The *weight of a directed cycle* is defined to be the product of the weights of the arcs of the cycle. The following characterization of sign-nonsingular matrices is due to Bassett, Maybee and Quirk[1968].

Theorem 7.5.8. *Let* $E = [e_{ij}]$ *be a* $(0, 1, -1)$-*matrix of order* n *with* $e_{ii} = -1, (i = 1, 2, \ldots, n)$. *Then* E *is sign-nonsingular if and only if the weight of each directed cycle of* $D_w(E)$ *equals* -1.

Proof. The matrix E is sign-nonsingular if and only if for each permutation $(\pi(1), \pi(2), \ldots, \pi(n))$ of $\{1, 2, \ldots, n\}$

$$e_{1\pi(1)}e_{2\pi(2)} \cdots e_{n\pi(n)} \neq 0 \text{ implies } \operatorname{sign}(\pi)e_{1\pi(1)}e_{2\pi(2)} \cdots e_{n\pi(n)} = (-1)^n.$$

Let $\gamma : i_1 \rightarrow i_2 \rightarrow \cdots \rightarrow i_k \rightarrow i_1$ be a directed cycle of $D_w(E)$. Then $\pi(i_j) = i_{j+1}$ for $j = 1, 2, \ldots, k$ and $\pi(i) = i$, otherwise defines a permutation of $\{1, 2, \ldots, n\}$ with $\operatorname{sign}(\pi)e_{1\pi(1)}e_{2\pi(2)} \cdots e_{n\pi(n)}$ equal to the $(-1)^{n-1}$ times the weight of γ. The theorem readily follows. $\square$

If in Theorem 7.5.8 we assume that all main diagonal elements of E equal 1, then the criterion for sign-nonsingularity is that the weight of each cycle have the opposite parity of its length.

Another characterization of $(0,1)$-matrices A for which there exists a $(0,1,-1)$-matrix E such that $|E| = A$ and $\operatorname{per}(A) = \det(E)$ is contained in the work of Seymour and Thomassen[1987] and we now describe this characterization.

Let D be a digraph which has no loops. A *splitting of an arc* (x, y) of D is the result of adjoining a new vertex z to D and replacing the arc (x, y) by the two arcs (x, z) and (z, y). A *subdivision* of D is a digraph obtained from D by successively splitting arcs (perhaps none). The digraph D is *even* provided every subdivision of D (including D itself) has a directed cycle of even length. The digraph is even if and only if for every weighting of the arcs of D with weights 1 and -1 there exists a directed cycle whose weight equals 1. A *splitting of a vertex* u of D is the result of adjoining a new vertex v and a new arc (u, v) and replacing each arc of the form (u, w) with an arc of the form (v, w). A *splitting* of the digraph D is a subdivision of a digraph obtained from D by splitting some (perhaps none) of its vertices.

Let k be an integer with $k \geq 3$ and let D_n^* denote the digraph which is obtained from an (undirected) cycle of length k by replacing each of its edges $\{x, y\}$ with two oppositely directed arcs (x, y) and (y, x). The following characterization of even digraphs is due to Seymour and Thomassen[1987].

Theorem 7.5.9.   *The digraph D is even if and only if it contains a splitting of D_k^* for some odd integer $k \geq 3$.*

Now let A be a $(0,1)$-matrix of order n with a nonzero permanent. Permuting lines if necessary, we may assume that all of the main diagonal elements of A equal 1. If there exists a $(0,1,-1)$-matrix E such that $|E| = A$ and $|\det(E)| = \text{per}(A)$ then we may assume that all of the main diagonal elements of E equal -1. It follows from Theorem 7.5.8 that $|\det(E)| = \text{per}(A)$ if and only if the digraph $D'(A)$ obtained from $D(A)$ by removing the loops at each of its vertices is even. We thus have the following characterization of coordinate subspaces on which the permanent can be converted into the determinant.

Theorem 7.5.10.   *Let A be a $(0, 1)$-matrix of order n with all of its main diagonal elements equal to 1. There exists a $(0, 1, -1)$-matrix E such that $|E| = A$ and $\text{per}(A) = \det(E)$ if and only if the digraph $D'(A)$ does not contain a splitting of D_k^* for any odd integer $k \geq 3$.*

The direct equivalence of the characterizations for converting the permanent into the determinant contained in Theorems 7.5.7 and 7.5.10 is discussed in Brualdi and Shader[1991].

The largest dimension of a coordinate subspace of $\mathcal{M}_n$ on which the permanent can be converted into the determinant is determined in the following theorem of Gibson[1971]. This result was independently obtained by Thomassen[1986] in the context of sign-nonsingular matrices. The description of the case of equality is due to Gibson.

Theorem 7.5.11.   *Let A be a $(0, 1)$-matrix of order n such that $\text{per}(A) \neq 0$ and suppose that there exists a $(0, 1, -1)$-matrix E such that $|E| = A$ and $\det(E) = \text{per}(A)$. Then the number of 1's of A is at most equal to $(n^2 + 3n - 2)/2$, with equality if and only if there exist permutation matrices P and Q such that*

$$PAQ = \begin{bmatrix} 1 & 1 & 0 & \cdots & 0 & 0 \\ 1 & 1 & 1 & \cdots & 0 & 0 \\ \vdots & \vdots & \vdots & \ddots & \vdots & \vdots \\ 1 & 1 & 1 & \cdots & 1 & 0 \\ 1 & 1 & 1 & \cdots & 1 & 1 \\ 1 & 1 & 1 & \cdots & 1 & 1 \end{bmatrix}.$$

Thomassen[1986] has observed that an algorithm which decides whether or not a square $(0,1)$-matrix A is sign-nonsingular, that is which satisfies $\text{per}(A) = \det(A)$, can be used to determine whether or not a square $(0, 1, -1)$-matrix is sign-nonsingular. The reason is as follows. Let B be a $(0, 1, -1)$-matrix and asssume without loss of generality that B has all 1's on its main diagonal. Let $D_w(B)$ be the weighted digraph associated with B. We construct an (unweighted) digraph $\hat{D}(B)$ by replacing each weighted arc $i \overset{-1}{\to} j$ with two arcs forming a path $i \to k_{ij} \to j$ of length 2, where k_{ij} is a new vertex. The number of vertices of $\hat{D}(B)$ is n plus the number of arcs of weight -1 of $D_w(B)$ (the number of -1's of B). Let $\hat{B}$ be the $(0,1)$-matrix with all 1's on its main diagonal which satisfies $D(\hat{B}) = \hat{D}(B)$. By Theorem 7.5.8 B is sign-nonsingular if and only if $\hat{D}(B) = D(\hat{B})$ has no directed cycle of even length. By Theorem 7.5.8 again, $D(\hat{B})$ has no directed cycle of even length if and only if the $(0,1)$-matrix $\hat{B}$ is sign-nonsingular.

We conclude this section by discussing the work of Valiant[1979] concerning the algorithmic complexity of computing the permanent of a $(0,1)$-matrix. We begin with a brief and informal discussion of the classes of decision problems known as **P** and **NP**.

A decision problem P has one of the two answers "yes" or "no." Assume that there is some "natural" way to measure the *size* of the problem P. For example, consider the decision problem, known as the *Hamilton cycle problem: Does a graph have a cycle whose length equals the order of the graph?* An instance of this problem is a specification of a graph, and we measure the size by the order n of the graph, that is by the number of its vertices. An algorithm to solve a decision problem P is a *nondeterministic polynomial algorithm* provided there exists a polynomial $p(n)$ such that if the solution to an instance of the problem of size n is "yes," then there is some guess which when input to the algorithm answers "yes" in a number of steps which is bounded by $p(n)$, while if the solution is "no," then for every guess either the algorithm answers "no" or does not halt. The set of decision problems which can be solved by a nondeterministic polynomial algorithm is denoted by **NP**. A nondeterministic algorithm can be converted to a deterministic algorithm by inputting all possible guesses. However, such an algorithm requires a number of steps which are not bounded by a polynomial in the size n.

The problems in **NP** have the property that they can be "certified" in a polynomial number of steps. The Hamilton cycle problem is in **NP**, since specifying a graph of order n and guessing a sequence of vertices γ, one can check in a number of steps bounded by a polynomial in n whether γ is a cycle of length n, that is whether γ is a cycle through all the vertices of the graph. Another problem in **NP** is the problem: *Is an integer N not a prime number?* A nondeterministic algorithm to solve this problem in a polynomial number of steps in the size n, the number of decimal digits in

N, can be described as follows. Specifying a factor F we divide N by F and obtain a remainder R. If $R = 0$, the answer is "yes"; if $R \neq 0$, the answer is "no." If N is not a prime, then a "certification" is a factor of N and such a factor can be checked in a polynomial number of steps.

Let **P** be the set of decision problems which can be solved deterministically in a number of steps bounded by a polynomial in the size of the problem. Thus for a problem P in **P** there exists an algorithm to determine (not just to check the correctness of a guess) whether the answer is "yes" or "no." The set **P** is contained in the set **NP** because given any guess, we set it aside and apply the polynomial deterministic algorithm. This results in a nondeterministic polynomial algorithm for the problem P. The question of whether **P**=**NP** has not yet been resolved, although the general belief is that **P** $\neq$ **NP**.

A decision problem P_1 is said to be *polynomially reducible* to another decision problem P_2 provided there is a function f from the set of inputs of P_1 to the set of inputs of P_2 such that the answer to the input I_1 for P_1 is "yes" if and only if the answer to the corresponding input $f(I_1)$ of P_2 is "yes," and there is a polynomially bounded algorithm to compute $f(I_1)$. If P_1 is polynomially reducible to P_2 and if P_2 is in **P**, then it follows that P_1 is also in **P**. A decision problem in **NP** is *NP-complete* provided every problem in **NP** can be polynomially reduced to it. It follows that a problem in **NP** to which an NP-complete problem can be polynomially reduced is also NP-complete. Also if some NP-complete problem belongs to **P** then **P** $=$ **NP**.

It was a fundamental contribution of Cook[1971] that there exist NP-complete problems. The problem that Cook established as NP-complete is the satisfiability problem SAT of conjunctive normal forms:

SAT

Given a finite set U of variables and a finite set C of clauses each of which is a disjunction of variables in U and their negations, is there a truth assignment for the variables in U for which all clauses in C are true?

There are many fundamental problems which are now known to belong to the class of NP-complete problems. These include the combinatorial problems of the existence of a coloring of the vertices of a graph with k colors, the existence of a complete subgraph of order k of a graph, and the existence of a Hamilton cycle in a graph. For more examples of NP-complete problems and a more formal description of the class **NP** we refer the reader to Karp[1972] and the books by Garey and Johnson[1979], Even[1979], and Wilf[1986].

Associated with a decision problem is the counting question: *How many solutions does the problem have?* The answer to the counting question is to be a number; a complete listing of all the solutions is not required. An NP-problem P is called *sharp P complete*, written *#P-complete*, provided the counting question for every other NP-complete problem is polynomially reducible to the counting question for P. Most "natural" NP-complete problems are also *#P*-complete (although it is not known whether this is always the case). The reason is that the reductions used to establish polynomial reducibility "preserve" the number of solutions. Thus, for instance, counting the number of satisfying assignments in SAT is *#P*-complete. The set of all problems in **NP** which are *#P*-complete is denoted by **#P**.

It was a fundamental contribution of Valiant[1979a,b] that there exist *#P*-complete problems for which the decision question can be decided in a number of steps which is bounded by a polynomial in the size of the problem, that is, there exist problems in **P** which are in **#P**. Thus unless **P = NP**, and perhaps even if **P ≠ NP**, there are NP-problems for which the decision question can be answered in a number of steps bounded by the size of the problem but the counting question cannot. The problem identified by Valiant is the problem of finding a system of distinct representatives:

$$SDR$$

Given a family $(X_1, X_2, \ldots, X_n)$ of subsets of the set $\{1, 2, \ldots, n\}$, does the family have a system of distinct representatives?

There are algorithms which show that SDR is in **P** (see, e.g., Even[1979]). A system of distinct representatives of $(A_1, A_2, \ldots, A_n)$ corresponds to a perfect matching in the associated bipartite graph of order $2n$.

Let A be the incidence matrix of order n of the family $(A_1, A_2, \ldots, A_n)$ of subsets of $\{1, 2, \ldots, n\}$. The problem SDR is equivalent to the problem

Is $\mathrm{per}(A) \neq 0$?

The counting question for SDR is: *How many systems of distinct representatives does $(A_1, A_2, \ldots, A_n)$ have?* In terms of the incidence matrix A, the counting question is

What is the value of $\mathrm{per}(A)$?

Thus determining the permanent of a square $(0,1)$-matrix is a *#P*-complete problem. It follows that the computation of the permanent of square matrices of 0's and 1's is a fundamental counting problem. The existence of an algorithm to compute the permanent of a $(0,1)$-matrix of order n in a number of steps bounded by a polynomial in n would imply the existence of polynomially bounded algorithms to compute the number of satisfying truth assignments in SAT, to compute the number of ways to color the vertices of a graph with k colors, to compute the number of Hamilton cycles in a graph and so on. However, *Valiant's hypothesis is that* *no polynomially bounded algorithm exists for computing the permanent of a $(0,1)$-matrix* (see also von zur Gathen[1987b].

Valiant showed that permanents of matrices $X = [x_{ij}]$ of order n can be calculated by determinants if one is allowed to increase the size of the matrix. Specifically he showed that it is possible to find an m by m matrix Y each of whose entries is either a constant or one of the elements $x_{11}, \ldots, x_{1n}, \ldots, x_{nn}$ of X such that $\operatorname{per}(X) = \det(Y)$. The size m of the matrix Y is, however, exponential in n, roughly a constant times $n^2 2^n$. Thus although the determinant of Y can be computed in a number of steps which is bounded by a polynomial in m, this number is exponential in n. It is shown in von zur Gathen[1987a] that every Y of the above type has order $m \geq \sqrt{2}n - 6\sqrt{n}$, even if one allows the elements of Y equal to negatives of elements of X. A matrix Y of order m where m is bounded by a polynomial in n would imply a polynomially bounded algorithm to compute the permanent of a matrix.

In closing we note that Everett and Stein[1973] have shown that the number of $(0,1)$-matrices of order n with zero permanent is asymptotic to $n \cdot 2^{n^2 - 2n + 1}$. Hence it follows that almost all $(0,1)$-matrices of order n have a nonzero permanent. Komlós[1967] has shown that almost all $(0,1)$-matrices of order n have a nonzero determinant.

Exercises

1. Determine whether the matrix

$$\begin{bmatrix} 1 & 0 & 1 & 1 & 0 & 0 \\ 1 & 1 & 1 & 0 & 0 & 0 \\ 0 & 1 & 1 & 0 & 0 & 1 \\ 0 & 1 & 0 & 1 & 0 & 0 \\ 1 & 0 & 0 & 0 & 1 & 0 \\ 0 & 0 & 0 & 0 & 1 & 1 \end{bmatrix}$$

is sign-nonsingular.
2. Show how to affix minus signs to some of the 1's of the matrix in Theorem 7.5.11 so that the determinant of the new matrix is the permanent of the original matrix.
3. Show that the incidence matrix A of the projective plane of order 2 given in (1.16) of Chapter 1 is a sign-nonsingular matrix (with no -1's). Show that each matrix obtained from A by replacing a 0 with ± 1 is not sign-nonsingular (thus A is a *maximal* sign-nonsingular matrix). (The matrix A, after line permutations, is the circulant matrix $I_7 + C_7 + C_7^3$.) (Brualdi and Shader[1991]).
4. Prove Theorem 7.5.6.
5. Prove that a digraph which contains a splitting of D_k^* for some odd integer $k \geq 3$ is even.

References

L. Bassett, J. Maybee and J. Quirk[1983], Qualitative economics and the scope of the correspondence principle, *Econometrica*, 26, pp. 544–563.

R.A. Brualdi[1988], Counting permutations with restricted positions: permanents of (0,1)-matrices. A tale in four parts. In The 1987 Utah State University Department of Mathematics Conference Report by L. Beasley and E.E. Underwood, *Linear Alg. Applics.*, 104, pp. 173–183.

R.A. Brualdi and B.L. Shader[1991], On converting the permanent into the determinant and sign-nonsingular matrices, *Applied Geometry and Discrete Mathematics* (P. Gritzmann and B. Sturmfels, eds.), Amer. Math. Soc., Providence, R.I.

S.A. Cook[1971], The complexity of theorem proving procedures, *Proc. 3rd ACM Symp. on Theory of Computing*, pp. 151–158.

S. Even[1979], *Graph Algorithms*, Computer Science Press, Potomac, Maryland.

C.J. Everett and P.R. Stein[1973], The asymptotic number of (0,1)-matrices with zero permanent, *Discrete Math.*, 6, pp. 29–34.

P. Gibson[1971], Conversion of the permanent into the determinant, *Proc. Amer. Math. Soc.*, 27, pp. 471–476.

R.M. Karp[1972], Reducibility among combinatorial problems, In *Complexity of Computer Calculations* (R.E. Miller and J.W. Thatcher, eds.), Plenum, New York, pp. 85–104.

V. Klee, R. Ladner and R. Manber[1983], Signsolvability revisited, *Linear Alg. Applics.*, 59, pp. 131–157.

J. Komlós[1967], On the determinant of (0,1)-matrices, *Studia Sci. Math. Hung.*, 2, pp. 7–21.

C.H.C. Little[1975], A characterization of convertible (0,1)-matrices, *J. Combin. Theory, Ser. B*, 18, pp. 187–208.

M. Marcus and H. Minc[1961], On the relation between the determinant and the permanent, *Illinois J. Math.*, 5, pp. 376–381.

J.S. Maybee[1981], Sign solvability, In *Computer assisted analysis and model simplification* (H. Greenberg and J. Maybee, eds.), Academic Press, New York.

G. Pólya[1913], Aufgabe 424, *Arch. Math. Phys.* (3), 20, p. 271.

P. Seymour and C. Thomassen[1987], Characterization of even directed graphs, *J. Combin. Theory, Ser. B*, 42, pp. 36–45.

G. Szegö[1913], Zu Aufgabe 424, *Arch. Math. Phys.* (3), 21, pp. 291–292.

C. Thomassen[1986], Sign-nonsingular matrices and even cycles in directed graphs, *Linear Alg. Applics.*, 75, pp. 27–41.

L.G. Valiant[1979], Completeness classes in algebra, *Proc. 11th Ann. ACM Symp. Theory of Computing*, pp. 249–261.

[1979], The complexity of computing the permanent, *Theoret. Comput. Sci.*, 8, pp. 189–201.

J. von zur Gathen[1987a], Permanent and determinant, *Linear Alg. Applics.*, 96, pp. 87–100.

[1987b], Feasible arithmetic computations: Valiant's hypothesis, *J. Symbolic Comp.*, 4, pp. 137–172.

8

Latin Squares

8.1 Latin Rectangles

Let S be a finite set with n elements. A *latin rectangle* based on S is an r by s matrix

$$A = [a_{ij}], \quad (i = 1, 2, \ldots, r; j = 1, 2, \ldots, s)$$

with the property that each row and each column of A contain distinct elements of S. The number r of rows and the number s of columns of the latin rectangle A satisfy $r \leq n$ and $s \leq n$. Usually the set S is chosen to be the set $\{1, 2, \ldots, n\}$ consisting of the first n positive integers. The matrix

$$\begin{bmatrix} 1 & 2 & 3 & 4 & 5 \\ 3 & 5 & 2 & 1 & 4 \\ 4 & 3 & 5 & 2 & 1 \end{bmatrix}$$

is a 3 by 5 latin rectangle based on $\{1, 2, 3, 4, 5\}$. If $s = n$, as it does in this example, then each row of the latin rectangle A contains a permutation of S and these permutations have the property that no column contains a repeated element. If the first row contains the permutation $(1, 2, \ldots, n)$, then the r by n latin rectangle is called *normalized*. An n by n latin rectangle based on the set S of n elements is a *latin square* of order n. Thus in a latin square each row and each column contains a permutation of S. The elements of S can always be labeled to normalize a latin square.

Let G be a group of order n whose set of elements in some order is $a_1, a_2, \ldots, a_n$, and let the binary operation of G be denoted by $*$. A *Cayley table* of G is the matrix $A = [a_{ij}]$ of order n in which

$$a_{ij} = a_i * a_j, \quad (i, j = 1, 2, \ldots, n).$$

The axioms for a group imply that A is a latin square of order n based on the set $\{a_1, a_2, \ldots, a_n\}$. Cyclic groups of order n give particularly simple examples of latin squares of order n. If P is a permutation matrix of order n, then PAP^T is also a Cayley table of G. If a_1 is the identity element of G, then the first row of A contains the permutation $a_1, a_2, \ldots, a_n$. Let P and Q be permutation matrices of order n. It is common to also call PAQ a Cayley table for G. Not every latin square is a Cayley table of a group because the axiom of associativity for a group imposes a further restriction on a Cayley table of a group. The algebraic systems whose Cayley tables are latin squares are called *quasigroups*. A Cayley table of a commutative group (or commutative quasigroup) is a symmetric latin square.

Let A be a latin square of order n based on $\{1, 2, \ldots, n\}$. Let P_i be the $(0,1)$-matrix of order n whose 1's are in those positions which in A are occupied by i. Then P_i is a permutation matrix of order n, $(i = 1, 2, \ldots, n)$. Moreover,

$$J_n = P_1 + P_2 + \cdots + P_n \tag{8.1}$$

and

$$A = 1P_1 + 2P_2 + \cdots + nP_n \tag{8.2}$$

are decompositions of the all 1's matrix J_n and A, respectively. Conversely, if (8.1) is a decomposition of J_n into n permutation matrices, then (8.2) defines a latin square A.

The matrix J_n is the reduced adjacency matrix of the complete bipartite graph $K_{n,n}$. The latin square A assigns a color from the color set $\{1, 2, \ldots, n\}$ to each of the edges of $K_{n,n}$ in such a way that adjacent edges are assigned different colors. Thus the set M_i of edges of color i is a perfect matching of $K_{n,n}$. The permutation matrix P_i is the reduced adjacency matrix of the spanning bipartite subgraph of $K_{n,n}$ whose set of edges is M_i. Conversely, an assignment of a color from the color set $\{1, 2, \ldots, n\}$ to each of the edges of $K_{n,n}$ produces a latin square of order n provided adjacent edges are assigned different colors.

Let $A = [a_{ij}]$ be a latin square of order n based on a set S. A *partial transversal* of size t of A is a set of t positions such that no two of the positions are on the same line and these positions are occupied in A by distinct elements. A *transversal* of A is a partial transversal of size n. Thus a transversal is a set of positions

$$\{(1, \sigma(1)), (2, \sigma(2)), \ldots, (n, \sigma(n))\}$$

where σ is a permutation of $\{1, 2, \ldots, n\}$ and

$$\{a_{1\sigma(1)}, a_{2\sigma(2)}, \ldots, a_{n\sigma(n)}\} = S.$$

Transversals are also known as *complete mappings* of quasigroups.

In terms of the decompositions (8.1) and (8.2) a transversal of A can be viewed as a permutation matrix Q of order n which for each $i = 1, 2, \ldots, n$ has exactly one 1 in common with P_i. Thus in a coloring of the edges of the complete bipartite graph $K_{n,n}$ with n colors, a transversal corresponds to a perfect matching with all edges colored differently.

Let A be a Cayley table for a group of odd order n with elements $a_1, a_2, \ldots, a_n$. Then each element of G has a unique square root and thus $\{a_1^2, a_2^2, \ldots, a_n^2\} = \{a_1, a_2, \ldots, a_n\}$. Hence the set of positions of the main diagonal is a transversal of A. A group of even order need not have a transversal as the group of order 2 shows. Conditions for the existence and the nonexistence of transversals in Cayley tables of groups of even order are discussed in Dénes and Keedwell[1974].

Let $A = [a_{ij}]$ be a latin square of order n based on the set $\{1, 2, \ldots, n\}$. A matrix obtained from A by row permutations and by column permutations is also a latin square as is the transposed matrix A^T. We may also apply a permutation to the elements of the set $\{1, 2, \ldots, n\}$ and obtain a latin square. There is one further basic transformation of latin squares which is less obvious and which we now discuss. To the latin square A of order n there corresponds a three-dimensional array

$$C = [c_{ijk}], \quad (i, j, k = 1, 2, \ldots, n) \tag{8.3}$$

of 0's and 1's in which $c_{ijk} = 1$ if and only if $a_{ij} = k$. A *line* of the array C is a set of positions (i, j, k) obtained by fixing two of i, j, and k and allowing the other index to vary from 1 to n. The array C obtained from the latin square of order n has the property that each line contains exactly one 1. Such an array is a 3-*dimensional line permutation matrix* of order n. Conversely, from a 3-dimensional line permutation matrix $C = [c_{ijk}]$ of order n one obtains a latin square $A = [a_{ij}]$ of order n by defining $a_{ij} = k$ if $c_{ijk} = 1$.

Let C be the 3-dimensional line permutation matrix of order n corresponding to the latin square A. Let (p, q, r) be a permutation of the three indices i, j and k of C and let $C_{(p,q,r)}$ be the 3-dimensional matrix obtained from C by taking the indices in the order p, q, r. Thus, for instance, $C_{(3,1,2)} = [x_{ijk}]$ where $x_{ijk} = c_{kij}$ for i, j and k between 1 and n. The matrix $C_{(p,q,r)}$ is a 3-dimensional line permutation matrix of order n and thus there corresponds a latin square $A_{(p,q,r)}$ of order n. We have $A_{(1,2,3)} = A$ and $A_{(2,1,3)} = A^T$. The latin square $A_{(3,2,1)}$ is the latin square obtained from A by interchanging row indices with elements. The (i, j)-entry of $A_{(3,1,2)}$ is k provided $a_{kj} = i$. Similarly, $A_{(1,3,2)}$ is the latin square obtained from A by interchanging column indices with elements. For example, let

$$A = \begin{bmatrix} 1 & 2 & 3 & 4 \\ 3 & 1 & 4 & 2 \\ 4 & 3 & 2 & 1 \\ 2 & 4 & 1 & 3 \end{bmatrix}. \tag{8.4}$$

Then

$$A_{(3,2,1)} = \begin{bmatrix} 1 & 2 & 4 & 3 \\ 4 & 1 & 3 & 2 \\ 2 & 3 & 1 & 4 \\ 3 & 4 & 2 & 1 \end{bmatrix}$$

and

$$A_{(1,3,2)} = \begin{bmatrix} 1 & 2 & 3 & 4 \\ 2 & 4 & 1 & 3 \\ 4 & 3 & 2 & 1 \\ 3 & 1 & 4 & 2 \end{bmatrix}.$$

Two latin squares A and B of order n based on $\{1, 2, \ldots, n\}$ are *equivalent* provided there is a permutation (p, q, r) of $\{1, 2, 3\}$ such that B can be obtained from $A_{(p,q,r)}$ by permutation of its rows, columns and elements. Any two latin squares of order 3 are equivalent. A latin square of order 4 is equivalent to one of the two latin squares

$$\begin{bmatrix} 1 & 2 & 3 & 4 \\ 2 & 1 & 4 & 3 \\ 3 & 4 & 1 & 2 \\ 4 & 3 & 2 & 1 \end{bmatrix} \quad \text{and} \quad \begin{bmatrix} 1 & 2 & 3 & 4 \\ 2 & 3 & 4 & 1 \\ 3 & 4 & 1 & 2 \\ 4 & 1 & 2 & 3 \end{bmatrix}. \tag{8.5}$$

The first of these latin squares C has the property that $C_{(p,q,r)} = C$ for each permutation (p, q, r) of $\{1, 2, 3\}$. More generally, let G be an elementary abelian 2-group with elements $\{a_1, a_2, \ldots, a_n\}$ where a_1 is the identity element. The Cayley table C of G has the property that $C_{(p,q,r)} = C$ for each permutation (p, q, r) of $\{1, 2, 3\}$.

Higher dimensional latin configurations are discussed by Jurkat and Ryser[1968].

Exercises

1. Find an example of a latin square A such that for no permutation matrices P and Q is PAQ a Cayley table of a group.
2. Find an example of a group of even order $n > 2$ whose Cayley table does not have a transversal.
3. Prove that all latin squares of order 3 are equivalent. Also prove that a latin square of order 4 is equivalent to one of the two latin squares given in (8.5).
4. Construct a Cayley table for the group of permutations of $\{1, 2, 3\}$ and obtain a latin square of order 6.
5. Let A be the latin square given in (8.4). Construct all the latin squares $A_{(p,q,r)}$.

References

J. Dénes and A.D. Keedwell[1974], *Latin Squares and Their Applications*, Academic Press, New York.

M. Hall, Jr.[1986], *Combinatorial Theory*, 2d edition, Wiley, New York.

W.B. Jurkat and H.J. Ryser[1968], Extremal configurations and decomposition theorems, *J. Algebra*, 8, pp. 194–222.

H.J. Ryser[1963], *Combinatorial Mathematics*, Carus Mathematical Monograph No. 14, Math. Assoc. of America, Washington, D.C.

8.2 Partial Transversals

Let $A = [a_{ij}]$ be a latin square of order n. If A has a transversal then each latin square equivalent to A also has a transversal. The following theorem of Mann[1942] shows in particular that a Cayley table of a cyclic group of even order (a circulant matrix of even order) does not have a transversal.

Theorem 8.2.1. *A Cayley table of an abelian group of even order with a unique element of order two does not have a transversal.*

Proof. Let n be an even integer and let G be an abelian group with elements $a_1, a_2, \ldots, a_n$. Assume that G has a unique element x of order 2. We have $x = a_1 * a_2 * \cdots * a_n$. Let $A = [a_{ij}]$ be the Cayley table of G in which $a_{ij} = a_i * a_j, (i, j = 1, 2, \ldots, n)$. Suppose that A has a transversal. Then there exists a permutation $(j_1, j_2, \ldots, j_n)$ of $\{1, 2, \ldots, n\}$ such that

$$\{a_1, a_2, \ldots, a_n\} = \{a_1 * a_{j_1}, a_2 * a_{j_2}, \ldots, a_n * a_{j_n}\}.$$

Since G is abelian, we obtain

$$x = (a_1 * a_{j_1}) * (a_2 * a_{j_2}) * \cdots * (a_n * a_{j_n})$$
$$= (a_1 * a_2 * \cdots a_n) * (a_{j_1} * a_{j_2} * \cdots * a_{j_n}) = x * x,$$

and this contradicts the fact that x has order 2. Therefore A does not have a transversal.   □

As noted in the previous section a Cayley table of a group of odd order always has a transversal. It has been proved by Paige[1947] that a Cayley table of an abelian group G of even order n has a transversal if G does not have a unique element of order two. Thus *a Cayley table of an abelian group of even order has a transversal if and only if it does not have a unique element of order two.* We now state without proof a more general theorem of Hall[1952].

Theorem 8.2.2. *Let G be an abelian group with elements $a_1, a_2, \ldots, a_n$ and let A be a Cayley table of G. Let $k_1, k_2, \ldots, k_n$ be a sequence of non-negative integers with $k_1 + k_2 + \cdots + k_n = n$. Then there exists in A n*

*positions no two on the same line such that in these positions a_i occurs
exactly k_i times if and only if $a_1^{k_1} * a_2^{k_2} * \cdots * a_n^{k_n}$ is the identity element
of G.*

If $k_1 = k_2 = \cdots = k_n = 1$, then Theorem 8.2.2 asserts that A has a
transversal if and only if

$$a_1 * a_2 * \cdots * a_n = 1, \qquad (8.6)$$

the identity element of G. If G has odd order, (8.6) is satisfied. If G has
even order, (8.6) is satisfied if and only if G has more than one element of
order two.

It has been conjectured by Ryser[1967] that every latin square of odd
order n has a transversal and by Brualdi (see Dénes and Keedwell[1974])
that every latin square of even order n has a partial transversal of size
$n - 1$. These conjectures remain unsettled (see Erdös et al.[1988]). The
remainder of this section concerns the progress that has been made toward
the resolution of these conjectures. First we remark that Theorem 8.2.2
implies at once that the Cayley table of an abelian group of order n has
a partial transversal of size $n - 1$ containing any specified subset of $n - 1$
elements of G.

Let A be a matrix of order n whose elements come from a set S. A
weak transversal of A is a set W of n positions of A no two from the same
line with the property that each element of S occurs at most twice in the
positions of W.

Theorem 8.2.3.   *A matrix of order n with no repeated element in a row
or in a column has a weak transversal. In particular, a latin square of order
n has a weak transversal.*

Proof. We prove the theorem by induction on n. If $n = 1$ the conclusion is trivial. Suppose that $n > 1$ and let A be a matrix of order n
such that each row and each column contains distinct elements. Let A'
be the matrix obtained from A by deleting the first row and the first
column. By the inductive hypothesis A' has a weak transversal. Without loss of generality we assume that the $n - 1$ diagonal positions of A'
form a weak transversal, and that the elements in these positions are
$1, 1, 2, 2, \ldots, r, r, r + 1, r + 2, \ldots, r + s$ where $2r + s = n - 1$. If the element
in position (1,1) does not equal any of $1, 2, \ldots, r$, then the n diagonal positions of A form a weak transversal. Otherwise we assume without loss
of generality that 1 is in position (1,1). Because row 1 of A contains distinct elements, row 1 contains $r + 1$ distinct elements $x_1, x_2, \ldots, x_r$ each of
which is different from $1, 2, \ldots, r$. Because column 1 has distinct elements,
there is an element x_i in row 1 such that the element y in column 1 which
is symmetrically opposite x_i is different from $1, 2, \ldots, r$. Let this x_i and

y occupy positions $(1, k)$ and $(k, 1)$, respectively. The set of n positions

$$\{(i, i) : i \neq 1, k\} \cup \{(1, k), (k, 1)\}$$

is a weak transversal of A. □

It is an immediate consequence of Theorem 8.2.3 that a latin square of order n has a partial transversal of size $\lceil n/2 \rceil$. Koksma[1969] proved that for $n \geq 3$ there is a partial transversal of size at least $\lceil (2n + 1)/3 \rceil$. Koksma's method was refined by de Vries and Wieringa[1978] who obtained the lower bound of $\lceil (4n - 3)/5 \rceil$ for the size of a partial transversal of a latin square of order n if $n \geq 12$. Woolbright[1978] and independently Brouwer, de Vries and Wieringa[1978] showed that a latin square of order n has a partial transversal of size at least $\lceil n - \sqrt{n} \rceil$. We rely on the proof of Brouwer et al.

Theorem 8.2.4. *A latin square of order n has a partial transversal of size t for some t satisfying the inequality $(n - t)(n - t + 1) \leq n$.*

Proof. Let $A = [a_{ij}]$ be a latin square of order n based on $\{1, 2, \ldots, n\}$ and let t be the largest size of a partial transversal of A. Without loss of generality we assume that $T = \{(1, 1), (2, 2), \ldots, (t, t)\}$ is a partial transversal of A and that $a_{kk} = k, (k = 1, 2, \ldots, t)$. Let $r = n - t$ and let $R = \{t + 1, t + 2, \ldots, n\}$. We inductively define subsets $W_0, W_1, \ldots, W_r$ of $\{1, \ldots, n\}$ as follows:

$$W_0 = \emptyset,$$

$$W_i = \{j : a_{j,t+i} \in W_{i-1} \cup R\}, \quad (i = 1, 2, \ldots, r).$$

Let

$$V_i = \{(j, t + i) : j \in W_i\}, \quad (i = 1, 2, \ldots, r)$$

and let

$$V = V_1 \cup V_2 \cup \cdots \cup V_r.$$

We define a digraph D with vertex set V by putting an arc from $(j, t + i)$ to $(k, t + l)$ if and only if $i < l$ and $j = a_{k,t+l}$.

Suppose that in D there is a path from a vertex $(j, t + i)$ to a vertex $(k, t + l)$ such that $a_{j,t+i} \in R$ and $k \in R$. Let

$$(j, t + i) = (j_0, t + i_0) \to (j_1, t + i_1) \to \cdots \to (j_p, t + i_p) = (k, t + l)$$

be such a path γ of smallest length. This path is pictured schematically in Figure 8.1, in which u denotes an element of R.

Let

$$T' = (T - \{(j_0, j_0), \ldots, (j_{p-1}, j_{p-1})\}) \cup \{(j_0, t+i_0), (j_1, t+i_1), \ldots, (j_p, t+i_p)\}.$$

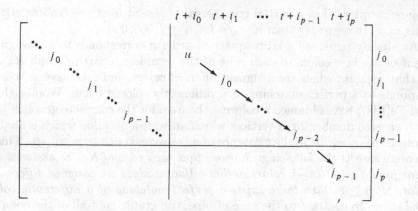

Figure 8.1

No two of the positions in T' belong to the same column. Suppose that two positions in T' belong to the same row. Then there are integers r and s with $0 \leq r < s \leq p$ and with $j_r = j_s$. If $s = p$ then $j_r = j_p = k$ and

$$(j_0, t + i_0) \to (j_1, t + i_1) \to \cdots \to (j_r, t + i_r)$$

is a path which contradicts the choice of γ. If $s < p$ then $j_r = j_s = a_{s+1,t+j_{s+1}}$ and

$$(j_0, t + i_0) \to \cdots \to (j_r, t + i_r) \to (j_{s+1}, t + j_{s+1}) \to \cdots \to (j_p, t + i_p)$$

is a path which contradicts the choice of γ. We therefore conclude that no two positions in T' belong to the same row. The elements in the $t + 1$ positions of T' are the numbers $1, 2, \ldots, t$ and u. Because u is an element of R, T' is a partial transversal of size $t + 1$, and this contradicts our choice of T. We now conclude that there is no path in D satisfying the conditions of the path γ. The definition of the sets W_i now implies that

$$W_0 \cup W_1 \cup \cdots \cup W_r \subseteq \{1, 2, \ldots, t\}$$

and

$$|W_i| = |W_{i-1}| + r, \quad (i = 1, 2, \ldots, r).$$

Hence $|W_r| = r^2$ and therefore

$$r^2 \leq t = n - r.$$

Because $r = n - t$ the theorem follows.   □

Theorem 8.2.4 has been extended by Csima[1979] to more general combinatorial configurations. In addition Shor[1982] has shown that every latin square of order n has a partial transversal of size at least $n - 5.53(\ell n\ n)^2$. This number is greater than $n - \sqrt{n}$ for $n \geq 2,000,000$.

As already remarked a latin square of order n corresponds to an assignment of one of n colors to each edge of the complete bipartite graph $K_{n,n}$ so that adjacent edges are assigned different colors, and a transversal corresponds to a perfect matching of n differently colored edges. Woolbright and Fu[1987] have obtained a coloring theorem for the complete graph K_{2n} with an even number $2n$ of vertices which answers a question which is analogous to the question of the existence of a transversal in a latin square. This theorem asserts the following: *Suppose that each edge of K_{2n} is assigned a color from a set of $2n-1$ colors so that adjacent edges are assigned different colors. If $n \geq 8$, then there exists a perfect matching of n differently colored edges.* In contrast to the case of bipartite graphs, not all of the colors appear as colors of the edges of the perfect matching.

Finally we remark that Ryser[1967] conjectured that the number of transversals of a latin square of order n is congruent to n modulo 2. If n is odd, then this conjecture is stronger than the conjecture that a latin square of odd order has a transversal. If n is even then the conjecture has been proved by Balasubramanian[1990], but notice that there is no implication concerning the existence of a transversal in a latin square of even order n. According to Parker [private communication], there are many latin squares of order 7, the number of whose transversals is an even positive integer. One such is

$$\begin{bmatrix} 1 & 2 & 3 & 4 & 5 & 6 & 7 \\ 2 & 3 & 6 & 7 & 4 & 5 & 1 \\ 3 & 4 & 7 & 5 & 1 & 2 & 6 \\ 4 & 6 & 1 & 2 & 3 & 7 & 5 \\ 5 & 7 & 2 & 1 & 6 & 4 & 3 \\ 6 & 5 & 4 & 3 & 7 & 1 & 2 \\ 7 & 1 & 5 & 6 & 2 & 3 & 4 \end{bmatrix}.$$

Exercises

1. Let A be a latin square of order n. Let t be the maximal size of a partial transversal of A. Prove that t is the maximal size of a partial transversal of each latin square equivalent to A.
2. Prove that the condition given in Theorem 8.2.2 is a necessary condition for there to exist n positions in a Cayley table of an abelian group with the stated properties.
3. Prove that Theorem 8.2.2 implies that a Cayley table of an abelian group G has a partial transversal of size $n - 1$ containing any specified subset of $n - 1$ elements of G.
4. Prove that the number of transversals of a latin square of even order is an even nonnegative integer (Balasubramanian[1990]).

References

K. Balasubramanian[1990], On transversals in latin squares, *Linear Alg. Applics.*, 131, pp. 125–129.

A.E. Brouwer, A.J. de Vries and R.M.A. Wieringa[1978], A lower bound for the length of partial transversals in a latin square, *Nieuw Archief voor Wiskunde* (3), XXVI, pp. 330–332.

J. Csima[1979], On the plane term rank of three dimensional matrices, *Discrete Math.*, 28, pp. 147–152.

J. Dénes and A.D. Keedwell[1974], *Latin Squares and Their Applications*, Academic Press, New York.

P. Erdös, D.R. Hickerson, D.A. Norton and S.K. Stein[1988], Has every latin square of order n a partial transversal of size $n-1$?, *Amer. Math. Monthly*, 95, pp. 428–430.

M. Hall, Jr.[1952], A combinatorial problem on abelian groups, *Proc. Amer. Math. Soc.*, 3, pp. 584–587.

K.K. Koksma[1969], A lower bound for the order of a partial transversal, *J. Combin. Theory*, 7, pp. 94–95.

H.B. Mann[1942], The construction of orthogonal latin squares, *Ann. Math. Statist.*, 13, pp. 418–423.

L.J. Paige[1947], A note on finite abelian groups, *Bull. Amer. Math. Soc.*, 53, pp. 590–593.

H.J. Ryser[1967], Neuere Probleme in der Kombinatorik (prepared by D.W. Miller), *Vorträge über Kombinatorik*, Oberwolfach, pp. 69–91.

P.W. Shor[1982], A lower bound for the length of a partial transversal in a latin square, *J. Combin. Theory, Ser. A*, 33, pp. 1–8.

A.J. de Vries and R.M.A. Wieringa, Een ondergrens voor de lengte van een partiele transversaal in een Latijns vierkant, preprint.

D.E. Woolbright[1978], An $n \times n$ latin square has a transversal with at least $n - \sqrt{n}$ distinct symbols, *J. Combin. Theory, Ser. A*, 24, pp. 235–237.

D.E. Woolbright and H.-L. Fu[1987], The rainbow theorem of 1-factorization, preprint.

8.3   Partial Latin Squares

Let S be the set $\{1, 2, \ldots, n\}$. We now consider matrices A of order n whose elements come from $S \cup \{\diamond\}$ where each occurrence of $\diamond$ is to be regarded as an unspecified element of A. Such a matrix A is called a *partial latin square* of order n provided that each element of S occurs at most once in each row and in each column. Notice that in a partial latin square there is no restriction on the number of times the symbol $\diamond$ may occur in a row or column. If each occurrence of the symbol $\diamond$ can be replaced by an element of S in such a way that the resulting matrix B is a latin square, then we say that the partial latin square A can be *completed to a latin square* and we call B a *completion* of A. A completion of A is a latin square which agrees with A in the elements specified.

Not every partial latin square of order n can be completed. Two simple examples with n specified elements which have no completion are

$$
\begin{bmatrix}
1 & \cdots & \diamond & \diamond \\
\vdots & \ddots & \vdots & \vdots \\
\diamond & \cdots & 1 & \diamond \\
\diamond & \cdots & \diamond & 2
\end{bmatrix}
\tag{8.7}
$$

and

$$
\begin{bmatrix}
1 & 2 & \cdots & n-1 & \diamond \\
\diamond & \diamond & \cdots & \diamond & n \\
\vdots & \vdots & \vdots & \vdots & \vdots \\
\diamond & \diamond & \cdots & \diamond & \diamond
\end{bmatrix}.
\tag{8.8}
$$

There does not exist a latin square of order n which agrees with (8.7) or with (8.8). Notice that if we extend the construction for latin squares which interchanges row indices and elements so that it applies to partial latin squares, then (8.8) can be obtained from (8.7) in this way.

We now regard an r by s latin rectangle based on the set $S = \{1, 2, \ldots, n\}$ as a partial latin square of order n in which only the elements in the r by s rectangle in the upper left corner have been specified. The following theorem is from Hall[1945].

Theorem 8.3.1.   *Let A be an r by n latin rectangle based on $\{1, 2, \ldots, n\}$. Then A can be completed to a latin square of order n.*

Proof. Let $C = [c_{ij}]$ be the (0,1)-matrix of order n in which $c_{ij} = 1$ if and only if the element i does not occur in column j of A, $(i, j = 1, 2, \ldots, n)$. Because A is an r by n latin rectangle each element of $\{1, 2, \ldots, n\}$ occurs once in each row of A and once in each of r different columns of A. It follows that each line sum of C equals $n - r$. We now apply Theorem 4.4.3 to C and conclude that there is a decomposition

$$
C = P_1 + P_2 + \cdots + P_{n-r}
$$

in which each P_i is a permutation matrix of order n. Let the 1's of P_1 occur in positions $(1, k_1), (2, k_2), \ldots, (n, k_n)$. Then $k_1, k_2, \ldots, k_n$ is a permutation of $\{1, 2, \ldots, n\}$ and we may adjoin this permutation to the r by n latin rectangle A to obtain an $r + 1$ by n latin rectangle. Repeating with $P_2, \ldots, P_{n-r}$ we obtain a latin square of order n.   □

Now suppose that A is a partial latin square of order n based on $\{1, 2, \ldots, n\}$ in which each specified element lies in one of rows $1, 2, \ldots, r+1$. Suppose further that all the elements in rows $1, 2, \ldots, r$ have been specified and that exactly d elements in row $r + 1$ are specified where $1 \le d \le n - 1$. Then Brualdi and Csima[1986] proved that for fixed n, r and d each partial latin square with the above properties can be completed to a latin square of order n if and only if either $d = 1, r = n - 1$ or $d \leq n - 2r$.

If $s < n$, an r by s latin rectangle A need not have a completion to a latin square of order n. For example, the 2 by 2 latin rectangle

$$\begin{bmatrix} 1 & 2 \\ 2 & 1 \end{bmatrix}$$

based on $\{1, 2, 3\}$ has no completion to a latin square of order 3. Ryser[1951] obtained necessary and sufficient conditions in order that an r by s latin rectangle have a completion to a latin square.

Theorem 8.3.2.   *Let A be an r by s latin rectangle based on $\{1, 2, \ldots, n\}$, and let $N(j)$ denote the number of times that the element j occurs in A, ($j = 1, 2, \ldots, n$). Then A can be completed to a latin square of order n if and only if*

$$N(j) \geq r + s - n, \quad (j = 1, 2, \ldots, n). \tag{8.9}$$

Proof. First suppose that there is a latin square B of order n which is a completion of A. Let j be an element of $\{1, 2, \ldots, n\}$. Then j occurs $n - r$ times in the last $n - r$ rows of B and $n - s$ times in the last $n - s$ columns, and hence at least $n - (n - r) - (n - s) = r + s - n$ times in the upper left r by s rectangle of B. Because this r by s rectangle is A (8.9) holds.

Now suppose that (8.9) is satisfied. It suffices to show that A may be extended to an r by n latin rectangle B based on $\{1, 2, \ldots, n\}$, for then we may apply Theorem 8.3.1 to B and obtain a completion of A. Let $D = [d_{ij}]$ be the r by n (0,1)-matrix in which $d_{ij} = 1$ if and only if the element j does not occur in row i of A, ($i = 1, 2, \ldots, r; j = 1, 2, \ldots, n$). Each row sum of D equals $n - s$. The sum of the entries in column j of D equals $r - N(j)$ which by (8.9) is at most equal to $n - s$. We now apply Theorem 4.4.3 to D and conclude that there is a decomposition

$$D = Q_1 + Q_2 + \cdots + Q_{n-s}$$

where each Q_i is an r by n subpermutation matrix. Since the number of 1's in D equals $r(n - s)$ each Q_i is a subpermutation matrix of rank r. Let the 1's of Q_i occur in positions $(1, j_{1i}), (2, j_{2i}), \ldots, (r, j_{ri})$. Then $j_{1i}, j_{2i}, \ldots, j_{ri}$ is an r-permutation of the set $\{1, 2, \ldots, n\}$. We now adjoin $j_{1i}, j_{2i}, \ldots, j_{ri}$ as a column to the r by s latin rectangle A for each $i = 1, 2, \ldots, n - s$ and obtain an r by n latin rectangle B. $\square$

Corollary 8.3.3.   *Let A be a partial latin square of order n whose specified elements all belong to the r by s rectangle L in its upper left corner. Assume that $r + s \leq n$. Then A can be completed to a latin square of order n.*

Proof. Suppose that the element in position (i,j) of A is unspecified where $1 \leq i \leq r$ and $1 \leq j \leq s$. The number of distinct integers that occur in row i or row j of A is at most $r + s - 2 \leq n - 2$. Hence there is a partial latin square of order n which can be obtained from A by specifying the element in position (i,j). Proceeding like this we conclude that there is a partial latin square B of order n which can be obtained from A by specifying all the unspecified elements in the r by s rectangle L. Because $r + s \leq n$ it now follows from Theorem 8.3.2 that B and hence A can be completed to a latin square of order n.   □

When it happens that a partial latin square of order n does not have a completion to a latin square of order n, there are two natural ways to proceed. One is to embed the partial latin square in a latin square of larger order. The other is to partition the partial latin square into parts each of which has a completion to a latin square of order n. The next two theorems address these two possibilities. The first is due to Evans[1960] and the second is due to Opencomb[1984].

Theorem 8.3.4.   *Let A be a partial latin square of order n. Then there is a latin square of order $2n$ which contains A in its upper left corner.*

Proof. We enlarge A to a partial latin square of order $2n$ in which all of the specified elements belong to the upper left n by n rectangle and then apply Corollary 8.3.3.   □

Theorem 8.3.5.   *Let $A = [a_{ij}]$ be a partial latin square of order n. Then there exist four latin squares B_1, B_2, B_3 and B_4 of order n such that for each specified element a_{ij} of A at least one of B_1, B_2, B_3 and B_4 has the property that the element in position (i,j) equals a_{ij}.*

Proof. First suppose that n is even, and let

$$A = \left[\begin{array}{cc} A_1 & A_2 \\ A_3 & A_4 \end{array} \right]$$

be a partition of A into four submatrices of order $n/2$. Let A_i' be the partial latin square of order n which is obtained from A by replacing all specified elements in A_1, A_2, A_3 and A_4 with $\diamond$ except for those specified in $A_i, (i = 1, 2, 3, 4)$. By Corollary 8.3.3 each A_i' can be completed to a latin square of order n.

Now assume that $n = 2m + 1$ is odd. If every element of A is specified then we let $B_1 = A$ and let B_2, B_3 and B_4 be arbitrary latin squares of order n. Now suppose that some element of A is unspecified. Without loss of generality we assume that the element in position $(m + 1, m + 1)$ is not specified. Except for this position, A can be partitioned into four submatrices A_1, A_2, A_3 and A_4 of sizes m by $m + 1$, $m + 1$ by m, m by

$m + 1$ and $m + 1$ by m, respectively. By Corollary 8.3.3 each A_i can be completed to a latin square of order n. □

It has been conjectured by Daykin and Häggvist[1981] that a partial latin square of order n can always be partitioned into two parts each of which can be completed to a latin square of order n.

The partial latin squares (8.7) and (8.8) of order n have n specified elements and cannot be completed to a latin square of order n. Evans[1960] conjectured that a partial latin square of order n with at most $n - 1$ specified elements can always be completed to a latin square of order n. Häggvist[1976] proved this conjecture provided $n \geq 1111$. The conjecture was proved in its entirety by Smetaniuk[1981] and Anderson and Hilton[1983]. The remainder of this section is devoted primarily to Smetaniuk's proof of the conjecture of Evans.

First we make the following definition. Let X be a matrix of order m. The set $\{(1, m), (2, m - 1), \ldots, (m, 1)\}$ of positions of X is called the *back diagonal* of X. Now let A be a latin square of order n based on the set $\{1, 2, \ldots, n\}$. We define a partial latin square $P(A)$ of order $n + 1$ based on $\{1, 2, \ldots, n + 1\}$ as follows. The elements on the back diagonal of $P(A)$ are all equal to $n + 1$. The triangular part of $P(A)$ *above* its back diagonal is the same as the triangular part of A on and above its back diagonal. All elements of $P(A)$ below its back diagonal are equal to ◇ and thus are unspecified. For example, if

$$A = \begin{bmatrix} 1 & 2 & 3 \\ 3 & 1 & 2 \\ 2 & 3 & 1 \end{bmatrix},$$

then

$$P(A) = \begin{bmatrix} 1 & 2 & 3 & 4 \\ 3 & 1 & 4 & \diamond \\ 2 & 4 & \diamond & \diamond \\ 4 & \diamond & \diamond & \diamond \end{bmatrix}.$$

Theorem 8.3.6. *Let $A = [a_{ij}]$ be a latin square of order n based on $\{1, 2, \ldots, n\}$. Then the partial latin square $P(A)$ can be completed to a latin square of order $n + 1$.*

Proof. We define inductively a sequence $L_1, L_2, \ldots, L_{n+1}$ of partial latin squares of order $n + 1$. First we let L_1 equal $P(A)$ and we observe that all the elements in column 1 are specified. For $k = 1, 2, \ldots, n$, L_{k+1} is to be obtained from L_k by specifying those elements in column $k + 1$ below its back diagonal. Thus L_{k+1} will have all its elements in columns $1, 2, \ldots, k + 1$ specified, and L_{n+1} will be a completion of $P(A)$ to a latin square of order $n + 1$.

Suppose that the partial latin squares $L_1, L_2, \ldots, L_k$ have been defined where $k \leq n-1$. Let i be one of the $k-1$ integers $n+2-k, n+3-k, \ldots, n$. We define the *deficiency* $d(i, L_k)$ of row i of L_k to be the set of those integers j such that j occurs in the first k positions of row i of A but not in row i of L_k. It follows from the definition of $P(A)$ that

$$d(n+2-k, L_k) = \{a_{n+2-k,k}\}, \quad (k = 2, 3, \ldots, n). \tag{8.10}$$

We now define L_2 by specifying the last element in column 2 of L_1 to be a_{n2}. We note that L_2 is a partial latin square. We assume inductively that

$$d(i, L_k) \text{ contains a unique element for each } i = n+2-k, \ldots, n \atop \text{and these } k-1 \text{ elements are distinct.} \tag{8.11}$$

If $d(i, L_k) = \{y\}$ we now write $d(i, L_k) = y$. We also assume inductively that

$$\text{the specified elements in row } n+1 \text{ of } L_k \text{ are} \atop \{n+1\} \cup \{d(j, L_k) : j = n+2-k, \ldots, n\}. \tag{8.12}$$

Properties (8.11) and (8.12) hold if $k = 2$. We now show how to define a partial latin square L_{k+1} by specifying the elements of column $k+1$ of L_k below its back diagonal in such a way that properties (8.11) and (8.12) hold with k replaced by $k+1$.

We begin with the element $a_{n+1-k,k+1}$ and determine the longest sequence of the form

$$a_{i_0,k+1}, d(i_1, L_k), a_{i_1,k+1}, d(i_2, L_k), a_{i_2,k+1}, \ldots, d(i_p, L_k), a_{i_p,k+1}, \tag{8.13}$$

where $i_0 = n+1-k, i_1, \ldots, i_p$ are distinct integers from the set $\{n+1-k, n+2-k, \ldots, n\}$ satisfying

$$\begin{aligned} d(i_1, L_k) &= a_{i_0,k+1}, \\ d(i_2, L_k) &= a_{i_1,k+1}, \\ &\cdots \\ d(i_p, L_k) &= a_{i_{p-1},k+1}. \end{aligned} \tag{8.14}$$

Because the sequence (8.13) is the longest sequence satisfying (8.14) it follows from the inductive property (8.11) that $a_{i_p,k+1}$ does not equal $d(j, L_k)$ for any integer j with $n+2-k \leq j \leq n$. We now specify the element below the back diagonal in row i of column $k+1$ of L_{k+1} to be

$$\begin{aligned} d(i, L_k), &\quad \text{if } i \text{ is one of } i_1, i_2, \ldots, i_p, \\ a_{i_p,k+1}, &\quad \text{if } i = n+1, \\ a_{i,k+1}, &\quad \text{if } i \text{ is not one of } i_1, i_2, \ldots, i_p, n+1. \end{aligned}$$

It follows from the construction and the inductive property (8.12) that L_{k+1} is a partial latin square of order $n+1$ with all of its elements in the first $k+1$ columns specified. Moreover,

$$d(i, L_{k+1}) = a_{i,k+1} \quad \text{if } i \text{ is one of } i_1, i_2, \ldots, i_k \text{ and}$$
$$d(i, L_{k+1}) = d(i, L_k) \quad \text{if } n+1-k \leq i \leq n \text{ and } i \text{ is not one}$$
$$\text{of } i_1, i_2, \ldots, i_p.$$

The inductive properties (8.11) and (8.12) now hold with k replaced by $k+1$. Hence we obtain a sequence $L_1, L_2, \ldots, L_n$ of partial latin squares of order $n+1$. The partial latin square L_n has all of its entries specified except for those below the back diagonal in column $n+1$. If we specify the element in row i of column $n+1$ of L_n to be the element in $\{1, 2, \ldots, n+1\}$ which does not yet appear in row i, $(i = 2, 3, \ldots, n+1)$, then we obtain a completion L_{n+1} of $P(A)$. $\square$

If A is a latin square of order n, then the latin square E_A of order $n+1$ obtained by completing $P(A)$ as in the proof of Theorem 8.3.6 is called the *enlargement* of A. As an illustration we construct the enlargement of the latin square

$$A = \begin{bmatrix} 3 & 5 & 1 & 7 & 6 & 2 & 4 \\ 6 & 7 & 3 & 2 & 4 & 1 & 5 \\ 5 & 2 & 4 & 6 & 1 & 7 & 3 \\ 1 & 6 & 5 & 4 & 7 & 3 & 2 \\ 2 & 1 & 6 & 5 & 3 & 4 & 7 \\ 4 & 3 & 7 & 1 & 2 & 5 & 6 \\ 7 & 4 & 2 & 3 & 5 & 6 & 1 \end{bmatrix}$$

for which

$$P(A) = \begin{bmatrix} 3 & 5 & 1 & 7 & 6 & 2 & 4 & 8 \\ 6 & 7 & 3 & 2 & 4 & 1 & 8 & \diamond \\ 5 & 2 & 4 & 6 & 1 & 8 & \diamond & \diamond \\ 1 & 6 & 5 & 4 & 8 & \diamond & \diamond & \diamond \\ 2 & 1 & 6 & 8 & \diamond & \diamond & \diamond & \diamond \\ 4 & 3 & 8 & \diamond & \diamond & \diamond & \diamond & \diamond \\ 7 & 8 & \diamond & \diamond & \diamond & \diamond & \diamond & \diamond \\ 8 & \diamond & \diamond & \diamond & \diamond & \diamond & \diamond & \diamond \end{bmatrix}$$

In the enlargement E_A of A below, the superscript preceding elements in positions (i, j) on and below the main diagonal equals the element in position (i, j) of $A, (1 \leq i \leq j \leq n)$. These elements are used in carrying out the construction of E_A.

$$
E_A = \begin{bmatrix}
3 & 5 & 1 & 7 & 6 & 2 & 4 & 8 \\
6 & 7 & 3 & 2 & 4 & 1 & {}^58 & 5 \\
5 & 2 & 4 & 6 & 1 & {}^78 & {}^37 & 3 \\
1 & 6 & 5 & 4 & {}^78 & {}^37 & {}^23 & 2 \\
2 & 1 & 6 & {}^58 & {}^33 & {}^44 & {}^75 & 7 \\
4 & 3 & {}^78 & {}^11 & {}^27 & {}^55 & {}^62 & 6 \\
7 & {}^48 & {}^22 & {}^33 & {}^55 & {}^66 & {}^11 & 4 \\
8 & 4 & 7 & 5 & 2 & 3 & 6 & 1
\end{bmatrix}
$$

If, for instance, $k = 6$ the sequence (8.13) in the proof of Theorem 8.3.6 is

$$5, d(5, L_6), 7, d(3, L_6), 3, d(4, L_6), 2, d(6, L_6), 6.$$

Theorem 8.3.6 provides the key step in an inductive proof of the Evans conjecture. The following lemma is also used in the inductive proof.

Lemma 8.3.7. *Let X be an n by n array with at most $n - 1$ of its elements specified, and let z be one of its specified elements. Then it is possible to permute the rows and the columns of X so that in the resulting array Y, z occurs on the back diagonal and the other specified elements of Y occur above the back diagonal.*

Proof. If $n = 2$ the lemma holds. We assume that $n > 2$ and proceed by induction on n. If all of the specified elements of X are in one row, we permute the rows of X so that the specified elements are in row 1, and then permute the columns so that z is in the last position of the first row. The resulting array Y satisfies the conclusions of the lemma. We now assume that there is an i such that row i does not contain z but contains at least one specified element. Since X has only $n - 1$ specified elements, there is a j such that column j of X has no specified element. We now permute the rows of X so that row i of X becomes the first row and we permute the columns so that column j becomes the last column. Let X' be the $n - 1$ by $n - 1$ array in the lower left corner. Then X' has at most $n - 2$ specified elements and one of these is z. Applying the inductive hypothesis to X' we complete the proof. □

The following theorem contains the solution to the Evans conjecture.

Theorem 8.3.8. *A partial latin square of order n with at most $n - 1$ specified elements can always be completed to a latin square of order n.*

Proof. Let A be a partial latin square of order n based on the set $\{1, 2, \ldots, n\}$ with at most $n - 1$ specified elements. If A has less than $n - 1$

specified elements, then it is possible to specify another element so as to obtain a partial latin square. Hence we assume that A has $n-1$ specified elements. We prove the theorem by induction on n. If $n=1$ or 2, the theorem holds. Now assume that $n>2$.

First suppose that some integer in $\{1,2,\ldots,n\}$ occurs exactly once in A. Without loss of generality we assume that n occurs exactly once in A. By Lemma 8.3.7 we also assume without loss of generality that n occurs on the back diagonal of A and that the other specified elements of A occur above the back diagonal. Let B be the $n-1$ by $n-1$ array obtained by deleting the last row and the last column of A and deleting the element n on A's back diagonal. Then B is a partial latin square of order $n-1$ based on the set $\{1,2,\ldots,n-1\}$ with $n-2$ prescribed elements. By the inductive hypothesis B can be completed to a latin square C of order $n-1$. By Theorem 8.3.6 $P(C)$ can be completed to a latin square E_C of order $n-1$ based on the set $\{1,2,\ldots,n\}$. The enlargement E_C has n throughout its back diagonal and hence is a completion of A to a latin square of order n.

Now suppose that every integer that occurs in A occurs at least twice. Thus the number of different integers that occur in A is at most $\lfloor (n-1)/2 \rfloor$. Suppose that there exists an integer i such that row i of A contains exactly one specified element. Then we replace A by the equivalent[1] partial latin square $A_{(3,2,1)}$ in which the row indices and the elements have been interchanged. The integer i occurs exactly once in $A_{(3,2,1)}$ and hence as proved above $A_{(3,2,1)}$ has a completion to a latin square U. The latin square $U_{(3,2,1)}$ is a completion of A. Thus we may now assume that no row of A contains exactly one specified element. Similarly, we may now also assume that no column of A contains exactly one specified element. Thus the specified elements of A lie in an r by s rectangle where $r \leq \lfloor (n-1)/2 \rfloor$ and $s \leq \lfloor (n-1)/2 \rfloor$. By Corollary 8.3.3 A can be completed to a latin square of order n. $\square$

Theorem 8.3.8 has an equivalent formulation in terms of $(0,1)$-matrices.

Theorem 8.3.9.   *Let B be a $(0,1)$-matrix with at most $n-1$ 1's, and let k be a positive integer with $k \leq n-1$. Assume that $B = Q_1 + Q_2 + \cdots + Q_k$ is a decomposition of B into subpermutation matrices $Q_1, Q_2, \ldots, Q_k$ of order n. Then there exist subpermutation matrices $P_1, P_2, \ldots, P_k$ of order n such that*

$$C = P_1 + P_2 + \cdots + P_k$$

is a $(0,1)$-matrix and $Q_i \leq P_i, (i=1,2,\ldots,k)$.

Proof. The array

$$A = 1Q_1 + 2Q_2 + \cdots + kQ_k$$

---

[1] We owe to B.L. Shader the observation that replacing A by an equivalent partial latin square enables one to avoid the theorem of Lindner[1970].

is a partial latin square of order n with at most $n-1$ specified elements. By Theorem 8.3.8 A can be completed to a latin square. This latin square is of the form

$$1P_1 + 2P_2 + \cdots + nP_n$$

where $P_1, P_2, \ldots, P_n$ are permutation matrices of order n and where $Q_i \leq P_i, (i = 1, 2, \ldots, k)$. The matrix

$$C = P_1 + P_2 + \cdots + P_k$$

is a (0,1)-matrix satisfying the conclusions of the theorem.   □

Finally we remark that Andersen and Hilton[1983] showed that a partial latin square A of order n with n prescribed elements can be completed to a latin square of order n if and only if A is not equivalent to

$$\begin{bmatrix} 1 & 2 & \cdots & k & \diamond & \cdots & \diamond \\ \diamond & \diamond & \cdots & \diamond & k+1 & \cdots & \diamond \\ & & \cdots & & \vdots & \ddots & \diamond \\ \diamond & \diamond & \cdots & \diamond & \diamond & \cdots & k+1 \\ \vdots & \vdots & \cdots & \vdots & \vdots & \cdots & \vdots \end{bmatrix} \tag{8.15}$$

for any $k = 1, 2, \ldots, n-1$. Damerell[1983] showed that Smetaniuk's proof of the Evans conjecture could be extended to yield a proof of this result as well.

Colburn[1984] has proved that the decision problem *Can a partial latin square of order n be completed to a latin square?* is an NP-complete problem.

Exercises

1. For each integer $n \geq 4$ show that there exists a partial latin square of order n which cannot be embedded in a latin square of any order strictly less than $2n$ (Evans[1960]).
2. Prove that the partial latin squares of order n given by (8.15) can be partitioned into two parts each of which can be completed to a latin square of order n.
3. Determine the enlargement of a Cayley table of a cyclic group of order 7.
4. Complete the partial latin square A given below to a latin square:

$$\begin{bmatrix} 1 & 2 & 3 & \diamond & \diamond & \diamond \\ \diamond & \diamond & \diamond & 4 & \diamond & \diamond \\ \diamond & \diamond & \diamond & \diamond & 5 & \diamond \\ \diamond & \diamond & \diamond & \diamond & \diamond & \diamond \\ \diamond & \diamond & \diamond & \diamond & \diamond & \diamond \\ \diamond & \diamond & \diamond & \diamond & \diamond & \diamond \end{bmatrix}.$$

5. Let n be a positive integer. Show that Lemma 8.3.7 does not hold in general if the array X has more than n specified elements.
6. Prove that a symmetric latin square of odd order has a transversal.
7. Let A be a partial latin square of order n whose specified elements all belong to an r by r square L in its upper left corner. Assume that each element of

L is specified and that L is symmetric. Prove that A can be completed to a *symmetric* latin square if and only if the number $N(j)$ of times the element j occurs in L satisfies the following two conditions:

(i) $N(j) \geq 2r - n, (j = 1, 2, \ldots, n)$;

(ii) $N(j) \equiv n \pmod 2$ for at least r of the integers $j = 1, 2, \ldots, n$ (Cruse[1974]).

8. Let A be a symmetric partial latin square of order n. Prove that there is a symmetric latin square of order $2n$ which contains A in its upper left corner. (A *symmetric partial latin square* is a partial latin square which, considered as a matrix with elements from $\{1, 2, \ldots, n, \diamond\}$, is symmetric.) For each integer $n \geq 4$ show that there exists a symmetric partial latin square of order n which cannot be embedded in this way in any symmetric latin square of order strictly less than $2n$(Cruse[1974]).

References

L.D. Andersen and A.J.W. Hilton[1983], Thanks Evans!, *Proc. London Math. Soc.* (3), 47, pp. 507–522.

R.A. Brualdi and J. Csima[1986], Extending subpermutation matrices in regular classes of matrices, *Discrete Math.*, 62, pp. 99–101.

C.C. Colburn[1984], The complexity of completing partial latin squares, *Discrete Applied Math.*, 8, pp. 25–30.

A.B. Cruse[1974], On embedding incomplete symmetric latin squares, *J. Combin. Theory, Ser. A*, 16, pp. 18–22.

R.M. Damerell[1983], On Smetaniuk's construction for latin squares and the Andersen-Hilton theorem, *Proc. London Math. Soc.* (3), 47, pp. 523–526.

D.E. Daykin and R. Häggvist[1981], Problem No. 6347, *Amer. Math. Monthly*, 88, p. 446.

T. Evans[1960], Embedding incomplete latin squares, *Amer. Math. Monthly*, 67, pp. 958–961.

R. Häggvist[1978], A solution to the Evans conjecture for latin squares of large size, *Combinatorics*, Proc. Conf. on Combinatorics, Kesthely (Hungary) 1976, János Bolyai Math. Soc. and North Holland, pp. 495–513.

M. Hall Jr.[1945], An existence theorem for latin squares, *Bull. Amer. Math. Soc.* 51, pp. 387–388.

C.C. Lindner[1970], On completing latin rectangles, *Canad. Math. Bull.*, 13, pp. 65–68.

W.E. Opencomb[1984], On the intricacy of combinatorial problems, *Discrete Math.*, 50, pp. 71–97.

H.J. Ryser[1951], A combinatorial theorem with an application to latin rectangles, *Proc. Amer. Math. Soc.*, 2, pp. 550–552.

B. Smetaniuk[1981], A new construction for latin squares I. Proof of the Evans conjecture, *Ars Combinatoria*, 11, pp. 155–172.

8.4 Orthogonal Latin Squares

Let A be a latin square of order n based on the set $S = \{1, 2, \ldots, n\}$. We let

$$X = \{(i, j) : i, j = 1, 2, \ldots, n\}$$

denote the set of n^2 positions of A, and we now call the elements of X *points*. The set of points

$$\{(i,1),(i,2),\ldots,(i,n)\}, \quad (i=1,2,\ldots,n)$$

in a row is a *horizontal line*. The set of points

$$\{(1,j),(2,j),\ldots,(n,j)\}, \quad (j=1,2,\ldots,n)$$

in a column is a *vertical line*. Each horizontal line contains n points, and the set $H(A)$ of horizontal lines partitions the set X of points. A similar statement holds for the set $V(A)$ of vertical lines. Because A is a latin square, there are permutation matrices $P_1, P_2, \ldots, P_n$ of order n such that

$$J_n = P_1 + P_2 + \cdots + P_n$$

and

$$A = 1P_1 + 2P_2 + \cdots + nP_n.$$

Each permutation matrix P_i determines a set of n points which we call a *latin line* of A. The set $L(A)$ of the n latin lines of A also partition the n^2 points of X. Unlike $H(A)$ and $V(A)$ the set $L(A)$ of latin lines depends on the elements of A. The latin squares obtained from A by permuting the lines within each of the three classes and by permuting the three classes are the latin squares which are equivalent to A.

We now seek to arrange the elements of S in an n by n array B in such a way that each line of each of the classes $H(A), V(A)$ and $L(A)$ contains each element of S exactly once. Such an array B is a latin square with the additional property that each latin line of A contains each element of S exactly once. For example, let

$$A = \begin{bmatrix} 1 & 2 & 3 & 4 \\ 3 & 4 & 1 & 2 \\ 4 & 3 & 2 & 1 \\ 2 & 1 & 4 & 3 \end{bmatrix}$$

$$= 1\begin{bmatrix} 1 & 0 & 0 & 0 \\ 0 & 0 & 1 & 0 \\ 0 & 0 & 0 & 1 \\ 0 & 1 & 0 & 0 \end{bmatrix} + 2\begin{bmatrix} 0 & 1 & 0 & 0 \\ 0 & 0 & 0 & 1 \\ 0 & 0 & 1 & 0 \\ 1 & 0 & 0 & 0 \end{bmatrix}$$

$$+ 3\begin{bmatrix} 0 & 0 & 1 & 0 \\ 1 & 0 & 0 & 0 \\ 0 & 1 & 0 & 0 \\ 0 & 0 & 0 & 1 \end{bmatrix} + 4\begin{bmatrix} 0 & 0 & 0 & 1 \\ 0 & 1 & 0 & 0 \\ 1 & 0 & 0 & 0 \\ 0 & 0 & 1 & 0 \end{bmatrix}.$$

The latin square

$$B = \begin{bmatrix} 1 & 2 & 3 & 4 \\ 2 & 1 & 4 & 3 \\ 3 & 4 & 1 & 2 \\ 4 & 3 & 2 & 1 \end{bmatrix}$$

$$= 1 \begin{bmatrix} 1 & 0 & 0 & 0 \\ 0 & 1 & 0 & 0 \\ 0 & 0 & 1 & 0 \\ 0 & 0 & 0 & 1 \end{bmatrix} + 2 \begin{bmatrix} 0 & 1 & 0 & 0 \\ 1 & 0 & 0 & 0 \\ 0 & 0 & 0 & 1 \\ 0 & 0 & 1 & 0 \end{bmatrix}$$

$$+ 3 \begin{bmatrix} 0 & 0 & 1 & 0 \\ 0 & 0 & 0 & 1 \\ 1 & 0 & 0 & 0 \\ 0 & 1 & 0 & 0 \end{bmatrix} + 4 \begin{bmatrix} 0 & 0 & 0 & 1 \\ 0 & 0 & 1 & 0 \\ 0 & 1 & 0 & 0 \\ 1 & 0 & 0 & 0 \end{bmatrix}$$

is of the desired type because each latin line of B has exactly one point in common with each latin line of A.

We define two latin squares A and B to be *orthogonal* provided each latin line of A and each latin line of B have exactly one point in common. Thus the latin squares $A = [a_{ij}]$ and $B = [b_{ij}]$ of order n are orthogonal if and only if the n^2 ordered pairs

$$(a_{ij}, b_{ij}), \quad (i, j = 1, 2, \dots, n)$$

are distinct. If A and B are orthogonal latin squares then each is an *orthogonal mate* of the other. The latin squares A and B of order 4 above are orthogonal latin squares. It follows from the definition that if A has an orthogonal mate, then A has a transversal and indeed the n^2 positions of A can be partitioned into n transversals (such a partition is the set of n latin lines of an orthogonal mate B). A latin square with no transversal cannot therefore have an orthogonal mate. Thus two latin squares of order 2 are never orthogonal. More generally, we conclude from Theorem 8.2.1 that a Cayley table of an abelian group of even order with a unique element of order two does not have an orthogonal mate.

The question of the existence of orthogonal latin squares was raised by Euler[1782] in the following *problem of the 36 officers*:

Is it possible to arrange 36 officers of 6 different ranks and from 6 different regiments in a square formation of size 6 by 6 so that each row and each column of this formation contains exactly one officer of each rank and exactly one officer from each regiment?

If we label the 6 ranks and the 6 regiments from 1 through 6, then this problem asks for two orthogonal latin squares of order 6. Euler was unable to find such a pair of latin squares and conjectured that no pair existed. Tarry[1901] verified that there was no pair of orthogonal latin squares of order 6. While the notion of orthogonal latin squares originated in this recreational problem of Euler, it has since become important in the design of statistical experiments (see, e.g., Raghavarao[1971] and Joshi[1987]). Orthogonal latin squares are also of fundamental importance in the study of finite projective planes.

We now extend our definition of orthogonality of latin squares to any finite set of latin squares of the same order. The latin squares $A_1, A_2, \ldots, A_t$ of order n are *mutually orthogonal* provided A_i and A_j are orthogonal for all i different from j. If $A_1, A_2, \ldots, A_t$ are mutually orthogonal latin squares of order n and A_i' is a latin square equivalent to $A_i, (i = 1, 2, \ldots, t)$, then $A_1', A_2', \ldots, A_t'$ are also mutually orthogonal latin squares.

Let $N(n)$ denote the largest number of mutually orthogonal latin squares of order n. Thus $N(1) = 2$ (because a latin square of order 1 is orthogonal to itself) and $N(2) = 1$. By Tarry's verification $N(6) = 1$. It is not difficult to show that $N(n) \geq 2$ for every odd integer $n \geq 3$.

Theorem 8.4.1. *If A is a Cayley table of an abelian group of odd order $n \geq 3$, then A has an orthogonal mate. In particular, $N(n) \geq 2$ for each odd integer $n \geq 3$.*

Proof. Let $a_1, a_2, \ldots, a_n$ be the elements of an (additive) abelian group of odd order $n \geq 3$. Let $A = [a_{ij}]$ be the Cayley table of G in which

$$a_{ij} = a_i + a_j, \quad (i, j = 1, 2, \ldots, n).$$

Let the matrix $B = [b_{ij}]$ of order n be defined by

$$b_{ij} = a_i - a_j, \quad (i, j = 1, 2, \ldots, n).$$

Then B is a latin square (B is a column permutation of a Cayley table for G). Because G has odd order each element of G can be written in the form $a_k + a_k$ for some integer $k = 1, 2, \ldots, n$. Let x and y be arbitrary elements of G. Then there exists i and j such that $a_i + a_i = x + y$ and $a_j + a_j = x - y$ implying that

$$a_i + a_j = x \qquad \text{and} \qquad a_i - a_j = y.$$

Therefore A and B are orthogonal latin squares. $\square$

We also have the following elementary result.

Theorem 8.4.2. *If $A_1, A_2, \ldots, A_t$ are mutually orthogonal latin squares of order $n \geq 2$, then $t \leq n - 1$. Thus $N(n) \leq n - 1$ for $n \geq 2$.*

Proof. Without loss of generality we assume that the first row of each of the latin squares A_i is $1, 2, \ldots, n$. Thus no A_i has a 1 in the $(2, 1)$ position nor can two of the latin squares have the same element in the $(2, 1)$ position. Hence $t \leq n - 1$.   □

The upper bound for $N(n)$ in Theorem 8.4.2 is attained if n is a power of a prime number. The verification of this statement uses the existence of the Galois fields $GF(p^\alpha)$ where p is a prime and α is a positive integer.

Theorem 8.4.3.   *Let $n = p^\alpha$ where p is a prime and α is a positive integer. Then there exist $n - 1$ mutually orthogonal latin squares of order n and hence $N(n) = n - 1$.*

Proof. Let the elements of the Galois field $GF(p^\alpha)$ be denoted by $a_1 = 0, a_2, \ldots, a_n$. For $k = 2, 3, \ldots, n$ we define n by n matrices

$$A^{(k)} = [a_{ij}^{(k)}], \quad (i, j = 1, \ldots, n)$$

where

$$a_{ij}^{(k)} = a_k a_i + a_j.$$

Suppose that $A^{(k)}$ has two equal elements in row i. Then there exist integers j and j' such that

$$a_k a_i + a_j = a_k a_i + a_{j'},$$

implying that $a_j = a_{j'}$ and hence $j = j'$. Now suppose that A has two equal elements in column j. Then there exist integers i and i' such that

$$a_k a_i + a_j = a_k a_{i'} + a_j,$$

which implies, because $a_k \neq 0$, that $a_i = a_{i'}$ and hence $i = i'$. It follows that each $A^{(k)}$ is a latin square.

Now let k and l be integers with $2 \leq k < l \leq n$. We show that $A^{(k)}$ and $A^{(l)}$ are orthogonal. Suppose that

$$\left(a_{ij}^{(k)}, a_{ij}^{(l)} \right) = \left(a_{i'j'}^{(k)}, a_{i'j'}^{(l)} \right).$$

Then

$$a_k a_i + a_j = a_k a_{i'} + a_{j'} \tag{8.16}$$

and

$$a_l a_i + a_j = a_l a_{i'} + a_{j'}. \tag{8.17}$$

Subtracting (8.17) from (8.16) we obtain

$$(a_l - a_k) a_i = (a_l - a_k) a_{i'}. \tag{8.18}$$

Because $a_l \neq a_k$, (8.18) implies that $a_i = a_{i'}$. Substituting into (8.16) we also get $a_j = a_{j'}$. Hence $i = i'$ and $j = j'$ and it follows that $A^{(k)}$ and $A^{(l)}$ are orthogonal.   □

We now show how to combine two pairs of orthogonal latin squares to obtain a pair of orthogonal latin squares of larger order. Let $X = [x_{ij}]$ and $Y = [y_{ij}]$ be matrices of orders m and n, respectively. The matrix

$$X \otimes Y = [(x_{ij}, y_{kl})], \quad (i, j = 1, 2, \ldots, m; k, l = 1, 2, \ldots, n)$$

of order mn is called the *symbolic direct product* of X and Y. Its rows and columns are indexed by the ordered pairs $(r, s), (r = 1, 2, \ldots, m; s = 1, 2, \ldots, n)$ in lexicographic order. The elements of $X \otimes Y$ are the ordered pairs of the elements of X with the elements of Y. The following theorem is from MacNeish[1922].

Theorem 8.4.4.   *Let $A = [a_{ij}]$ and $B = [b_{ij}]$ be orthogonal latin squares of order m, and let $C = [c_{kl}]$ and $D = [d_{kl}]$ be orthogonal latin squares of order n. Then $A \otimes C$ and $B \otimes D$ are orthogonal latin squares of order mn.*

Proof. Let the latin squares A and B be based on the set $S = \{1, 2, \ldots, m\}$ and let the latin squares C and D be based on the set $T = \{1, 2, \ldots, n\}$. Then $A \otimes C$ and $B \otimes D$ are latin squares based on the cartesian product $S \times T$ with rows and columns indexed by the elements of $S \times T$. We now show that $A \otimes C$ and $B \otimes D$ are orthogonal.

Suppose that

$$((a_{ij}, c_{kl}), (b_{ij}, d_{kl})) = ((a_{i'j'}, c_{k'l'}), (b_{i'j'}, d_{k'l'})).$$

Then

$$(a_{ij}, c_{kl}) = (a_{i'j'}, c_{k'l'}) \quad \text{and} \quad (b_{ij}, d_{kl}) = (b_{i'j'}, d_{k'l'}),$$

and hence

$$a_{ij} = a_{i'j'} \quad \text{and} \quad b_{ij} = b_{i'j'} \tag{8.19}$$

and

$$c_{kl} = c_{k'l'} \quad \text{and} \quad d_{kl} = d_{k'l'}. \tag{8.20}$$

Because A and B are orthogonal, (8.19) implies that $i = i'$ and $j = j'$. Because C and D are orthogonal, (8.20) implies that $k = k'$ and $l = l'$. Hence $(i, j) = (i', j')$ and $(k, l) = (k', l')$, and it follows that $A \otimes C$ and $B \otimes D$ are orthogonal.   □

Corollary 8.4.5.   *Let $m_1, m_2, \ldots, m_k$ be positive integers. Then*

$$N(m_1 m_2 \cdots m_k) \geq \min\{N(m_i) : i = 1, 2, \ldots, k\}. \tag{8.21}$$

Proof. The inequality (8.21) follows from repeated application of Theorem 8.4.4.   □

By applying Theorem 8.4.3 and Corollary 8.4.5 we obtain the following lower bound for the number $N(n)$ of mutually orthogonal latin squares of order n.

Corollary 8.4.6.   *Let $n \geq 2$ be an integer and let $n = p_1^{\alpha_1} p_2^{\alpha_2} \cdots p_k^{\alpha_k}$ where $p_1, p_2, \ldots, p_k$ are distinct primes. Then*

$$N(n) \geq \min\{p_i^{\alpha_i} - 1 : i = 1, 2, \ldots, k\}. \tag{8.22}$$

In particular, if n is not of the form $4m + 2$, then $N(n) \geq 2$.

It was conjectured by MacNeish[1922] that equality holds in (8.22) for every integer $n \geq 2$. By Theorem 8.4.3 MacNeish's conjecture holds if n is a power of a prime. MacNeish's conjecture has its origin in a conjecture of Euler[1782] which asserts that there does not exist a pair of orthogonal latin squares of order n for any integer n which is twice an odd number. Euler's conjecture thus asserts that if $n \geq 6$ and n is of the form $4m + 2$, then $N(n) = 1$, which is in agreement with MacNeish's conjecture for these integers n. While Tarry [1901] verified the validity of Euler's conjecture for $n = 6$, the combined efforts of Bose, Shrikhande and Parker[1960] disproved Euler's conjecture in *all* other cases.

Theorem 8.4.7.   *Let $n > 6$ be an integer of the form $4m + 2$. Then there exists a pair of orthogonal latin squares of order n.*

We shall not give a complete proof of Theorem 8.4.7. We shall only discuss some of the techniques of construction which are used in its proof.

Parker[1959a] disproved MacNeish's conjecture by showing that $N(21) \geq 3$. Then Bose and Shrikhande[1959, 1960b] disproved Euler's conjecture by showing that $N(50) \geq 5$ and $N(22) \geq 2$. Parker[1959b] constructed a pair of orthogonal latin squares of order 10 thereby showing that $N(10) \geq 2$ and thus settling negatively the smallest unsolved case of the Euler conjecture since Tarry[1901] had verified that $N(6) = 1$.

The following two orthogonal latin squares of order 10 based on the set $\{0, 1, 2, \ldots, 9\}$ are those constructed by Parker[1959b].

$$
\begin{bmatrix}
0 & 6 & 5 & 4 & 7 & 8 & 9 & 1 & 2 & 3 \\
9 & 1 & 0 & 6 & 5 & 7 & 8 & 2 & 3 & 4 \\
8 & 9 & 2 & 1 & 0 & 6 & 7 & 3 & 4 & 5 \\
7 & 8 & 9 & 3 & 2 & 1 & 0 & 4 & 5 & 6 \\
1 & 7 & 8 & 9 & 4 & 3 & 2 & 5 & 6 & 0 \\
3 & 2 & 7 & 8 & 9 & 5 & 4 & 6 & 0 & 1 \\
5 & 4 & 3 & 7 & 8 & 9 & 6 & 0 & 1 & 2 \\
2 & 3 & 4 & 5 & 6 & 0 & 1 & 7 & 8 & 9 \\
4 & 5 & 6 & 0 & 1 & 2 & 3 & 9 & 7 & 8 \\
6 & 0 & 1 & 2 & 3 & 4 & 5 & 8 & 9 & 7
\end{bmatrix}
\quad
\begin{bmatrix}
0 & 9 & 8 & 7 & 1 & 3 & 5 & 2 & 4 & 6 \\
6 & 1 & 9 & 8 & 7 & 2 & 4 & 3 & 5 & 0 \\
5 & 0 & 2 & 9 & 8 & 7 & 3 & 4 & 6 & 1 \\
4 & 6 & 1 & 3 & 9 & 8 & 7 & 5 & 0 & 2 \\
7 & 5 & 0 & 2 & 4 & 9 & 8 & 6 & 1 & 3 \\
8 & 7 & 6 & 1 & 3 & 5 & 9 & 0 & 2 & 4 \\
9 & 8 & 7 & 0 & 2 & 4 & 6 & 1 & 3 & 5 \\
1 & 2 & 3 & 4 & 5 & 6 & 0 & 7 & 8 & 9 \\
2 & 3 & 4 & 5 & 6 & 0 & 1 & 8 & 9 & 7 \\
3 & 4 & 5 & 6 & 0 & 1 & 2 & 9 & 7 & 8
\end{bmatrix}
$$

This construction for a pair of orthogonal latin squares of order 10 was generalized by Bose, Shrikhande and Parker[1960] and by Menon[1961] to give a pair of orthogonal latin squares of order $n = 3m + 1$ for every $m \geq 3$ satisfying $N(m) \geq 2$.

Theorem 8.4.8.   *Let m be an integer for which there exists a pair of orthogonal latin squares of order m. Then there exists a pair of orthogonal latin squares of order $n = 3m + 1$.*

Proof. Let σ be the permutation of $\{0, 1, \ldots, 2m\}$ defined by $\sigma(i) \equiv i + 1 \pmod{2m + 1}$. Let P denote the permutation matrix of order $2m + 1$ corresponding to σ. Then P^k is the permutation matrix corresponding to $\sigma^k, (k = 0, 1, \ldots, 2m)$. We first construct a matrix $X = [x_{ij}]$ of order $2m + 1$ as follows. The elements on the main diagonal of X (the positions in which $P^0 = I_{2m+1}$ has 1's) are $0, 1, \ldots, 2m$. For $k = 1, 2, \ldots, m$ the elements of X in those positions in which P^k has 1's are $2m - (k-1), 2m - (k-2), \ldots, 0, 1, \ldots, 2m - k \pmod{2m + 1}$ in the order of their row index. For $k = m + 1, m + 2, \ldots, 2m$ the elements of X in those positions in which P^k has 1's are $2m + 1, 2m + 2, \ldots, 3m$ again in the order of their row index.

Next we construct a $2m + 1$ by m matrix Y as follows. The elements in the first column of Y are $1, 2, \ldots, 2m, 0$ in the order of their row index. The elements in column k of Y are obtained by cyclically shifting upwards by one row the elements in column $k - 1$ of $Y, (k = 2, 3, \ldots, m)$.

Finally we construct an m by $2m + 1$ matrix Z. The elements in the first row of Z are $2, 3, \ldots, 2m, 0, 1$ in the order of their column index. The elements in row k of Z are obtained by cyclically shifting to the left by two columns the elements in row $k - 1$ of $Z, (k = 2, 3, \ldots, m)$.

Let U and V be orthogonal latin squares of order m based on the m-element set $\{2m + 1, 2m + 2, \ldots, 3m\}$. Then it can be verified that

$$\begin{bmatrix} X & Y \\ Z & U \end{bmatrix} \quad \text{and} \quad \begin{bmatrix} X^T & Z^T \\ Y^T & V \end{bmatrix}$$

are orthogonal latin squares of order $3m + 1$ based on $\{0, 1, 2, \ldots, 3m\}$. $\square$

Corollary 8.4.9.   *Let n be a positive integer satisfying $n \equiv 10 \pmod{12}$. Then there exists a pair of orthogonal latin squares of order n.*

Proof. Let n be an integer satisfying the hypothesis. Then there exists a nonnegative integer k such that $n = 3(4k + 3) + 1$. By Theorem 8.4.1 there exists a pair of orthogonal latin squares of order $4k + 3$. We now apply Theorem 8.4.8 with $m = 4k + 3$. $\square$

To further discuss constructions of orthogonal latin squares, we continue with the geometric interpretation of latin squares given at the beginning of the section. Let $A_1, A_2, \ldots, A_t$ be a set of t mutually orthogonal latin squares of order n based on the set $\{1, 2, \ldots, n\}$, and let L_i denote the set of latin lines of $A_i, (i = 1, 2, \ldots, t)$. Let H denote the set of horizontal lines (of each A_i) and V the set of vertical lines. Thus each of the $t + 2$ sets

$$H, V, L_1, L_2, \ldots, L_t$$

consists of n pairwise disjoint lines of n points each of which partition the point set $X = \{(i, j) : i, j = 1, 2, \ldots, n\}$. We call each of these sets a *parallel class* of lines. Because $A_1, A_2, \ldots, A_t$ are mutually orthogonal, lines from different parallel classes have exactly one point in common. We are thus led to make the following general definition (Bruck[1951,1963]).

Let $n \geq 2$ be an integer. Let X be a set of n^2 elements called *points*, and let $\mathcal{B} = \{T_1, T_2, \ldots, T_r\}$ where

$$T_i = T_i^1, T_i^2, \ldots, T_i^n$$

is a partition of X into n sets of n points each. The sets $T_i^j, (j = 1, 2, \ldots, n)$ in T_i are called *lines* and each T_i is a *parallel class* of lines. The pair $(X, \mathcal{B})$ is an (n, r)-*net* provided that lines in different parallel classes intersect in exactly one point. It follows from our previous discussion that if there exists a set of t mutually orthogonal latin squares of order n, then there exists an $(n, t + 2)$-net. The converse also holds.

Theorem 8.4.10.   *Let $n \geq 2$ and $r \geq 2$ be integers. Then $N(n) \geq r - 2$ if and only if an (n, r)-net exists.*

Proof. If $N(n) \geq r - 2$, then an (n, r)-net exists. Now suppose that $(X, \mathcal{B})$ is an (n, r)-net where $\mathcal{B} = \{T_1, T_2, \ldots, T_r\}$. We may label the n^2 points of X so that

$$X = \{(i, j) : i, j = 1, 2, \ldots, n\},$$
$$T_1^i = \{\{(i, j) : j = 1, 2, \ldots, n\}\}, \quad (1 \leq i \leq n)$$

and

$$T_2^j = \{\{(i, j) : i = 1, 2, \ldots, n\}\}, \quad (i \leq j \leq n).$$

Let

$$P_k^{(l)}, \quad (k = 3, 4, \ldots, r; l = 1, 2, \ldots, n)$$

be the $(0,1)$-matrix of order n whose 1's are in those positions (i, j) for which (i, j) is a point of the lth line of T_k. Then each $P_k^{(l)}$ is a permutation matrix of order n and the $r - 2$ matrices

$$A_k = 1P_k^{(1)} + 2P_k^{(2)} + \cdots + nP_k^{(n)}, \quad (k = 3, 4, \ldots, r),$$

are mutually orthogonal latin squares of order n. $\square$

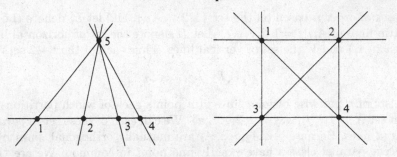

Figure 8.2

Bose and Shrikhande[1960a] defined another combinatorial configuration as follows. Let X be a finite set of elements called *points*, and let $\mathcal{A}$ be a collection of subsets of X called *lines* (or *blocks*). Then the pair $(X, \mathcal{A})$ is a *pairwise balanced design* provided:

(L1) *Each line in $\mathcal{A}$ contains at least two points.*

and

(L2) *For each pair x, y of distinct points in X, there is exactly one line in $\mathcal{A}$ containing both x and y.*

A simple example of a pairwise balanced design is obtained by letting

$$X = \{1, 2, 3, 4, 5\}$$

and

$$\mathcal{A} = \{\{1, 2, 3, 4\}, \{1, 5\}, \{2, 5\}, \{3, 5\}, \{4, 5\}\}.$$

Another simple example is obtained by letting $X = \{1, 2, 3, 4\}$ and $\mathcal{A} = \mathcal{P}_2(X)$, the collection of all two element subsets of X. These two examples are pictured geometrically in Figure 8.2.

An *affine plane* is a pairwise balanced design $(X, \mathcal{A})$ which satisfies the additional properties:

(L3) (Playfair's axiom or the parallel postulate) *For each point x in X and each line A in $\mathcal{A}$ not containing x, there is a unique line B in $\mathcal{A}$ which contains x but has no points in common with A.*

and

(L4) (A nondegeneracy condition) *There exist four points in X no three of which are together in a line of $\mathcal{A}$.*

The pairwise balanced design in Figure 8.2 which has four points is an affine plane.

Let p be a prime number and let α be a positive integer. The finite field $GF(p^\alpha)$ with $n = p^\alpha$ elements can be used to construct an affine plane. This construction, which mimics the analytic representation of the real Euclidean plane, proceeds as follows.

Let the set of n^2 points be

$$X = \{(a, b) : a, b \text{ in } GF(p^\alpha)\}.$$

A line in $\mathcal{A}$ is defined to be a set of points which satisfies a linear equation $ax + by + c = 0$ where a, b and c are elements of $GF(p^\alpha)$ with not both a and b equal to 0. Thus the lines are in one-to-one correspondence with the set of $n^2 + n$ linear equations of the form

$$\begin{aligned} x &= a & (a \text{ in } GF(p^\alpha)) \\ y &= mx + b & (m, b \text{ in } GF(p^\alpha)) \end{aligned}$$

It is now entirely straightforward to verify that the pair $(X, \mathcal{A})$ is an affine plane. We call a plane constructed in this way a *Galois affine plane* and denote it by $AP(p^\alpha)$. The affine plane $AP(p^\alpha)$ has n^2 points and $n^2 + n$ lines; in addition, each line contains n points and each point is contained in $n + 1$ lines. In the next theorem we show that arithmetic conditions like these hold for affine planes in general.

Let $(X, \mathcal{A})$ be an affine plane. Two lines in $\mathcal{A}$ are called *parallel* provided the lines are identical or have no point in common. It follows from the definition of an affine plane that the relation of parallelism is an equivalence relation on $\mathcal{A}$. Hence the lines in $\mathcal{A}$ are partitioned into parallel classes of lines. Two lines in the same class are parallel while lines from each of two different classes have exactly one point in common.

Theorem 8.4.11.   *Let $(X, \mathcal{A})$ be an affine plane. Then there exists an integer $n \geq 2$ for which the following properties hold:*

$$X \text{ has } n^2 \text{ points.} \tag{8.23}$$

$$\mathcal{A} \text{ has } n^2 + n \text{ lines.} \tag{8.24}$$

$$\text{Each line in } \mathcal{A} \text{ contains } n \text{ points.} \tag{8.25}$$

$$\text{Each point in } X \text{ is contained in } n + 1 \text{ lines.} \tag{8.26}$$

$$\text{There are } n + 1 \text{ parallel classes of lines.} \tag{8.27}$$

$$\text{Each parallel class contains } n \text{ lines.} \tag{8.28}$$

Proof. We first show that (8.25) holds. Let B and B' be distinct lines in $\mathcal{A}$. First suppose that B and B' have a point x in common. By (L1) and (L2) there exist a point $a \neq x$ on B and a point $b \neq x$ on B', and a line B'' containing both a and b. By Playfair's axiom (L3) there exist a line containing x which is parallel to B'' and hence there exists a point z

contained in neither B nor B'. It follows from (L2) and Playfair's axiom
(L3) that exactly one line containing z intersects B but not B' and exactly
one line containing z intersects B' but not B. We thus conclude that B
and B' have the same number of points. Now suppose that B and B' are
parallel. Let c be a point on B and let d be a point on B'. Then there
is a line B^* containing both c and d. As above B and B^* have the same
number of points, and B' and B^* have the same number of points. Thus in
this case also B and B' have the same number of points. Therefore there
exists an integer $n \geq 2$ such that (8.25) holds. By (L2) and Playfair's axiom
(L3), (8.26) also holds. From (L2) and (8.26) we obtain (8.23). From (8.26)
we conclude that there are $n + 1$ parallel classes of lines and each of these
parallel classes contains n lines. Hence (8.24), (8.27) and (8.28) also hold.

$\square$

Let $(X, \mathcal{A})$ be an affine plane. The integer $n \geq 2$ satisfying the conclusions of Theorem 8.4.12 is called the *order* of the affine plane. The Galois
affine plane $AP(p^\alpha)$ has order $n = p^\alpha$. It is not known whether there exist affine planes whose orders do not equal a prime power, although there
exist affine planes which are not Galois affine planes.

Theorem 8.4.12.   *Let $n \geq 2$ be an integer. Then the following are equivalent:*

    (i) There is a set of $n - 1$ mutually orthogonal latin squares of order n.
    (ii) There is an $(n, n + 1)$-net.
    (iii) There is an affine plane of order n.

Proof. The equivalence of (i) and (ii) follows from Theorem 8.4.10. It
follows from Theorem 8.4.11 that (iii) implies (ii). Now suppppose that (ii)
holds and let $(X, \mathcal{B})$ be an $(n, n + 1)$-net. Let $\mathcal{A}$ be the set of all lines in the
net. One may directly show that $(X, \mathcal{A})$ is an affine plane of order n. $\square$

A *projective plane* is a pairwise balanced design $(X, \mathcal{A})$ which satisfies
the nondegeneracy condition (L4) and the additional property

(L5) (The no-parallels postulate) *Two distinct lines in $\mathcal{A}$ have exactly
one point in common.*

Let the parallel classes of lines of an affine plane $(X, \mathcal{A})$ be $T_1, T_2, \ldots, T_{n+1}$.
Let $Y = \{y_1, y_2, \ldots, y_{n+1}\}$ be a set of $n + 1$ points which is disjoint from
X. By adjoining y_i to each line of T_i, $(i = 1, 2, \ldots, n + 1)$ and adjoining Y
as a new line to $\mathcal{A}$ we obtain a projective plane $(X \cup Y, \mathcal{A}')$ with $n^2 + n + 1$
points and $n^2 + n + 1$ lines. Each line of $\mathcal{A}'$ contains exactly $n + 1$ points of
$X \cup Y$, and each point of $X \cup Y$ is contained in exactly $n + 1$ lines of $\mathcal{A}'$.
Conversely, upon removing from a projective plane the set of points on any
prescribed line B and removing the line B we obtain an affine plane. The

order of the projective plane is defined to be the order of the resulting affine plane. Hence by Theorem 8.4.12 *a projective plane of order $n \geq 2$ exists if and only if there exist $n - 1$ mutually orthogonal latin squares of order n.* These connections between orthogonal latin squares and projective and affine planes are due to Bose[1938].

We now describe two constructions of Bose, Shrikhande and Parker which combine a set of mutually orthogonal latin squares and a pairwise balanced design in order to obtain another set of mutually orthogonal latin squares.

A latin square of order n based on an n-set S is *idempotent* provided its main diagonal is a transversal. A latin square with a transversal is equivalent (under row and column permutations only) to an idempotent latin square. Let $A_0, A_1, \ldots, A_k$ be $k+1$ mutually orthogonal latin squares of order n. Without loss of generality, we assume that the elements on the main diagonal of A_0 are identical. Then for $i \geq 1$ the main diagonal of A_i is a transversal of A_i, and hence A_i is an idempotent latin square. Thus from the set $A_0, A_1, \ldots, A_k$ of $k+1$ mutually orthogonal latin squares of order n we obtain a set $A_1', A_2', \ldots, A_k'$ of k mutually orthogonal idempotent latin squares of order n.

Theorem 8.4.13.   *Let $(X, \mathcal{A})$ be a pairwise balanced design with n points and m blocks $B_1, B_2, \ldots, B_m$. Let n_i denote the number of points in $B_i, (i = 1, 2, \ldots, m)$. If there exists a set of $t \geq 2$ mutually orthogonal idempotent latin squares of order n_i for each $i = 1, 2, \ldots, m$, then there exists a set of t mutually orthogonal idempotent latin squares of order n.*

Proof. Without loss of generality we assume that the set of points is $X = \{1, 2, \ldots, n\}$. Suppose that

$$A_i^{(1)}, A_i^{(2)}, \ldots, A_i^{(t)}, \quad (i = 1, 2, \ldots, m) \tag{8.29}$$

is a set of t mutually orthogonal idempotent latin squares of order n_i based on the set B_i of n_i points. We assume that the rows and columns of each $A_i^{(r)}$ are indexed by the elements of B_i in the same order and that for each x in B_i the element x occurs in the main diagonal position of $A_i^{(r)}$ corresponding to x. For each $k = 1, 2, \ldots, t$ we define a matrix $A^{(k)}$ of order n by

the element in position (p, p) of $A^{(k)}$ is p, $\quad (p = 1, 2, \ldots, n)$

and

the element in position (p, q) of $A^{(k)}$ equals the element in position (p, q) of $A_i^{(k)}$ where i is chosen so that B_i is the unique line in $\mathcal{A}$ containing both p and q.

We first show that each $A^{(k)}$ is a latin square, and hence an idempotent latin square. Suppose that the element u occurs in both the (p, q_1)

and (p, q_2) positions of $A^{(k)}$ where $q_1 \neq q_2$. The idempotence of the latin squares in (8.29) implies that none of q_1, q_2 and u equals p. Thus there exists a unique block of $\mathcal{A}$ which contains both p and q_1 and a unique block containing both p and q_2, and these two blocks also contain u. Hence p and u are in two distinct blocks contradicting the axiom (L2) for a pairwise balanced block design. Thus no element occurs twice in a row of $A^{(k)}$, and in a similar way one shows that no element occurs twice in a column. Hence $A^{(k)}$ is a latin square.

We now show that $A^{(k)}$ and $A^{(l)}$ are orthogonal for $k \neq l$. Let i and j be points in X. We show that there exist p and q such that the element in position (p, q) of $A^{(k)}$ is i and the element in position (p, q) of $A^{(l)}$ is j. If $j = i$, then we choose p and q equal to i. Now suppose that $i \neq j$. Then there exists a unique block B_r containing both i and j. Because $A_r^{(k)}$ and $A_r^{(l)}$ are orthogonal, there exist p and q in B_r such that the element in position (p, q) of $A_r^{(k)}$ is i and the element in position (p, q) of $A_r^{(l)}$ is j. It now follows that the element in position (p, q) of $A^{(k)}$ is i and the element in position (p, q) of $A^{(l)}$ is j. Hence $A^{(k)}$ and $A^{(l)}$ are orthogonal if $k \neq l$.

$\square$

We illustrate the application of Theorem 8.4.13 in obtaining mutually orthogonal latin squares.

Let $(X, \mathcal{A})$ be a projective plane of order 4. Thus X has 21 points and $\mathcal{A}$ has 21 lines, and each line contains 5 points. Since 5 is a prime number, there exists an affine plane of order 5. Therefore by Theorem 8.4.12 there exist four mutually orthogonal latin squares of order 5. Hence there exist three mutually orthogonal idempotent latin squares of order 5. Applying Theorem 8.4.13 we obtain three mutually orthogonal latin squares of order 21. Thus $N(21) \geq 3$. We observe that MacNeish's conjecture asserted that $N(21) = 2$.

Now let $(X, \mathcal{A})$ be a projective plane of order 8. Thus X has 73 points and 73 lines, and each line contains 9 points. Let a, b and c be three points of X not all on the same line. Let X' be obtained from X by removing the three points a, b and c, and let $\mathcal{A}'$ be obtained from $\mathcal{A}$ by removing the points a, b and c from those lines containing them. Then the pair $(X, \mathcal{A}')$ is a pairwise balanced design with 70 points and 73 blocks, and each block contains 7, 8 or 9 points. Because $N(9) = 8 > N(8) = 7 > N(7) = 6$, there exist 5 mutually orthogonal idempotent latin squares of each of the orders 7, 8 and 9. Applying Theorem 8.4.13 we conclude that $N(70) \geq 5$. We observe that Euler's conjecture asserted that $N(70) = 1$.

Theorem 8.4.14.   *Let $(X, \mathcal{A})$ be a pairwise balanced design with n points and m blocks $B_1, B_2, \ldots, B_m$. Let n_i equal the number of points in B_i, $(i = 1, 2, \ldots, m)$. Suppose that for some positive integer s with $s \leq m$, the blocks $B_1, B_2, \ldots, B_s$ are pairwise disjoint. If there exists a set of $t \geq 2$ mutually* *orthogonal latin squares of order n_i for each $i = 1, 2, \ldots, s$ and there exists a set of t mutually orthogonal idempotent latin squares of order n_i for each $i = s+1, s+2, \ldots, m$, then there exists a set of t mutually orthogonal latin squares of order n.*

Proof. We continue with the notation used in the proof of Theorem 8.4.13, but now the mutually orthogonal latin squares

$$A_i^{(1)}, A_i^{(2)}, \ldots, A_i^{(t)}$$

of order n_i based on the set B_i of points are assumed to be idempotent only for $i = s+1, s+2, \ldots, m$. The matrices $A^{(k)}$ are defined as in the proof of Theorem 8.4.13 with, however, the following change: If p is a point which belongs to B_r for some $r = 1, 2, \ldots, s$ the element in position (p, p) of $A^{(k)}$ equals the element in position (p, p) of $A_r^{(k)}$. One may modify the proof of Theorem 8.4.13 to show that $A^{(1)}, A^{(2)}, \ldots, A^{(k)}$ are orthogonal latin squares of order n. □

The pairwise balanced design $(X', \mathcal{A}')$ with 70 points and 73 lines defined in the paragraph immediately preceding Theorem 8.4.14 has exactly three blocks with 7 points and these blocks are pairwise disjoint. The other lines have 8 or 9 points. Hence Theorem 8.4.14 implies that $N(70) \geq 6$. Applying Theorem 8.4.14 to the pairwise balanced design obtained by removing three noncollinear points from a projective plane of order 4 we obtain $N(18) \geq 2$.

A "complete disproof" of the Euler conjecture can be found in Dénes and Keedwell[1974], Hall[1986], Raghavarao[1971] and Zhu[1982].

Exercises

1. Construct a pair of orthogonal latin squares of order 9.
2. Construct a pair of orthogonal latin squares of order 12.
3. Use the construction in the proof of Theorem 8.4.8 in order to obtain a pair of orthogonal latin squares of order 16.
4. Prove that there exists a symmetric, idempotent latin square of order n if and only if n is odd.
5. Prove that a set of $n - 2$ mutually orthogonal latin squares of order n can be extended to a set of $n - 1$ mutually orthogonal latin squares of order n. (In fact Shrikhande[1961] has shown that a set of $n - 3$ mutually orthogonal latin squares of order n can be extended to a set of $n - 1$ mutually orthogonal latin squares of order n.)
6. Show that the latin squares of order $3m + 1$ constructed in the proof of Theorem 8.4.8 are indeed orthogonal.
7. Let $(X, \mathcal{A})$ be a projective plane of order n. Let Y be obtained from X by removing four points, no three of which are together on a line in $\mathcal{A}$ and let $\mathcal{A}'$ be obtained by intersecting the lines in $\mathcal{A}$ with Y. Show that $(Y, \mathcal{A}')$ is a pairwise balanced design in which each line contains $n - 1$, n or $n + 1$ points.
8. Use Exercise 7 and Theorem 8.4.13 to show that $N(69) \geq 5$.
9. Let $N_I(n)$ denote the largest number of mutually orthogonal idempotent latin

squares of order n. Prove that for all positive integers m and n, $N_I(mn) \geq N_I(m)N_I(n)$.

10. Complete the proof of Theorem 8.4.14.

References

R.C. Bose[1938], On the application of the properties of Galois fields to the problem of construction of hyper-Graeco-Latin squares, *Sankhyā*, 3, pp. 323–338.

R.C. Bose, S.S. Shrikhande and E.T. Parker[1960], Further results on the construction of mutually orthogonal Latin squares and the falsity of Euler's conjecture, *Canad. J. Math.*, 12, pp. 189–203.

R.C. Bose and S.S. Shrikhande[1959], On the falsity of Euler's conjecture about the non-existence of two orthogonal Latin squares of order $4t+2$, *Proc. Nat. Acad. Sci. U.S.A.*, 45, pp. 734–737.

[1960a], On the composition of balanced incomplete block designs, *Canad. J. Math.*, 12, pp. 177–188.

[1960b], On the construction of sets of mutually orthogonal Latin squares and the falsity of a conjecture of Euler, *Trans. Amer. Math. Soc.*, 95, pp. 191–209.

R.H. Bruck[1951], Finite nets. I. Numerical invariants, *Canad. J. Math.*, 3, pp. 94–107.

[1963], Finite nets. II. Uniqueness and imbedding, *Pac. J. Math.*, 13, pp. 421–457.

J. Dénes and A.D. Keedwell[1974], *Latin Squares and Their Application*, Academic Press, New York.

M. Hall Jr.[1986], *Combinatorial Theory*, 2d edition, Wiley, New York.

D.D. Joshi[1987], *Linear Estimation and Design of Experiments*, Wiley, New York.

H.F. MacNeish[1922], Euler squares, *Ann. Math.*, 23, pp. 221–227.

P.K. Menon[1961], Method of constructing two mutually orthogonal latin squares of order $3n+1$, *Sankhyā* A, 23, pp. 281–282.

E.T. Parker[1959a], Construction of some sets of mutually orthogonal Latin squares, *Proc. Amer. Math. Soc.*, 10, pp. 946–949.

[1959b], Orthogonal Latin squares, *Proc. Nat. Acad. Sci.*, 45, pp. 859–862.

D. Raghavarao[1971], *Construction and combinatorial problems in design of experiments*, Wiley, New York (reprinted[1988] by Dover, Mineola, NY).

S.S. Shrikhande[1961], A note on mutually orthogonal latin squares, *Sankhyā* A, 23, pp. 115–116.

G. Tarry[1900,1901], Le problème de 36 officeurs, *Compte Rendu de l'Association Française pour l'Avancement de Science Naturel*, 1, pp. 122–123, and 2, pp. 170–203.

L. Zhu[1982], A short disproof of Euler's conjecture concerning orthogonal latin squares, *Ars Combinatoria*, 14, pp. 47–55.

8.5   Enumeration and Self-Orthogonality

In this final section we discuss without proof some results concerning the enumeration of latin squares, and latin squares which are orthogonal to their transpose.

One of the major unsolved problems in the theory of latin squares is the determination of the number L_n of distinct latin squares of order n based on the set $S = \{1, 2, \ldots, n\}$, and more generally the number $L_{r,n}$ of distinct r by n latin rectangles based on S.

The number ℓ_n of normalized latin squares and the number $\ell_{r,n}$ of normalized latin rectangles satisfy, respectively,

$$L_n = n!\ell_n \quad \text{and} \quad L_{r,n} = n!\ell_{r,n}.$$

We have $\ell_{1,n} = 1$ and $\ell_{2,n} = D_n$ where

$$D_n = n!\left(1 - \frac{1}{1!} + \frac{1}{2!} - \cdots + (-1)^n\frac{1}{n!}\right)$$

is the number of derangements of $\{1, 2, \ldots, n\}$. The formula of Riordan[1946]

$$\ell_{3,n} = \sum_{k=0}^{\lfloor n/2 \rfloor} \binom{n}{k} D_{n-k} D_k U_{n-2k}$$

gives the number of 3 by n normalized latin rectangles in terms of binomial coefficients, derangements and ménage numbers U_{n-2k}. The number $\ell_{3,n}$ is also known (see Goulden and Jackson[1983]) to be the coefficient of $x^n/n!$ in

$$e^{2x} \sum_{n=0}^{\infty} n!\frac{x^n}{(1+x)^{3n+3}}.$$

This exponential generating function for $\ell_{3,n}$ was generalized by Gessel [1985]. An explicit formula for $L(4, n)$, and thus for $\ell(4, n)$, is given in Athreya, Pranesacher and Singhi[1980] and Pranesachar[1981]. Nechvatal [1981], Athreya, Pranesachar and Singhi[1980], Pranesachar[1981] and Gessel[1987] obtained formulas for $L(r, n)$ in terms of the Möbius function for the partitions of a set. Formulas for the numbers $\ell_n, (n = 1, 2, \ldots, 9)$ are given in Ryser[1963] and Bammell and Rothstein[1975].

Hall[1948] proved that

$$L_{r+1,n} \geq (n - r)!L_{r,n}, \quad (r = 1, 2, \ldots, n - 1) \tag{8.30}$$

from which it follows that

$$L_n \geq n!(n - 1)! \cdots 2!1!. \tag{8.31}$$

The inequality (8.30) is a consequence of Theorem 7.4.1. Smetaniuk[1982] improved (8.31) by showing that in fact

$$L_n \geq n!L_{n-1}, \quad (n \geq 2). \tag{8.32}$$

It follows from (8.32) that L_n is a strictly increasing function of n. Jucys showed that L_n is a structure constant of an algebra defined on magic squares of order n from which he was then able to express L_n in terms

of the eigenvalues of a certain element of the algebra. Euler, Burkhard
and Grommes[1986] investigated the facial structure of a certain polytope
associated with latin squares of order n.

It follows from results in section 8.4 that the number $N(n)$ of mutually
orthogonal latin squares of order n satisfies

$$N(n) \le n - 1, \quad (n \ge 2)$$

with equality if and only if there exists a projective (or affine) plane of
order n. In particular, $N(n) = n - 1$ if n is a power of a prime number.
Thus we have

$$N(n) = n - 1 \text{ if } n = 2, 3, 4, 5, 7, 8, 9$$

and by Tarry's verification we also have $N(6) = 1$. The first undecided
value of $N(n)$ is $N(10)$. From the construction of Parker we know that
$N(10) \ge 2$. As noted in section 1.3 it has recently been concluded by an
extensive computer calculation that there does not exist a projective plane
of order 10. Hence it follows that $N(10) \le 8$. But a little more can be said.

Theorem 8.5.1.   *Let $A_1, A_2, \ldots, A_{n-2}$ be $n-2$ mutually orthogonal latin
squares of order $n \ge 3$. Then there exists a latin square A_{n-1} of order n
such that $A_1, A_2, \ldots, A_{n-2}, A_{n-1}$ are mutually orthogonal latin squares of
order n. In particular, $N(n) \ne n - 2$ for all $n \ge 3$.*

Proof. Let

$$A^{(t)} = \left[a_{ij}^{(t)} : i, j = 1, 2, \ldots, n \right], \quad (t = 1, 2, \ldots, n - 2).$$

Let $L_1, L_2, \ldots, L_{n-2}$ be the sets of latin lines of these latin squares, and let
H and V be, respectively, the set of horizontal and the set of vertical lines.
Each point belongs to one line of each of these n classes, each class consists
of n pairwise disjoint lines and each line contains n points. It follows that
the relation defined on points by

$$(i, j) \sim (i', j')$$

if and only if $i = i'$ and $j = j'$, or there does not exist a line containing
both (i, j) and (i', j') is an equivalence relation. This equivalence relation
partitions the set $\{(i, j) : i, j = 1, 2, \ldots, n\}$ of points into n equivalence
classes

$$\ell^1, \ell^2, \ldots, \ell^n \tag{8.33}$$

and each equivalence class ℓ^j contains n points. The equivalence classes
(8.33) are the latin lines of a latin square

$$A^{(n-1)} = \left[a_{ij}^{(n-1)} \right]$$

orthogonal to each of $A^{(1)}, A^{(2)}, \ldots, A^{(n-2)}$. More specifically, if we define

$$a_{ij}^{(n-1)} = k$$

if (i, j) belongs to ℓ^k, $(i, j = 1, 2, \ldots, n)$, then A^{n-1} is a latin square which is orthogonal to each of $A^{(1)}, A^{(2)}, \ldots, A^{(n-2)}$. $\square$

We therefore have that $N(10) \leq 7$ and more generally that $N(n) \leq n-3$ whenever there does not exist a projective (affine) plane of order n. We remark that the inequality $N(10) \geq 2$ has never been improved. However, the general inequality $N(n) \geq 2, (n > 6)$ has been improved. For example, Guérin has proved that $N(n) \geq 4, (n \geq 53)$, Hanani[1970] has proved that $N(n) \geq 5, (n \geq 63)$ and Wilson[1974] has proved that $N(n) \geq 6, (n \geq 91)$. In addition, Beth[1983], improving a result of Wilson[1974], has shown that

$$N(n) \geq n^{1/14.8} - 2$$

for all n sufficiently large.

A latin square A is called *self-orthogonal* provided A is orthogonal to its transpose A^T. The main diagonal of a self-orthogonal latin square is necessarily a transversal. Because there does not exist a pair of orthogonal latin squares of either of the orders 2 and 6, no latin square of order 2 or 6 can be orthogonal to its transpose. A simple examination reveals that there does not exist a self-orthogonal latin square of order 3. Modifying existing techniques and using some special constructions, Brayton, Coppersmith and Hoffman[1974] have shown how to construct a self-orthogonal latin square of order n for each positive integer n different from 2, 3 and 6. It follows that for those integers n for which there exists a pair of orthogonal latin squares of order n, there exists a self-orthogonal latin square of order n except if $n = 3$.

We now describe a construction of Mendelsohn for a self-orthogonal latin square of order n for all prime powers $n \neq 2, 3$.

We begin with the Galois field $GF(q)$ where $q \neq 2, 3$. There exists an element λ in this field with λ different from $0, 1$ and 2^{-1}. Let the elements of $GF(q)$ be denoted by $a_1, a_2, \ldots, a_q$. We define a matrix $A = [a_{ij}]$ of order q by

$$a_{ij} = \lambda a_i + (1 - \lambda)a_j, \quad (i, j = 1, 2, \ldots, n).$$

Because $\lambda \neq 0, 1$, A is a latin square of order q based on the set of elements of $GF(q)$. We now show that A is orthogonal to A^T. If not then there exist i, j, k and l such that

$$\lambda a_i + (1 - \lambda)a_j = \lambda a_k + (1 - \lambda)a_l \tag{8.34}$$

and

$$\lambda a_j + (1 - \lambda)a_i = \lambda a_l + (1 - \lambda)a_k. \tag{8.35}$$

Adding (8.34) and (8.35) we obtain

$$a_i + a_j = a_k + a_l. \tag{8.36}$$

Substituting (8.36) into (8.34) and using the fact that $\lambda \neq 2^{-1}$, we get $a_i = a_k$ and $a_j = a_l$. Hence $i = k$ and $j = l$, and A is orthogonal to A^T. A self-orthogonal latin square of order 5 constructed in this way using the field $GF(5) = \{0, 1, 2, 3, 4\}$ of integers modulo 5 and $\lambda = 2$ is

$$A = [a_{ij}] = \begin{bmatrix} 0 & 4 & 3 & 2 & 1 \\ 2 & 1 & 0 & 4 & 3 \\ 4 & 3 & 2 & 1 & 0 \\ 1 & 0 & 4 & 3 & 2 \\ 3 & 2 & 1 & 0 & 4 \end{bmatrix}.$$

Here $a_{ij} = 2i - j \bmod 5, (i, j = 0, 1, 2, 3, 4)$.

The symbolic direct product $X \otimes Y$ of matrices X and Y of orders m and n, respectively, satisfies $(X \otimes Y)^T = X^T \otimes Y^T$. Hence the preceding construction and Theorem 8.4.4 can be used to obtain a self-orthogonal latin square of order n for each integer $n > 1$ whose prime factorization does not contain exactly one 2 or exactly one 3.

We conclude with an application of self-orthogonal latin squares to *spouse-avoiding mixed doubles round robin tournaments* as described by Brayton, Coppersmith and Hoffman[1974].

It is desired to arrange a schedule of mixed doubles matches for n married couples which has the following four properties:

  (i) In each match two teams, each composed of one man and one woman, compete.
  (ii) A husband and wife pair never appear in the same match, as either partners or opponents.
  (iii) Two players of the same sex oppose each other exactly once.
  (iv) Two players of the opposite sex, if not married to each other, play in exactly one match as partners and in exactly one match as opponents.

In such a tournament each person plays $n - 1$ matches and hence there are a total of $n(n-1)/2$ matches.

Let the couples be labeled $1, 2, \ldots, n$ and let the husband and wife of couple i be M_i and $W_i, (i = 1, 2, \ldots, n)$. Let $A = [a_{ij}]$ be a self-orthogonal latin square of order n based on the set $\{1, 2, \ldots, n\}$. The main diagonal of A is a transversal and without loss of generality we assume that $a_{ii} = i$, $(i = 1, 2, \ldots, n)$. The $n(n-1)/2$ matches

$$\{M_i, W_{a_{ij}}\} \text{ versus } \{M_j, W_{a_{ji}}\}, \quad (1 \leq i < j \leq n)$$

determine a spouse-avoiding mixed doubles round robin tournament. Conditions (i)–(iv) follow from the fact that A is a self-orthogonal latin square of order n with main diagonal $1, 2, \ldots, n$.

Conversely, given a spouse-avoiding mixed doubles round robin tournament, a self-orthogonal latin square $A = [a_{ij}]$ of order n is obtained by defining a_{ii} to be $i, (i = 1, 2, \ldots, n)$ and defining a_{ij} to be k provided W_k is the partner of M_i in the match in which M_i opposes M_j.

Exercises

1. Show that (8.30) is a consequence of Theorem 7.4.1.
2. Construct a self-orthogonal latin square of order 7.
3. Construct a self-orthogonal latin square of order 20.
4. Verify that a spouse-avoiding mixed doubles round robin tournament with n couples gives a self-orthogonal latin square as described at the end of this section.

References

K.B. Athreya, C.B. Pranesachar and N.M. Singhi[1980], On the number of latin rectangles and chromatic polynomial of $L(K_{r,s})$, *Europ. J. Combinatorics*, 1, pp. 9–17.

S.E. Bammel and J. Rothstein[1975], The number of 9×9 latin squares, *Discrete Math.*, 11, pp. 93–95.

T. Beth[1983], Eine Bemerkung zur Abschätzung der Anzahl orthogonaler lateinischer Quadrate mittels Siebverfahren, *Abh. Math. Sem. Hamburg*, 53, pp. 284–288.

R.K. Brayton, D. Coppersmith and A.J. Hoffman[1974], Self-orthogonal latin squares of all orders $n \neq 2, 3, 6$, *Bull. Amer. Math. Soc.*, 80, pp. 116–118.

R. Euler, R.E. Burkhard and R. Grommes[1986], On latin squares and the facial structure of related polytopes, *Discrete Math.*, 62, pp. 155–181.

I. Gessel[1985], Counting three-line latin rectangles, *Proc. Colloque de Combinatoire Énumérative*, UQAM, pp. 106–111.

[1987], Counting latin rectangles, *Bull. Amer. Math. Soc. (new Series)*, 16, pp. 79–82.

R. Guérin[1968], Existence et propriétés des carrés latins orthogonaux II, *Publ. Inst. Statist. Univ. Paris*, 15, pp. 215–293.

M. Hall Jr.[1948], Distinct representatives of subsets, *Bull. Amer. Math. Soc.*, 54, pp. 958–961.

H. Hanani[1970], On the number of orthogonal latin squares, *J. Combin. Theory*, 8, pp. 247–271.

A.-A.A. Jucys[1976], The number of distinct latin squares as a group-theoretical constant, *J. Combin. Theory, Ser. A*, 20, pp. 265–272.

J.R. Nechvatal[1981], Asymptotic enumeration of generalized latin rectangles, *Utilitas Math.*, 20, pp. 273–292.

C.R. Pranesachar, Enumeration of latin rectangles via SDR's, *Combinatorics and Graph Theory* (S.B. Rao, ed.), Lecture Notes in Math., 885, Springer-Verlag, Berlin and New York, pp. 380–390.

J. Riordan[1946], Three-line latin rectangles - II, *Amer. Math. Monthly*, 53, pp. 18–20.

H.J. Ryser[1963], *Combinatorial Mathematics*, Carus Mathematical Monograph
   No. 14, Math. Assoc. of America, Washington, D.C.
B. Smetaniuk[1982], A new construction on latin squares - II: The number of
   latin squares is strictly increasing, *Ars Combinatoria*, 14, pp. 131–145.
R.M. Wilson[1974], Concerning the number of mutually orthogonal latin squares,
   Discrete Math., 9, pp. 181–198.
   [1974], A few more squares, *Congressus Numerantium*, No. X, pp. 675–680.

9

Combinatorial Matrix Algebra

9.1 The Determinant

We begin this chapter by examining in more detail the definition of the determinant function from a combinatorial point of view.

Let

$$A = [a_{ij}], \quad (i, j = 1, 2, \ldots, n)$$

be a matrix of order n. Then the determinant of A is defined by the formula

$$\det(A) = \sum_{\pi} (\operatorname{sign} \pi) a_{1\pi(1)} a_{2\pi(2)} \cdots a_{n\pi(n)} \tag{9.1}$$

where the summation extends over all permutations π of $\{1, 2, \ldots, n\}$. Suppose that the permutation π consists of k permutation cycles of sizes $\ell_1, \ell_2, \ldots, \ell_k$, respectively, where $\ell_1 + \ell_2 + \cdots + \ell_k = n$. Then $\operatorname{sign} \pi$ can be computed by

$$\operatorname{sign} \pi = (-1)^{\ell_1 - 1 + \ell_2 - 1 + \cdots + \ell_k - 1} = (-1)^{n-k} = (-1)^n (-1)^k. \tag{9.2}$$

Let D_n be the *complete digraph* of order n with vertices $\{1, 2, \ldots, n\}$ in which each ordered pair (i, j) of vertices forms an arc of D_n. We assign to each arc (i, j) of D_n the weight a_{ij} and thereby obtain a weighted digraph. The *weight of a directed cycle*

$$\gamma : i_1 \to i_2 \to \cdots \to i_t \to i_1$$

is defined to be

$$-a_{i_1 i_2} \cdots a_{i_{t-1} i_t} a_{i_t i_1},$$

the negative of the product of the weights of its arcs.

Let π be a permutation of $\{1, 2, \ldots, n\}$. The *permutation digraph* $D(\pi)$ is the digraph with vertices $\{1, 2, \ldots, n\}$ and with the n arcs

$$\{(i, \pi(i)) : i = 1, 2, \ldots, n\}.$$

The digraph $D(\pi)$ is a spanning subdigraph of the complete digraph D_n. The directed cycles of $D(\pi)$ are in one-to-one correspondence with the permutation cycles of π and the arc sets of these directed cycles partition the set of arcs of $D(\pi)$. The *weight* $\mathrm{wt}(D(\pi))$ *of the permutation digraph* $D(\pi)$ is defined to be the product of the weights of its directed cycles. Hence if π has k permutation cycles,

$$\mathrm{wt}(D(\pi)) = (-1)^k a_{1\pi(1)} a_{2\pi(2)} \cdots a_{\pi(n)}.$$

Using (9.1) and (9.2) we obtain

$$\det(-A) = \sum \mathrm{wt}(D(\pi)), \tag{9.3}$$

where the summation extends over all permutation digraphs of order n. Let σ denote a permutation of a subset X of $\{1, 2, \ldots, n\}$. The digraph $D(\sigma)$ is a permutation digraph with vertex set X and is a (not necessarily spanning) subdigraph of D_n with weight equal to the product of the weights of its cycles. (If $X = \emptyset$, the weight is defined to be 1.) Since $\det(I_n - A)$ is the sum of the determinants of all principal submatrices of $-A$, we also have

$$\det(I_n - A) = \sum \mathrm{wt}(D(\sigma)), \tag{9.4}$$

where the summation extends over all permutation digraphs whose vertices form a subset of $\{1, 2, \ldots, n\}$.

Similar formulas hold for the permanent. If we define the weight of a directed cycle to be the product of the weights of its arcs, and define the weight $\mathrm{wt}'(D(\sigma))$ of a permutation digraph to be the product of the weights of its directed cycles, then we have

$$\mathrm{per}(A) = \sum \mathrm{wt}'(D(\pi)),$$

and

$$\mathrm{per}(I_n + A) = \sum \mathrm{wt}'(D(\sigma)).$$

We remark that the fact that the determinant of a matrix is equal to the determinant of its transpose is a direct consequence of the formula (9.3) for the determinant. This is because assigning weights to the arcs of D_n using the elements of A^T is equivalent to assigning weights using the elements of A and then reversing the direction of all arcs. The resulting involution on the permutation digraphs of D_n is weight-preserving and hence $\det(A) = \det(A^T)$.

Exercises

1. Let A be the tridiagonal matrix of order n

$$\begin{bmatrix} a & b & 0 & \cdots & 0 & 0 \\ c & a & b & \cdots & 0 & 0 \\ 0 & c & a & \cdots & 0 & 0 \\ \vdots & \vdots & \vdots & \ddots & \vdots & \vdots \\ 0 & 0 & 0 & \cdots & a & b \\ 0 & 0 & 0 & \cdots & c & a \end{bmatrix}$$

Verify that

$$\det(A) = \sum_{k=0}^{\lfloor n/2 \rfloor} \binom{n-k}{k} a^{n-2k} (-bc)^k.$$

2. Let A and B be two matrices of order n. Prove that

$$\det(AB) = \det(A)\det(B).$$

3. Prove that the determinant of a matrix with two identical rows equals 0.

References

C.L. Coates[1959], Flow graph solutions of linear algebraic equations, *IREE Trans. Circuit Theory*, CT-6, pp. 170–187.

D.M. Cvetković[1975], The determinant concept defined by means of graph theory, *Mat. Vesnik*, 12, pp. 333–336.

M. Doob[1984], Applications of graph theory in linear algebra, *Math. Mag.*, 57, pp. 67–76.

F. Harary[1962], The determinant of the adjacency matrix of a graph, *SIAM Review*, 4, pp. 202–210.

D. König[1916], Über Graphen and ihre Anwendung auf Determinanttheorie und Mengenlehre, *Math. Ann.*, 77, pp. 453–465.

D. Zeilberger[1985], A combinatorial approach to matrix algebra, *Discrete Math.*, 56, pp. 61–72.

9.2 The Formal Incidence Matrix

Let

$$X = \{x_1, x_2, \ldots, x_n\}$$

be a nonempty set of n elements, and let

$$X_1, X_2, \ldots, X_m \tag{9.5}$$

be a configuration of m not necessarily distinct subsets of the n-set X. We recall from section 1.1 that the incidence matrix for these subsets is the $(0, 1)$-matrix

$$A = [a_{ij}], \quad (i = 1, 2, \ldots, m; j = 1, 2, \ldots, n)$$

of size m by n with the property that row i of A displays the subset X_i and column j displays the occurrences of the element x_j among the subsets. The matrix A may be regarded as a matrix over any field F, although we generally think of A as a matrix over the rational or real field.

A more general incidence matrix for our configuration may be obtained by replacing the 1's in A by not necessarily identical nonzero elements of the field F. Thus the nonzero elements of A now play the role of the identity element of F. In this generality every m by n matrix A over F may serve as an incidence matrix of some configuration of m subsets of the n-set X. In the discussion that follows A is an arbitrary matrix over F of size m by n.

Let

$$Z = [z_{ij}], \quad (i = 1, 2, \ldots, m; j = 1, 2, \ldots, n)$$

be a matrix of size m by n whose elements are mn independent indeterminates z_{ij} over the field F. The Hadamard product

$$M = A * Z = [a_{ij} z_{ij}], \quad (i = 1, 2, \ldots, m; j = 1, 2, \ldots, n)$$

is called a *formal incidence matrix* of the configuration (9.5). The elements of M belong to the polynomial ring

$$F^* = F[z_{11}, z_{12}, \ldots, z_{mn}],$$

and the nonzero elements of M are independent indeterminates over F.

We now call an m by n matrix

$$M = [m_{ij}], \quad (i = 1, 2, \ldots, m; j = 1, 2, \ldots, n)$$

a *generic matrix*, with respect to the field F, provided its nonzero elements are independent indeterminates over the field F. Every generic matrix with respect to F may serve as a formal incidence matrix. In that which follows we frame our discussion in terms of generic matrices with respect to a field F and work in the polynomial ring F^* obtained by adjoining the nonzero elements of M to F.

The term rank of the $(0,1)$-matrix A has been defined as the maximal number of 1's in A with no two of the 1's on a line. This important combinatorial invariant is equal to an algebraic invariant of the formal incidence matrix $A * Z$. The rank of $A * Z$ equals the maximal order of a square submatrix with a nonzero determinant. A square submatrix of $A * Z$ of order r has a nonzero determinant if and only if the corresponding submatrix of A has term rank r. Thus the term rank of A equals the rank of $A * Z$. (The term rank of an arbitrary matrix is defined as the maximal number of its nonzero elements with no two of the elements on a line.) The above argument implies the following observation of Edmonds[1967].

Theorem 9.2.1. *The term rank of a generic matrix equals its rank.*

The next theorem, already evident in the work of Frobenius[1912,1917] (see also Ryser[1973] and Schneider[1977]), shows that the combinatorial property of full indecomposability of a matrix also has an algebraic characterization. We first prove two simple lemmas about polynomials.

Lemma 9.2.2.   *Let $u_1, u_2, \ldots, u_k$ be k independent indeterminates over a field F. Let $p(u_1, u_2, \ldots, u_k)$ be a polynomial in $F[u_1, u_2, \ldots, u_k]$ which is linear in each of the indeterminates $u_1, u_2, \ldots, u_k$. Suppose that there is a factorization*

$$p(u_1, u_2, \ldots, u_k) = p_1(u_1, u_2, \ldots, u_k) p_2(u_1, u_2, \ldots, u_k)$$

where $p_1(u_1, u_2, \ldots, u_k)$ and $p_2(u_1, u_2, \ldots, u_k)$ are polynomials of positive degree in $F(u_1, u_2, \ldots, u_k)$. Then there is a partition

$$\{u_1, u_2, \ldots, u_k\} = \{u_{i_1}, \ldots, u_{i_r}\} \cup \{u_{j_1}, \ldots, u_{j_{k-r}}\}$$

of the set of indeterminates into two nonempty sets such that p_1 is a polynomial in the indeterminates $\{u_{i_1}, \ldots, u_{i_r}\}$, p_2 is a polynomial in the indeterminates $\{u_{j_1}, \ldots, u_{j_{k-r}}\}$ and each of the polynomials p_1 and p_2 is linear in its indeterminates.

Proof. For each $i = 1, 2, \ldots, k$ we may regard the polynomials p, p_1 and p_2 as polynomials over the integral domain $F[u_1, \ldots, u_{i-1}, u_{i+1}, \ldots, u_k]$ in the indeterminate u_i. These polynomials so regarded are linear polynomials and the conclusions follow.   □

Let $M = [m_{ij}]$ be a generic matrix of order n. A polynomial with coefficients in F of the form

$$f(M) = \sum c_\pi m_{1\pi(1)} m_{2\pi(2)} \cdots m_{n\pi(n)}$$

where the summation is over all permutations π of $\{1, 2, \ldots, n\}$ is called a *generalized matrix function* over F. A nonzero generalized matrix function is linear in each of its indeterminates and is homogeneous of degree n. The determinant and the permanent are examples of generalized matrix functions.

Lemma 9.2.3.   *Let $M = [m_{ij}]$ be a generic matrix of order n and let $f(M)$ be a nonzero generalized matrix function. Suppose that there is a factorization in F^* of the form*

$$f(M) = pq$$

where p and q are polynomials in the nonzero indeterminates m_{ij} with positive degrees r and $n - r$, respectively. Then there are complementary square submatrices M_r and M_{n-r} of M of orders r and $n - r$, respectively, such that p is a generalized matrix function of M_r and q is a generalized matrix function of M_{n-r}.

Proof. It follows from Lemma 9.2.2 that p and q are homogeneous polynomials of degree r and $n - r$, respectively, in distinct indeterminates. Let p_i be a typical nonzero term that occurs in p. Then, apart from a scalar factor, p_i is a product of r elements of M no two from the same line, and hence a product of r elements no two from the same line from a submatrix M_r of M of order r. It follows from the factorization $f = pq$ that a typical nonzero term q_j that occurs in q is, apart from a scalar, a product of $n - r$ elements no two on a line from the complementary submatrix M_{n-r}. Hence q is a generalized matrix function of M_{n-r}. Returning to p we now also see that p is a generalized matrix function of M_r.   □

Theorem 9.2.4.   *Let M be a generic matrix of order n. Then M is fully indecomposable if and only if $\det(M)$ is an irreducible polynomial in the polynomial ring F^*.*

Proof. First assume that M is not fully indecomposable. Then there exist positive integers r and $n - r$ and permutation matrices P and Q of order n such that

$$PMQ = \left[\begin{array}{cc} M_r & O \\ * & M_{n-r} \end{array} \right]$$

where M_r and M_{n-r} are square matrices of order r and $n - r$, respectively. Hence

$$\det(M) = \pm \det(M_r) \det(M_{n-r})$$

and it follows that $\det(M)$ is not an irreducible polynomial.

Now assume that M is fully indecomposable. Suppose that $\det(M)$ is reducible and thus admits a factorization

$$\det(M) = pq \tag{9.6}$$

into two polynomials p and q of positive degree. By Lemma 9.2.3 there exist complementary submatrices M_r and M_{n-r} of M of orders r and $n - r$, respectively, such that p is a generalized matrix function of M_r and q is a generalized matrix function of M_{n-r}. Let P and Q be permutation matrices such that

$$PMQ = \left[\begin{array}{cc} M_r & * \\ * & M_{n-r} \end{array} \right]. \tag{9.7}$$

Since M is fully indecomposable it follows (see Theorem 4.2.2) that each nonzero element of M appears in $\det(M)$. Since p and q are generalized matrix functions of M_r and M_{n-r}, respectively, it now follows from the factorization (9.6) that the asterisks in (9.7) correspond to zero matrices and this contradicts the assumption that M is fully indecomposable.   □

Theorem 9.2.4 remains true if we replace the determinant function with the permanent function. Almost no change is required in its proof.

It was established in Theorem 4.2.6 that the fully indecomposable components of a matrix of order n with term rank equal to n are uniquely determined to within arbitrary permutations of their lines. As noted by Ryser[1973] this fact follows from the algebraic characterization of full indecomposability given in Theorem 9.2.4. The following corollary is equivalent to the uniqueness conclusion of Theorem 4.2.6.

Corollary 9.2.5. *Let M be a generic matrix of order n such that $\det(M) \neq 0$. Then the set of fully indecomposable components of M is uniquely determined apart from arbitrary permutations of the lines of each component.*

Proof. Suppose that $E_1, E_2, \ldots, E_r$ and $F_1, F_2, \ldots, F_s$ are two sets of fully indecomposable components of M. Then by Theorem 9.2.4 we have that

$$\det(M) = \pm p_1 p_2 \cdots p_r = \pm q_1 q_2 \cdots q_s$$

where the p_i and the q_j are irreducible polynomials equal to the determinant of appropriate submatrices of M. Since F^* is a unique factorization domain, we conclude that $r = s$ and that the p_i and the q_j are the same apart from order and scalar factors. Hence the fully indecomposable components are the same apart from order and line permutations.   □

The property of irreducibility of a matrix also admits an algebraic characterization. We now let

$$Y = \begin{bmatrix} y_1 & 0 & \cdots & 0 \\ 0 & y_2 & \cdots & 0 \\ \vdots & \vdots & \ddots & \vdots \\ 0 & 0 & \cdots & y_n \end{bmatrix}$$

be the diagonal matrix of order n whose main diagonal elements are independent indeterminates over the field F. The polynomial $\det(A + Y)$ is a polynomial in the polynomial ring $F^* = F[y_1, y_2, \ldots, y_n]$ which is linear in each of the indeterminates $y_1, y_2, \ldots, y_n$. If f is a polynomial in F^* which is linear in the indeterminates, then for $\alpha \subseteq \{1, 2, \ldots, n\}$ the coefficient of $\prod_{i \in \alpha} y_i$ in f is denoted by f_α. The following lemma and theorem is from Schneider[1977].

Lemma 9.2.6.   *Let $f(y_1, y_2, \ldots, y_n)$ be a polynomial in F^* which is linear in each of the indeterminates. Suppose that for some integer r with $1 \leq r < n$*

$$f(y_1, y_2, \ldots, y_n) = p(y_1, \ldots, y_r) q(y_{r+1}, \ldots, y_n)$$

where p and q are polynomials that are linear in the indeterminates which they contain. Then

$$f_\emptyset f_{\{1,2,\ldots,n\}} = f_{\{1,\ldots,r\}} f_{\{r+1,\ldots,n\}}.$$

Proof. We have

$$f_{\{1,2,\ldots,n\}} = p_{\{1,\ldots,r\}} q_{\{r+1,\ldots,n\}} \qquad \text{and} \qquad f_\emptyset = p_\emptyset q_\emptyset$$

and

$$f_{\{1,\ldots,r\}} = p_{\{1,\ldots,r\}} q_\emptyset \qquad \text{and} \qquad f_{\{r+1,\ldots,n\}} = p_\emptyset q_{\{r+1,\ldots,n\}}. \qquad \square$$

Theorem 9.2.7. *Let A be a matrix of order n over the field F. Then A is irreducible if and only if $\det(A + Y)$ is an irreducible polynomial in the polynomial ring $F^\star$.*

Proof. The polynomial $\det(A + Y)$ contains the nonzero term $y_1 y_2 \cdots y_n$ and hence is a nonzero polynomial. First assume that A is reducible. Then there exist positive integers r and $n - r$ and a permutation matrix P of order n such that

$$P(A + Y)P^T = \begin{bmatrix} A_1 + Y_1 & O \\ * & A_2 + Y_2 \end{bmatrix}$$

where A_1 and A_2 are square matrices of order r and $n - r$, respectively, and Y_1 and Y_2 are diagonal matrices the union of whose diagonal elements is $\{y_1, y_2, \ldots, y_n\}$. The factorization

$$\det(A + Y) = \det(A_1 + Y_1) \det(A_2 + Y_2)$$

of $\det(A + Y)$ into two nonconstant polynomials in $F^\star$ implies that the polynomial $\det(A + Y)$ is reducible.

Now assume that the matrix $A = [a_{ij}]$ is irreducible. Suppose that $\det(A + Y)$ is reducible. Then it follows from Lemma 9.2.2 that there is a partition of the set of indeterminates $\{y_1, y_2, \ldots, y_n\}$ into two nonempty sets U and V and a factorization

$$\det(A + Y) = pq \tag{9.8}$$

where p is a polynomial in the indeterminates of U of positive degree r and q is a polynomial in the indeterminates of V of positive degree $n - r$. The polynomials p and q are linear in the indeterminates of U and V, respectively. Without loss of generality we assume that

$$U = \{y_1, \ldots, y_r\} \qquad \text{and} \qquad V = \{y_{r+1}, \ldots, y_n\}$$

and hence that

$$p = p(y_1, \ldots, y_r) \qquad \text{and} \qquad q = q(y_{r+1}, \ldots, y_n).$$

We now consider the digraph $D_0(A)$ associated with the off-diagonal elements of A. The set of vertices of $D_0(A)$ is taken to be $\{y_1, y_2, \ldots, y_n\}$ and there is an arc from y_i to y_j if and only if $i \neq j$ and $a_{ij} \neq 0$. Since A is irreducible, the digraph $D_0(A)$ is strongly connected. It follows that there is a directed cycle γ containing at least one vertex from $\{y_1, y_2, \ldots, y_r\}$ and at least one vertex from $\{y_{r+1}, y_{r+2}, \ldots, y_n\}$. We choose such a cycle γ of minimal length. Let α be the set of vertices of γ, and let $\alpha_1 = \alpha \cap \{y_1, y_2, \ldots, y_r\}$ and $\alpha_2 = \alpha \cap \{y_{r+1}, y_{r+2}, \ldots, y_n\}$. The minimality of γ implies that the principal submatrix of $A + Y$ determined by the indices in α satisfies

$$\det(A[\alpha] + Y[\alpha]) = \det(A[\alpha_1] + Y[\alpha_1])\det(A[\alpha_2] + Y[\alpha_2]) + c \quad (9.9)$$

where c is a nonzero scalar. Let

$$\det(A[\alpha] + Y[\alpha]) = \sum_{\beta \subseteq \alpha} c_\beta \prod_{i \in \beta} y_i.$$

It follows from (9.8) that

$$\det(A[\alpha] + Y[\alpha]) = p'q'$$

where p' is the coefficient of $\prod_{i \in \{1,2,\ldots,r\}-\alpha_1} y_i$ in p and q' is the coefficient of $\prod_{j \in \{r+1,r+2,\ldots,n\}-\alpha_2} y_j$ in q. By Lemma 9.2.6 we have

$$c_\emptyset = c_\emptyset c_\alpha = c_{\alpha_1} c_{\alpha_2}.$$

By (9.9) we may also apply Lemma 9.2.6 to $\det(A[\alpha] + Y[\alpha]) - c$ and conclude that

$$c_\emptyset - c = c_{\alpha_1} c_{\alpha_2}.$$

Combining these last two equations we obtain $c = 0$, a contradiction. Hence $\det(A + Y)$ is irreducible. $\qquad\square$

Ryser[1975] has obtained a different algebraic characterization of irreducible matrices.

The following corollary gives another algebraic characterization of fully indecomposable matrices. It is a direct consequence of Theorem 9.2.7 and Theorem 4.2.3.

Corollary 9.2.8.   *Let A be a matrix of order n over F and assume that the main diagonal of A contains no 0's. Then A is a fully indecomposable matrix if and only if $\det(A + Y)$ is an irreducible polynomial in F^*.*

We now show how the fundamental theorem of Hall[1935](Theorem 1.2.1) on systems of distinct representatives (SDR's) can be proved algebraically. The proof is taken from Edmonds[1967] (see also Mirsky[1971]). As already noted in section 1.2, the fundamental minimax theorem of König[1936]

(Theorem 1.2.1) can be derived from Hall's theorem. A direct algebraic proof of König's theorem is given at the end of this section.

Theorem 9.2.9.   *The subsets $X_1, X_2, \ldots, X_m$ of the n-set X have an SDR if and only if the set union $X_{i_1} \cup X_{i_2} \cup \cdots \cup X_{i_k}$ contains at least k elements for $k = 1, 2, \ldots, m$ and for all k-subsets $\{i_1, i_2, \ldots, i_k\}$ of the integers $1, 2, \ldots, m$.*

Proof. The necessity part of the theorem is obvious, and we turn to the sufficiency of Hall's condition for an SDR. Let M be a generic matrix of size m by n associated with the given subsets of X. The existence of an SDR is equivalent to M having term rank equal to m. By Theorem 9.2.1 it suffices to show that the rows of M are linearly independent over the polynomial ring F^* (equivalently, its quotient field) obtained by adjoining to the field F the nonzero elements of M.

Suppose that the rows of M are linearly dependent. We choose a minimal set of linearly dependent rows, which we assume without loss of generality are the first k rows of M. By Hall's condition M has no zero row and hence $k > 1$. The submatrix of M consisting of its first k rows has rank $k - 1$ and hence contains $k - 1$ linearly independent columns. Without loss of generality we assume that

$$M = \left[\begin{array}{cc} M_1 & M_2 \\ M_3 & M_4 \end{array} \right]$$

where M_1 is a k by $k - 1$ matrix of rank $k - 1$. The rows of M_1 are linearly dependent and hence there exists a row vector $u = (u_1, u_2, \ldots, u_k)$ whose elements are polynomials in the indeterminates of M_1 such that

$$uM_1 = 0.$$

Since the columns of M_2 are linear combinations of the columns of M_1, we also have

$$u \left[\begin{array}{cc} M_1 & M_2 \end{array} \right] = 0.$$

Since the first k rows of M form a minimal linearly dependent set of rows, each element of u is different from 0. Let

$$\left[\begin{array}{c} c_1 \\ c_2 \\ \vdots \\ c_k \end{array} \right]$$

be any column of M_2. Then

$$u_1 c_1 + u_2 c_2 + \cdots + u_k c_k = 0.$$

Since the u_i are polynomials in the indeterminates of M_1 and since the nonzero elements of M are independent indeterminates, we now conclude that each of the c_i's equals 0 and hence that $M_2 = O$. This means that $A_1 \cup A_2 \cup \cdots \cup A_k$ contains at most $k - 1$ elements, contradicting Hall's condition.   □

We conclude this section by giving the proof of Kung[1984] of the König theorem. This proof makes use of a classical determinantal identity of Jacobi[1841]. We include a proof of Jacobi's identity. The *adjugate* of a square matrix A is denoted by $\text{adj}(A)$.

Lemma 9.2.10.   *Let A be a nonsingular matrix of order n with elements in a field F. Let B be a submatrix of order r of $\text{adj}(A)$ and let C be the complementary submatrix of order $n - r$ in A. Then*

$$\det(B) = \det(A)^{r-1} \det(C).$$

In particular, B is nonsingular if and only if C is nonsingular.

Proof. Without loss of generality we assume that A and $\text{adj}(A)$ are in the form

$$A = \begin{bmatrix} * & * \\ * & C \end{bmatrix}, \; \text{adj}(A) = \begin{bmatrix} B & F \\ * & * \end{bmatrix}.$$

We now form

$$\begin{bmatrix} B & F \\ O & I_{n-r} \end{bmatrix} A = \begin{bmatrix} \det(A)I_r & O \\ * & C \end{bmatrix}.$$

Taking determinants we obtain

$$\det(B) \det(A) = \det(A)^r \det(C).$$ □

We now prove the theorem of König in the following equivalent form.

Theorem 9.2.11.   *Let M be a generic matrix of size m by n. Then the minimal number of lines in M that cover all the indeterminates of M equals the maximal number of indeterminates of M with no two on the same line.*

Proof. By Theorem 9.2.1 the rank of M equals its term rank, and we let this number be r. Let N be a submatrix of M of order and rank r. Without loss of generality we may assume that N occurs in the upper left corner of M. We permute the first r rows and the first r columns of M so that M assumes the form

$$M = \begin{bmatrix} * & * & M_1 \\ * & U & O \\ M_2 & O & O \end{bmatrix},$$

where the matrix M_1 of size e by $n - r$ has nonzero rows and the matrix M_2 of size $m - r$ by f has nonzero columns.

We first deal with certain degeneracies. In case $r = \min\{m, n\}$ the theorem is clearly valid so that we may take

$$r < \min\{m, n\}.$$

In case $e = 0$ we have zero columns in M and we may cover M with the first r columns. A similar situation holds for $f = 0$. Consider the case in which $e = r$. Suppose that $f > 0$. There is some submatrix of N of order and rank $r - 1$ within its last $r - 1$ rows. This submatrix along with a nonzero element in column 1 of M_2 and a nonzero element in an appropriate row of M_1 implies that M has rank greater than r. Thus if $e = r$ then $f = 0$, and similarly if $f = r$ then $e = 0$. Hence we may now assume that all nine blocks displayed in our decomposition of M are actually present.

We next write $\text{adj}(N)$ in the form

$$\text{adj}(N) = \begin{bmatrix} W & * \\ * & * \end{bmatrix}$$

where the matrix W is of size e by f. We assert that

$$W = [w_{ij}] = O.$$

Suppose that some $w_{ij} \neq 0$. Now $\text{adj}(N)$ is nonsingular, and by Lemma 9.2.10, $w_{ij} \neq 0$ implies that the submatrix N_{ij} of order $r - 1$ of N in the rows complementary to i and the columns complementary to j is nonsingular. But now we may take $r - 1$ nonzero elements no two on a line in N_{ij} and add two nonzero elements from M_1 and M_2. We thus increase the term rank of M and this is a contradiction. Hence $W = O$.

We note that

$$e + f \leq r.$$

For if $e + f > r$ then the nonsingular matrix $\text{adj}(N)$ of order r has a zero submatrix of e by f with $e + f > r$, and this contradicts the nonsingularity of $\text{adj}(N)$.

We now prove that

$$\text{rank}(U) \leq r - (e + f).$$

We deny this and suppose that

$$\text{rank}(U) \geq g \text{ where } g = r - (e + f) + 1.$$

This means that U contains a nonsingular submatrix U' of order g. We return to the nonsingular matrix $\text{adj}(N)$. The complementary matrix of U' in $\text{adj}(N)$ is a matrix V of order $e + f - 1$. But by Lemma 9.2.10 V is nonsingular and at the same time contains a zero submatrix of size e by f. This is again a contradiction.

The theorem now follows by induction on the (term) rank of M. By the induction hypothesis we may cover U with $r - (e + f)$ lines and hence M with r lines.   □

We remark that Perfect[1966] has given an algebraic proof of a symmetrized form of Theorem 9.2.11 due to Dulmage and Mendelsohn[1958]. A generalization of Theorem 9.2.4 to mixed matrices, matrices whose nonzero elements are either indeterminates or scalars, has been given by Murota [1989]. Hartfiel and Loewy[1984] show that a singular mixed matrix contains a submatrix with special properties.

Exercises

1. Let A and B be fully indecomposable matrices of order n with elements from a field F. Let $M = A * Z$ and $N = B * Z$ be formal incidence matrices associated with A and B, respectively. Suppose that $\det(M) = c \det(N) \neq 0$ where c is a scalar in F. Use Lemma 7.5.3 and prove that there exist diagonal matrices D_1 and D_2 with elements from F such that $D_1 A D_2 = B$ (Ryser[1973]).

2. Let A be an m by n matrix. Let r and s be integers with $1 \leq r \leq m$ and $1 \leq s \leq n$. Assume that the r by n submatrix of A determined by the first r rows has term rank r and that the m by s submatrix determined by the first s columns has term rank s. Prove that there exists a set T of nonzero elements of A, no two on the same line, such that T contains one element from each of the first r rows of A and one element from each of the first s columns of A (Dulmage and Mendelsohn[1958] and Perfect[1966]).

References

A.L. Dulmage and N.S. Mendelsohn[1958], Some generalizations of the problem of distinct representatives, *Canad. J. Math.*, 10, pp. 230–241.

J. Edmonds[1967], Systems of distinct representatives and linear algebra, *J. Res. Nat. Bur. Standards*, 71B, pp. 241–245.

G. Frobenius[1912], Über Matrizen aus nicht negativen Elementen, *Sitzungsber. Preuss. Akad. Wiss.*, pp. 456–477.

[1917], Über zerlegbare Determinanten, *Sitzungsber. Preuss. Akad. Wiss.*, pp. 274–277.

D.J. Hartfiel and R. Loewy[1984], A determinantal version of the Frobenius–König theorem, *Linear Multilin. Alg.*, 16, pp. 155–165.

C.G. Jacobi[1841], De determinantibus functionalibus, *J. Reine Angew. Math.*, 22, pp. 319–359.

J.P.S. Kung[1984], Jacobi's identity and the König-Egerváry theorem, *Discrete Math.*, 49, pp. 75–77.

L. Mirsky[1971], *Transversal Theory*, Academic Press, New York.

K. Murota[1989], On the irreducibility of layered mixed matrices, *Linear Multilin. Alg.*, 24, pp. 273–288.

H. Perfect1966], Symmetrized form of P. Hall's theorem on distinct representatives, *Quart. J. Math. Oxford* (2), 17, pp. 303–306.

H.J. Ryser[1973], Indeterminates and incidence matrices, *Linear Multilin. Alg.*, 1, pp. 149–157.

[1975], The formal incidence matrix, *Linear Multilin. Alg.*, 3, pp. 99–104.

H. Schneider[1977], The concepts of irreducibility and full indecomposability of a matrix in the works of Frobenius, König and Markov, *Linear Alg. Applics.*, 18, pp. 139–162.

9.3   The Formal Intersection Matrix

We again let

$$\{x_1, x_2, \ldots, x_n\} \tag{9.10}$$

be a nonempty set of n elements. Now we consider two configurations

$$X_1, X_2, \ldots, X_m \tag{9.11}$$

and

$$Y_1, Y_2, \ldots, Y_p \tag{9.12}$$

of subsets of (9.10). Let F be a field. Let A be a general m by n incidence matrix for the subsets (9.11) and let B be a general p by n incidence matrix for the subsets (9.12). Thus the elements of A and B come from the field F, and there is a nonzero element in position (i, j) if and only if x_j is a member of the ith set of the configuration. We now regard the elements $x_1, x_2, \ldots, x_n$ as indeterminates over the field F and consider the diagonal matrix

$$X = \begin{bmatrix} x_1 & 0 & \cdots & 0 \\ 0 & x_2 & \cdots & 0 \\ \vdots & \vdots & \ddots & \vdots \\ 0 & 0 & \cdots & x_n \end{bmatrix} \tag{9.13}$$

of order n. We then form the matrix product

$$Y = Y(x_1, x_2, \ldots, x_n) = AXB^T. \tag{9.14}$$

The matrix Y is of size m by p and we know the structure of Y quite explicitly. Indeed the matrix Y has in its (i, j) position a linear form

$$c_1 x_1 + c_2 x_2 + \cdots + c_n x_n$$

in the indeterminates $x_1, x_2, \ldots, x_n$ in which the coefficient c_k of x_k is nonzero if and only if x_k is in the intersection $X_i \cap Y_j$. The matrix Y is a *formal intersection matrix* of the configurations (9.11) and (9.12).

Two special cases of (9.14) are of considerable importance in their own right. We set $p = m$ and assume that the configurations (9.11) and (9.12) are identical. Then $B = A$ and (9.14) becomes

$$Y = AXA^T. \tag{9.15}$$

The matrix Y in (9.15) is now a symmetric matrix of order m and has in its (i, j) position a linear form in the indeterminates in the set intersection

$$X_i \cap X_j$$

in which each of these indeterminates appears with a nonzero coefficient. In this case Y is called a *formal symmetric intersection matrix* of the configuration (9.11).

Now let (9.12) be the complementary configuration

$$\overline{X}_1, \overline{X}_2, \ldots, \overline{X}_m$$

whose sets are the complements of the sets of the configuration (9.11). The matrix Y now has in its (i, j) position a linear form in the indeterminates in the set difference

$$X_i - X_j$$

in which each of these indeterminates appears with a nonzero coefficient. In this case we call Y a *formal set difference matrix* of the configuration (9.11).

Now suppose that A and B are the $(0,1)$-adjacency matrices of the configurations (9.11) and (9.12), respectively. Then the element in the (i, j) position of Y is

$$\sum_{x_k \in X_i \cap X_j} x_k,$$

the sum of the indeterminates in the intersection of X_i and X_j. If we set

$$x_1 = x_2 = \cdots = x_n = 1,$$

then we obtain the basic equation

$$Y(1, 1, \ldots, 1) = AB^T, \tag{9.16}$$

which reveals the cardinalities of the set intersections $X_i \cap X_j$. If Y is the formal symmetric intersection matrix of the configuration, then (9.16) reveals the cardinalities of the set intersections $X_i \cap X_j$. If Y is the formal set difference matrix, then (9.16) reveals the cardinalities of the set differences $X_i - X_j$.

The basic matrix equations (9.14), (9.15) and (9.16) allow one to apply the methods of matrix theory to the study of the intersection pattern of a configuration of sets. The vast area of combinatorial designs is primarily concerned with these equations. We make no attempt in this volume to study this equation in general or to develop the basic theory of combinatorial designs. These topics will be included in a subsequent volume to be entitled *Combinatorial Matrix Classes*. We briefly discuss some applications of these matrix equations.

An application of standard theorems on rank to the formal symmetric intersection matrix Y in (9.15) yields that

$$\text{rank}(AA^T) = \text{rank}(Y(1,1,\ldots,1)) \le \text{rank}(Y(x_1,x_2,\ldots,x_n))$$
$$\le \text{rank}(A) = \text{rank}(AA^T).$$

Hence in the case of a formal symmetric intersection matrix we have

$$\text{rank}(Y) = \text{rank}(A).$$

The following two theorems are from Ryser[1972].

Theorem 9.3.1.   *Let $X_1, X_2, \ldots, X_m$ be a configuration of subsets of the n-set*

$$\{x_1, x_2, \ldots, x_n\}$$

and let A be the m by n incidence matrix for this configuration. Assume that the number of distinct nonempty set intersections

$$X_i \cap X_j, \quad (i \ne j)$$

is strictly less than n. Then there exists an integral nonzero diagonal matrix D such that

$$ADA^T = E$$

is a diagonal matrix.

Proof. We consider the formal symmetric intersection matrix

$$AXA^T = Y.$$

Let t denote the number of distinct nonzero elements occurring in the positions of Y not on the main diagonal. The assumption in the theorem implies that $t < n$. We equate to zero these t nonzero elements and this gives us a homogeneous system of t equations in the n unknowns $x_1, x_2, \ldots, x_n$. This system of equations has a nonzero rational solution and hence an integral solution $d_1, d_2, \ldots, d_n$ with at least one d_i different from zero. The desired matrix D is the diagonal matrix whose main diagonal elements are $d_1, d_2, \ldots, d_n$.   □

Theorem 9.3.2.   *Let A be a $(0,1)$-matrix of order n. Assume that A satisfies the matrix equation*

$$ADA^T = E \tag{9.17}$$

where D and E are complex diagonal matrices and E is nonsingular. Then A is a permutation matrix of order n.

Proof. The assumption that E is nonsingular implies that both A and D are nonsingular matrices. The equation (9.17) implies that we may write

$$ADA^T E^{-1} = DA^T E^{-1} A = I_n.$$

Hence it follows that

$$A^T E^{-1} A = D^{-1}. \tag{9.18}$$

Let the elements on the main diagonal of D and E be $d_1, d_2, \ldots, d_n$ and $e_1, e_2, \ldots, e_n$, respectively. Then inspecting the main diagonal of (9.18) and using the fact that A is a (0,1)-matrix we see that

$$A^T \begin{bmatrix} e_1^{-1} \\ e_2^{-1} \\ \vdots \\ e_n^{-1} \end{bmatrix} = \begin{bmatrix} d_1^{-1} \\ d_2^{-1} \\ \vdots \\ d_n^{-1} \end{bmatrix}. \tag{9.19}$$

We now multiply (9.19) by AD and this gives

$$A \begin{bmatrix} 1 \\ 1 \\ \vdots \\ 1 \end{bmatrix} = \begin{bmatrix} 1 \\ 1 \\ \vdots \\ 1 \end{bmatrix}.$$

Thus each of the row sums of A equals 1. Because A is a nonsingular (0,1)-matrix, A is a permutation matrix. $\qquad\qquad\square$

We now consider set differences and first prove a theorem of Marica and Schönheim[1969].

Theorem 9.3.3. *Let $X_1, X_2, \ldots, X_m$ be a configuration of m distinct subsets of $\{x_1, x_2, \ldots, x_n\}$. Then the number of distinct differences*

$$X_i - X_j, \quad (i, j = 1, 2, \ldots, m)$$

is at least m.

Proof. We prove the theorem by induction on m. If $m \leq 2$, the theorem clearly holds. Now let $m \geq 3$. We denote our configuration by $\mathcal{C}$, and we let

$$\Delta(\mathcal{C}) = \{X_i - X_j : i, j = 1, 2, \ldots, m\}$$

be the collection of differences of the configuration $\mathcal{C}$. Let

$$k = \min_{i \neq j}\{|X_i \cap X_j|\}.$$

Without loss of generality we assume that $X_1 \cap X_2 = F$ where $|F| = k$. We partition $\mathcal{C}$ into three configurations:

(i) $\mathcal{C}_1$ consists of those X_i satisfying $F \not\subseteq X_i$.

The remaining sets of the configuration $\mathcal{C}$ contain F and for these sets we write $X_i' = X_i - F$.

(ii) $\mathcal{C}_2$ consists of those X_i not in $\mathcal{C}_1$ for which

$$X_i' \cap X_j' \neq \emptyset \text{ for all } X_j'.$$

(iii) $\mathcal{C}_3$ consists of those X_i not in $\mathcal{C}_1$ for which

$$X_i' \cap X_j' = \emptyset \text{ for some } X_j'.$$

The configuration $\mathcal{C}_3$ contains $t \geq 2$ sets, since X_1 and X_2 are in $\mathcal{C}_3$.

Let X_i be a set in $\mathcal{C}_3$ and let X_j be a set not in $\mathcal{C}_1$ satisfying $X_i' \cap X_j' = \emptyset$. Then X_j is also in $\mathcal{C}_3$. Since $X_i - X_j = X_i$ it follows that for each set X_i in $\mathcal{C}_3$, X_i' is a difference in $\Delta(\mathcal{C}_3)$. We now show that no such X_i' is a difference of the configuration $\mathcal{C}_1 \cup \mathcal{C}_2$ consisting of the sets in $\mathcal{C}_1$ and the sets in $\mathcal{C}_2$. It follows from the definition of $\mathcal{C}_2$ that each set in $\mathcal{C}_2$ has a nonempty intersection with X_i'. Since a set in $\mathcal{C}_1$ does not contain F but has at least k elements in common with X_i, it follows that each set in $\mathcal{C}_1$ also has a nonempty intersection with X_i'. Therefore the difference of two sets in $\mathcal{C}_1 \cup \mathcal{C}_2$ cannot equal X_i'. We thus have the inequality

$$|\Delta(\mathcal{C})| \geq |\Delta(\mathcal{C}_1 \cup \mathcal{C}_2)| + t.$$

Applying the induction hypothesis to the configuration $\mathcal{C}_1 \cup \mathcal{C}_2$ of $m - t \leq m - 2$ sets we complete the proof.   □

The following theorem of Ryser[1984] demonstrates how an assumption on the rank of a formal set intersection matrix implies a stronger conclusion than that given in Theorem 9.3.3.

Theorem 9.3.4.   *Let*

$$Y = Y(x_1, x_2, \ldots, x_n) = A X B^T$$

be a formal set intersection matrix of size m by p, and assume that the rank of Y equals m. Then Y contains m distinct nonzero elements, one in each of the m rows of Y.

Proof. Let R_i denote the set consisting of the nonzero elements in row i of the matrix $Y, (i = 1, 2, \ldots, m)$. The theorem asserts that the sets $R_1, R_2, \ldots, R_m$ have a system of distinct representatives. By Theorem 9.2.9 it suffices to prove that every k rows of Y contain in their union at least k distinct nonzero elements.

Assume to the contrary that there is an integer k with $1 \leq k \leq m$ and a k by p matrix Y_k of k rows of Y with the property that Y_k contains fewer than k distinct nonzero elements. Let the distinct nonzero elements in Y_k be denoted by $\alpha_1, \alpha_2, \ldots, \alpha_s$ where $s < k$. Suppose that the element α_i of Y_k occurs exactly e_{ij} times in row j of Y_k. We consider the system of s linear homogeneous equations

$$e_{i1}z_1 + e_{i2}z_2 + \cdots + e_{ik}z_k = 0, \quad (i = 1, 2, \ldots, s) \qquad (9.20)$$

in the unknowns $z_1, z_2, \ldots, z_k$. Since $s < k$ this system has an integral solution with not all of the $z_j = 0$. We let $z_1, z_2, \ldots, z_k$ denote such a solution and we multiply (9.20) by α_i and obtain

$$e_{i1}z_1\alpha_i + e_{i2}z_2\alpha_i + \cdots + e_{ik}z_k\alpha_i = 0, \quad (i = 1, 2, \ldots, s). \qquad (9.21)$$

We return to the matrix Y_k and multiply row j of Y_k by z_j, $(i = 1, 2, \ldots, k)$. We call the resulting matrix Z_k. It follows from (9.21) that the sum of all of the elements of Z_k equals 0. Because each α_i is a sum of certain of the indeterminates $x_1, x_2, \ldots, x_n$, all terms of the form $z_j x_i$ in Z_k involving a particular indeterminate x_i must also sum to 0.

We now consider the location of these terms within Z_k. It follows from the nature of the formal set intersection matrix that the element x_i is confined to a certain submatrix of Y and appears in every position of that submatrix. (The possibility that x_i does not appear in Y is not excluded.) An entirely similar situation holds concerning the location of all terms of the form $z_j x_i$ in Z_k involving a particular indeterminate x_i. Now we have noted that all such terms within Z_k sum to 0. The location of these terms within Z_k implies that all such terms within a particular column of Z_k must also sum to 0. Hence all of the column sums of Z_k are 0. This implies that there is a dependence relation among the rows of Y_k and this contradicts the hypothesis that Y is of rank m. $\square$

Daykin and Lovász[1976] have refined Theorem 9.3.3 by proving that if in the formal set difference matrix $Y = AX(J - A)^T$ of order $m \geq 2$, the incidence matrix A has distinct rows, then Y has m distinct nondiagonal elements with no two of the nondiagonal elements on a line. Ahlswede and Daykin[1979] (see also Baston[1982]) have extended Theorem 9.3.3 to include more general intersection matrices.

Ryser[1982] has studied the determinant and characteristic polynomial of the formal symmetric intersection matrix. In addition, Ryser[1973] has solved the "inverse problem" for the formal intersection matrix.

Theorem 9.3.5.   *Let Y be a matrix of order $n \geq 3$, and assume that each element of Y is a linear form in the n indeterminates $x_1, x_2, \ldots, x_n$ with respect to a field F. Let X be the diagonal matrix of order n whose elements* *on the main diagonal are $x_1, x_2, \ldots, x_n$. Assume that the determinant of Y satisfies*

$$\det(Y) = c x_1 x_2 \cdots x_n$$

where c is a nonzero element of F. Assume further that each element of Y^{-1} is a linear form in $x_1^{-1}, x_2^{-1}, \ldots, x_n^{-1}$ with respect to the field F. Then there exist matrices A and B of order n with elements in F such that

$$Y = AXB.$$

Exercise

1. Let $Y = Y(x_1, x_2, \ldots, x_n)$ be a matrix of size m by p such that every element of Y is a linear form in the indeterminates $x_1, x_2, \ldots, x_n$ over a field F. Assume that the rank of the matrix $Y(0, \ldots, 0, x_i, 0, \ldots, 0)$ equals 0 or 1 for each $i = 1, 2, \ldots, n$. Prove that Y is a formal intersection matrix, that is, prove that there exist matrices A and B of sizes m by n and p by n, respectively, with elements from the field F such that $Y = AXB^T$ (Ryser[1973]).

References

R. Ahlswede and D.E. Daykin, The number of values of combinatorial functions, *Bull. London Math. Soc.*, 11, pp. 49–51.

V.J. Baston[1982], A Marica-Schönheim theorem for an infinite sequence of finite sets, *Math. Z.*, 179, pp. 531–533.

D.E. Daykin and L. Lovász[1976], The number of values of a Boolean function, *J. London Math. Soc.*, 12 (2), pp. 225–230.

J. Marica and J. Schönheim[1969], Differences of sets and a problem of Graham, *Canad. Math. Bull.*, 12, pp. 635–637.

H.J. Ryser[1972], A fundamental matrix equation for finite sets, *Proc. Amer. Math. Soc.*, 34, pp. 332–336.

[1973], Analogs of a theorem of Schur on matrix transformations, *J. Algebra*, 25, pp. 176–184.

[1974], Indeterminates and incidence matrices, in *Combinatorics*, Math. Centre Tracts Vol. 55, Mathematisch Centrum, Amsterdam, pp. 3–17.

[1982], Set intersection matrices, *J. Combin. Theory, Ser. A*, 32, pp. 162–177.

[1984], Matrices and set differences, *Discrete Math.*, 49, pp. 169–173.

9.4 MacMahon's Master Theorem

Let $A = [a_{ij}]$, $(i, j = 1, 2, \ldots, n)$ be a matrix of order n over a field F. As in section 9.2 we let

$$Y = \begin{bmatrix} y_1 & 0 & \cdots & 0 \\ 0 & y_2 & \cdots & 0 \\ \vdots & \vdots & \ddots & \vdots \\ 0 & 0 & \cdots & y_n \end{bmatrix}$$

be the diagonal matrix of order n whose main diagonal elements are independent indeterminates over the field F, and we let $F^\star$ denote the polynomial ring $F[y_1, y_2, \ldots, y_n]$. The *Master Theorem for Permutations* of MacMahon[1915] identifies the coefficients $A(m_1, m_2, \ldots, m_n)$ in the expansion

$$\det(I_n - AY)^{-1} = \sum_{(m_1, m_2, \ldots, m_n)} A(m_1, m_2, \ldots, m_n) y_1^{m_1} y_2^{m_2} \cdots y_n^{m_n}$$

(9.22)

where the summation extends over all n-tuples $(m_1, m_2, \ldots, m_n)$ of nonnegative integers. The combinatorial proof of the Master Theorem given below is due to Foata[1965] (see also Cartier and Foata[1969]), as it is described by Zeilberger[1985].

Theorem 9.4.1.   *The coefficient $A(m_1, m_2, \ldots, m_n)$ in (9.22) equals the coefficent of $y_1^{m_1} y_2^{m_2} \cdots y_n^{m_n}$ in the product*

$$\prod_{i=1}^{n} (a_{i1}y_1 + a_{i2}y_2 + \cdots + a_{in}y_n)^{m_i}.$$

(9.23)

Proof. We consider general digraphs whose vertex sets are $\{1, 2, \ldots, n\}$, and we denote by $\mathcal{D}$ the collection of all general digraphs D of order n for which each vertex has the same indegree as outdegree. We assume that the arcs of D with the same initial vertex have been linearly ordered. We denote by $\mathcal{H}$ the collection of all digraphs H of order n for which each vertex has the same indegree as outdegree and this common value is either 0 or 1 (H consists of a number of pairwise disjoint directed cycles and hence is a permutation digraph on a subset of $\{1, 2, \ldots, n\}$).

The weight of an arc (i, j) is defined by

$$\text{wt}(i, j) = a_{ij}y_j.$$

The weight $\text{wt}(D)$ of a digraph D in $\mathcal{D}$ is defined to be the product of the weights of its arcs. The weight $\text{wt}(H)$ of a digraph H in $\mathcal{H}$ is defined to be

$$(-1)^{c(H)}(\text{the product of the weights of its arcs}),$$

where $c(H)$ equals the number of directed cycles of H. Notice that a digraph in $\mathcal{H}$ is also in $\mathcal{D}$, and thus the computation of its weight depends on whether it is being regarded as a member of $\mathcal{D}$ or of $\mathcal{H}$. We define

$$\text{wt}\mathcal{D} = \sum_{D \in \mathcal{D}} \text{wt}(D)$$

and

$$\text{wt}\mathcal{H} = \sum_{H \in \mathcal{H}} \text{wt}(H).$$

We consider the cartesian product

$$\mathcal{G} = \mathcal{D} \times \mathcal{H} = \{(D, H) : D \in \mathcal{D}, H \in \mathcal{H}\}$$

and define the weight of the pair (D, H) by

$$\mathrm{wt}(D, H) = \mathrm{wt}(D)\mathrm{wt}(H).$$

The weight of $\mathcal{G}$ is

$$\mathrm{wt}\mathcal{G} = \sum_{(D,H)\in\mathcal{G}} \mathrm{wt}(D, H) = (\mathrm{wt}\mathcal{D})(\mathrm{wt}\mathcal{H}).$$

It follows from section 9.1 that

$$\mathrm{wt}\mathcal{H} = \det(I_n - AY).$$

Let $B(m_1, m_2, \ldots, m_n)$ denote the coefficient of $y_1^{m_1} y_2^{m_2} \cdots y_n^{m_n}$ in the product (9.23). We now show that

$$\mathrm{wt}\mathcal{D} = \sum_{(m_1,m_2,\ldots,m_n)} B(m_1, m_2, \ldots, m_n) y_1^{m_1} y_2^{m_2} \cdots y_n^{m_n}, \qquad (9.24)$$

where the summation is over all n-tuples of nonnegative integers $(m_1, m_2, \ldots, m_n)$.

Let $\mathcal{D}_{(m_1,m_2,\ldots,m_n)}$ be the subset of $\mathcal{D}$ consisting of those general digraphs in which vertex i has outdegree (and hence indegree) $m_i, (i = 1, 2, \ldots, n)$. Each term in the expanded product (9.23) is the weight of a general digraph with outdegrees $m_1, m_2, \ldots, m_n$. In order that the general digraph have indegrees equal to $m_1, m_2, \ldots, m_n$ as well, we take only those terms containing $y_1^{m_1} y_2^{m_2} \cdots y_n^{m_n}$. It follows that the weight of $\mathcal{D}_{(m_1,m_2,\ldots,m_n)}$ equals

$$B(m_1, m_2, \ldots, m_n) y_1^{m_1} y_2^{m_2} \cdots y_n^{m_n}.$$

Therefore (9.24) holds and hence

$$\mathrm{wt}\mathcal{G} = \left(\sum_{(m_1,m_2,\ldots,m_n)} B(m_1, m_2, \ldots, m_n) y_1^{m_1} y_2^{m_2} \cdots y_n^{m_n} \right) \det(I_n - AY).$$

To complete the proof we show that $\mathrm{wt}\mathcal{G} = 1$.

We now define a mapping

$$\sigma : \mathcal{G} - \{(\emptyset, \emptyset)\} \to \mathcal{G} - \{(\emptyset, \emptyset)\},$$

where $\emptyset$ denotes the digraph with vertices $\{1, 2, \ldots, n\}$ and with an empty set of edges. Given a pair $(D, H) \neq (\emptyset, \emptyset)$, we determine the first vertex u whose outdegree in either D or H is positive. Beginning at that vertex u

we walk along the arcs of D, always choosing the topmost arc, until one of the following occurs:

(i) We encounter a previously visited vertex (and thus have located a directed cycle γ of D).

(ii) We encounter a vertex which has positive outdegree in H (and thus is a vertex on a directed cycle δ of H).

We note that if u is a vertex with positive outdegree in H then we are immediately in case (ii). We also note that cases (i) and (ii) cannot occur simultaneously. If case (i) occurs, we remove γ from D and put it in H. If case (ii) occurs, we remove δ from H and put it in D in such a way that each arc of γ is put in front of (in the linear order) those with the same initial vertex. Let (D', H') be the pair obtained in this way. Then D' is in $\mathcal{D}$ and H' is in $\mathcal{H}$, and hence (D', H') is in $\mathcal{G}$. Moreover, since the number of directed cycles in H' differs from the number in H by one, it follows that $\mathrm{wt}(D', H') = -\mathrm{wt}(D, H)$. We define $\sigma(D, H) = (D', H')$ and note that $\sigma(D', H') = (D, H)$. Thus $\sigma : \mathcal{G} - \{(\emptyset, \emptyset)\} \to \mathcal{G} - \{(\emptyset, \emptyset)\}$ is an involution which is sign-reversing on the weight. It follows that

$$\mathrm{wt}\mathcal{G} = \mathrm{wt}(\emptyset, \emptyset) = 1,$$

and the proof of the theorem is complete.   □

By Theorem 9.4.1 the coefficient $B(m_1, m_2, \ldots, m_n)$ of $y_1^{m_1} y_2^{m_2} \cdots y_n^{m_n}$ in the product (9.23) equals $A(m_1, m_2, \ldots, m_n)$. We now show that this coefficient can be expressed in terms of permanents. Let $\mathcal{R}$ denote the set of all nonnegative integral matrices $R = [r_{ij}]$ of order n with row sum vector and column sum vector equal to $(m_1, m_2, \ldots, m_n)$. Thinking of r_{ij} as the number of times the y_j term is selected from the factor $(a_{i1}y_1 + a_{i2}y_2 + \cdots + a_{in}y_n)^{m_i}$ in (9.23), we see that

$$B(m_1, m_2, \ldots, m_n) = \sum_{R \in \mathcal{R}} \left(\prod_{i=1}^{n} \frac{m_i!}{r_{i1}! r_{i2}! \cdots r_{in}!} \right) \prod_{i,j=1}^{n} a_{ij}^{r_{ij}}. \qquad (9.25)$$

The quantity on the right-hand side of (9.25) is easily seen to be equal to

$$\frac{1}{m_1! m_2! \cdots m_n!} \mathrm{per}_{(m_1, m_2, \ldots, m_n)}(A)$$

where $\mathrm{per}_{(m_1, m_2, \ldots, m_n)}(A)$ equals the permanent of the matrix of order $m_1 + m_2 + \cdots + m_n$ obtained from A by replacing each element a_{ij} by the constant matrix of size m_i by m_j each of whose elements equals a_{ij}. We thus obtain the following identity of Vere-Jones[1984].

Corollary 9.4.2.

$$\det(I_n - AY)^{-1} = \sum \frac{1}{m_1! m_2! \cdots m_n!} per_{(m_1, m_2, \ldots, m_n)}(A) y_1^{m_1} y_2^{m_2} \cdots y_n^{m_n}$$

where the summation extends over all n-tuples $(m_1, m_2, \ldots, m_n)$ of non-negative integers.

We conclude this section by presenting an expansion for the determinant of a matrix in terms of the permanents of its principal submatrices and an expansion for the permanent in terms of the determinants of its principal submatrices.

Let S be a subset of the set $\{1, 2, \ldots, n\}$ and as usual let $A[S]$ denote the principal submatrix of A with rows and columns indexed by the elements of S. The set of ordered partitions of S into nonempty sets is denoted by Λ_S. Let $\alpha = (\alpha_1, \alpha_2, \ldots, \alpha_j)$ be in Λ_S. The number j of nonempty parts of α is denoted by p_α. We define $\det_\alpha(A)$ and $\operatorname{per}_\alpha(A)$ by

$$\det{}_\alpha(A) = \prod_{i=1}^{p_\alpha} \det(A[\alpha_i])$$

and

$$\operatorname{per}{}_\alpha(A) = \prod_{i=1}^{p_\alpha} \operatorname{per}(A[\alpha_i]).$$

The following identity is from Chu[1989].

Theorem 9.4.3. *The matrix A and the diagonal matrix Y of order n satisfy*

$$\det(A + Y) = \sum_{S \subseteq \{1,2,\ldots,n\}} \sum_{\alpha \in \Lambda_S} (-1)^{|S|+p_\alpha} \left(\prod_{i \in \overline{S}} y_i \right) \operatorname{per}{}_\alpha(A). \qquad (9.26)$$

Proof. The equation (9.26) asserts that for each subset S of $\{1, 2, \ldots, n\}$ we have

$$\det(A[S]) = \sum_{\alpha \in \Lambda_S} (-1)^{|S|+p_\alpha} \operatorname{per}{}_\alpha(A). \qquad (9.27)$$

Clearly it suffices to prove (9.27) in the case that $S = \{1, 2, \ldots, n\}$, that is to prove that

$$\det(A) = \sum_{\alpha \in \Lambda_{\{1,2,\ldots,n\}}} (-1)^{n+p_\alpha} \operatorname{per}{}_\alpha(A). \qquad (9.28)$$

Let σ be a permutation of $, 2, \ldots, n$ and suppose that σ has k permutation cycles in its cycle decomposition. We verify (9.28) by showing that the coefficient $c(\sigma)$ of

$$a_{1\sigma(1)} a_{2\sigma(2)} \cdots a_{n\sigma(n)}$$

in the expression on the right-hand side is the same as its coefficient sign $\sigma = (-1)^{n+k}$ in $\det(A)$.

Let $\alpha \in \Lambda_{\{1,2,\ldots,n\}}$. If there is some cycle of σ which is not wholly contained in a part of α, then $\text{per}_\alpha(A)$ does not contribute to the coefficient $c(\sigma)$. Therefore

$$c(\sigma) = \sum_{j=0}^{k} (-1)^{n+j} p(k,j)$$

where $p(k,j)$ equals the number of partitions of a set of k distinct elements (the k cycles of σ) into j distinguishable boxes (the parts of α) with no empty box. It follows from the inclusion-exclusion principle (see, e.g., Brualdi[1977] or Stanley[1986]) that

$$p(k,j) = \sum_{i=0}^{j} (-1)^{j-i} \binom{j}{i} i^k.$$

Hence

$$\begin{aligned}
c(\sigma) &= \sum_{j=0}^{k} (-1)^{n+j} \sum_{i=0}^{j} (-1)^{j-i} \binom{j}{i} i^k \\
&= \sum_{i=0}^{k} (-1)^{n-i} i^k \sum_{j=0}^{k} \binom{j}{i} \\
&= \sum_{i=0}^{k} (-1)^{n-i} i^k \binom{k+1}{i+1} = (-1)^{n+k}.
\end{aligned}$$

The penultimate equation follows from a basic identity for the binomial coefficients. The last equation follows from the identity for divided differences (see, e.g., Stanley[1986]):

$$\sum_{i=0}^{k+1} (-1)^{k+1-i} \binom{k+1}{i} f(i) = 0$$

for a polynomial of degree at most k, in particular for the polynomial $f(x) = (x-1)^k$. □

The next identity, obtained by interchanging the determinant and permanent in (9.26), can be proved in a similar way.

Theorem 9.4.4. *The matrix A and diagonal matrix Y of order n satisfy*

$$\text{per}(A+Y) = \sum_{S \subseteq \{1,2,\ldots,n\}} \sum_{\alpha \in \Lambda_S} (-1)^{|S|+p_\alpha} \left(\prod_{i \in \overline{S}} y_i \right) \det_\alpha(A). \qquad (9.29)$$

The following recurrence relation involving both the determinant and the permanent was established by Muir[1897].

Theorem 9.4.5. *Let A be a matrix of order n. Then*

$$\sum_{S\subseteq\{1,2,\ldots,n\}} (-1)^{|S|}\,\mathrm{per}\,(A[\overline{S}])\det(A[S]) = 0. \qquad (9.30)$$

Proof. Using the identity (9.27) in the left-hand side of (9.30), we obtain

$$\sum_{S\subseteq\{1,2,\ldots,n\}}\ \sum_{\alpha\in\Lambda_S} (-1)^{p_\alpha}\mathrm{per}(A[\overline{S}])\mathrm{per}_\alpha(A)$$

$$= \sum_{\alpha\in\Lambda_{\{1,2,\ldots,n\}}} (-1)^{p_\alpha}\mathrm{per}_\alpha(A) + \sum_{S\subset\{1,2,\ldots,n\}}\ \sum_{\alpha\in\Lambda_S}(-1)^{p_\alpha}\mathrm{per}(A[\overline{S}])\mathrm{per}_\alpha(A)$$

$$= \sum_{\alpha\in\Lambda_{\{1,2,\ldots,n\}}} (-1)^{p_\alpha}\mathrm{per}_\alpha(A) + \sum_{\alpha\in\Lambda_{\{1,2,\ldots,n\}}}(-1)^{p_\alpha-1}\mathrm{per}_\alpha(A) = 0. \qquad \square$$

Further identities for $\mathrm{per}(I_n - AY)^{-1}$ and $\det(I_n - AY)^{-1}$ and their relation to MacMahon's master theorem can be found in Chu[1989]. We also mention the identity of Vere-Jones[1988] for arbitrary powers of $\det(I_n-AY)$.

Exercises

1. Prove Theorem 9.4.4.
2. Let A and B be two matrices of order n. Prove that

$$\mathrm{per}(A)\det(B) = \sum_\sigma \mathrm{sign}(\sigma)\det(A * (P_\sigma B))$$

where the sum extends over all permutations σ of $\{1,2,\ldots,n\}$ and P_σ denotes the permutation matrix of order n corresponding to σ (Muir[1882]).

References

R.A. Brualdi[1977], *Introductory Combinatorics*, Elsevier/North-Holland, New York.

P. Cartier and D. Foata[1969], *Problèmes Combinatoires de Commutation et Réarrangement*, Lec. Notes in Math. No. 85, Springer, Berlin.

W.C. Chu[1985], Some algebraic identities concerning determinants and permanents, *Linear Alg. Applics.*, 116, pp. 35–40.

P.A. MacMahon[1915], *Combinatory Analysis*, vol. 1, Cambridge University Press, Cambridge.

T. Muir[1882], On a class of permanent symmetric functions, *Proc. Roy. Soc. Edinburgh*, 11, pp. 409–418.

[1897], A relation between permanents and determinants, *Proc. Roy. Soc. Edinburgh*, 22, pp. 134–136.

R.P. Stanley[1986], *Enumerative Combinatorics*, vol. I, Wadsworth & Brooks/ Cole, Monterrey.

D. Vere-Jones[1984], An identity involving permanents, *Linear Alg. Applics.*, 63,
    pp. 267–270.
    [1988], A generalization of permanents and determinants, *Linear Alg. Applics.*,
    111, pp. 119–124.
D. Zeilberger[1985], A combinatorial approach to matrix algebra, *Discrete Math.*,
    56, pp. 61–72.

9.5   The Formal Adjacency Matrix

Let G denote a graph of order n with vertex set

$$V = \{1, 2, \ldots, n\}. \tag{9.31}$$

As defined in section 2.2 the adjacency matrix $A = [a_{ij}], (i, j = 1, 2, \ldots, n)$
of G is a symmetric (0,1)-matrix of order n of trace zero. Row i of A displays
the vertices which are adjacent to vertex i. We now let

$$Z = [z_{ij}], \quad (i, j = 1, 2, \ldots, n)$$

be a skew-symmetric matrix of order n whose elements $z_{ij}, (1 \leq i < j \leq n)$
above the main diagonal are independent indeterminates over the field F.
Since Z is skew-symmetric we have $z_{ii} = 0, (i = 1, 2, \ldots, n)$ and $z_{ij} =
-z_{ji}, (i, j = 1, 2, \ldots, n)$. The Hadamard product

$$M = A * Z = [a_{ij}z_{ij}], \quad (i, j = 1, 2, \ldots, n)$$

is called the *formal adjacency matrix* of the graph G. The nonzero elements
of the skew-symmetric matrix M above the main diagonal are independent
indeterminates over F. We call such a matrix a *generic skew-symmetric
matrix* with respect to the field F. Every generic skew-symmetric matrix is
the formal adjacency matrix of a graph. We let F^* denote the polynomial
ring obtained by adjoining the nonzero elements above the main diagonal
of M to F.

If G is a bipartite graph, the formal adjacency matrix of G can be taken
in the form

$$M = \begin{bmatrix} O & M_1 \\ -M_1^T & O \end{bmatrix}. \tag{9.32}$$

The matrix M_1 is a generic matrix of size k by l for some integers k and
l with $k + l = n$. Conversely, a generic matrix M_1 determines a formal
adjacency matrix of a bipartite graph by means of the equation (9.32).

Let $M = [m_{ij}]. (i, j = 1, 2, \ldots, n)$ be a generic skew-symmetric matrix of
order n. Then $M^T = -M$ and hence $\det(M) = (-1)^n \det(M)$. Hence if n
is odd then $\det(M) = 0$ and M is a singular matrix. We henceforth assume
that n is even.

Let K_n denote the complete graph with vertex set (9.31). Let $\alpha = \{i, j\}$ be an edge of K_n where $i < j$. We define the *weight* of α by

$$\text{wt } \alpha = \text{wt}\{i, j\} = m_{ij}, \quad (1 \le i < j \le n).$$

Let

$$L = \{\{i_1, i_2\}, \{i_3, i_4\}, \ldots, \{i_{n-1}, i_n\}\}$$

be a perfect matching (1-factor) of K_n. To standardize the notation, we assume that

$$i_1 < i_2, i_3 < i_4, \ldots, i_{n-1} < i_n; i_1 < i_3 < \cdots < i_{n-1}.$$

To the perfect matching L we let correspond the permutation

$$\pi_L = (i_1, i_2, i_3, i_4, \ldots, i_{n-1}, i_n)$$

of $\{1, 2, \ldots, n\}$. The weight of the perfect matching (1-factor) L is defined to be

$$\text{wt}(L) = (\text{sign}\pi_L) \prod_{\alpha \in L} \text{wt}(\alpha),$$

the signed product of the weights of its edges. Since M is skew-symmetric, the weight of a perfect matching L depends only on L and not on the order in which the vertices are listed in the edges or the order in which its edges are listed. We denote the set of perfect matchings in K_n by $\mathcal{F}$. The size of the set $\mathcal{F}$ is $1 \cdot 3 \cdot 5 \cdots (n - 1)$.

The classical definition of the *pfaffian* of the matrix M is equivalent to the following:

$$\text{pf}(M) = \sum_{L \in \mathcal{F}} \text{wt}(L), \tag{9.33}$$

the sum of the weights of the perfect matchings of K_n. The pfaffian of M is a linear form in the indeterminates m_{ij} that appear in M and is homogeneous of degree $n/2$. Since the weight of a perfect matching in $\mathcal{F}$ is zero if it contains an edge not belonging to the graph G, it suffices in (9.33) to sum only over the perfect matchings in G. The following theorem is now a consequence of the definitions involved.

Theorem 9.5.1.   *The graph G has a perfect matching if and only if the pfaffian of its formal adjacency matrix is not identically zero.*

Pfaffians occupy an obscure position in matrix theory. The pfaffian of the skew-symmetric matrix $M = [m_{ij}]$ of order $n = 4$ is

$$m_{12}m_{34} - m_{13}m_{24} + m_{14}m_{23},$$

the three terms corresponding, respectively, to the three perfect matchings of K_n

$$\{\{1,2\},\{3,4\}\},\{\{1,3\},\{2,4\}\},\{\{1,4\},\{2,3\}\}.$$

Unlike determinants, pfaffians lack a simple multiplication formula. Their usefulness in combinatorial problems is demonstrated by the original proof of Tutte[1947] of his theorem for the existence of a perfect matching in a graph (cf. Theorem 2.6.1) and the solution by Kasteleyn[1961] of the dimer problem.

We now present the proof of Kasteleyn[1967] of the theorem of Cayley[1854] that the determinant of a skew-symmetric matrix is the square of its pfaffian.

Theorem 9.5.2.   *The skew-symmetric matrix M of even order n satisfies*

$$\det(M) = (\mathrm{pf}(M))^2.$$

Proof. Let $\mathcal{I}$ denote the set of permutations of $\{1, 2, \ldots, n\}$ which contain at least one (permutation) cycle of odd length, and let $\mathcal{J}$ denote the set of permutations of $\{1, 2, \ldots, n\}$ each of whose cycles has even length. Since n is even, (9.3) implies that

$$\det(M) = \sum_{\pi \in \mathcal{I}} \mathrm{wt}\,(D(\pi)) + \sum_{\pi \in \mathcal{J}} \mathrm{wt}\,(D(\pi)). \qquad (9.34)$$

We show that the first summation in (9.34) equals zero and the second summation equals $(\mathrm{pf}(M))^2$.

We define a mapping

$$\sigma : \mathcal{I} \to \mathcal{I}$$

as follows. Let π be a permutation in $\mathcal{I}$, and consider the cycle of π of odd length which contains the smallest integer $i \in \{1, 2, \ldots, n\}$. Let $\sigma(\pi)$ be the permutation obtained from π by reversing the direction of that cycle. Then $\sigma(\pi)$ is in $\mathcal{I}$. Moreover, since M is skew-symmetric, it follows that

$$\mathrm{wt}\,(D(\pi)) = -\mathrm{wt}\,(D(\sigma(\pi))).$$

We have $\sigma(\sigma(\pi)) = \pi$, and hence σ is an involution on $\mathcal{I}$ which is sign-reversing on the weight. [Note that if the cycle defined above has length one, then $\sigma(\pi) = \pi$ and $\mathrm{wt}\,D((\pi)) = 0$ follows since the main diagonal elements of M are all equal to zero.] We conclude that

$$\sum_{\pi \in \mathcal{I}} \mathrm{wt}\,(D(\pi)) = 0.$$

We now define a mapping

$$\tau : \mathcal{F} \times \mathcal{F} \to \mathcal{J}$$

for which

$$\text{wt } (L_1)\text{wt } (L_2) = \text{wt } (\tau((L_1, L_2))), (L_1, L_2 \in \mathcal{F}).$$

Let L_1 and L_2 be perfect matchings in $\mathcal{F}$. We define a multigraph $G(L_1, L_2)$ with vertex set $\{1, 2, \ldots, n\}$ in which $\{i, j\}$ is an edge of multiplicity 1 or 2 according as $\{i, j\}$ is an edge of exactly one or both of L_1 and L_2. Each vertex of the multigraph $G(L_1, L_2)$ has degree equal to 2, and, since L_1 and L_2 are perfect matchings, $G(L_1, L_2)$ decomposes into $t \geq 1$ cycles γ_j of even length $k_j, (j = 1, 2, \ldots, t)$ no two of which pass through the same vertex. This same collection of cycles results from 2^t pairs (L_1, L_2) of perfect matchings in $\mathcal{F}$. Each of the cycles γ_j can be oriented in two ways and this gives us 2^t permutations σ in $\mathcal{J}$ each of which has sign equal to $(-1)^t$. Each permutation in $\mathcal{J}$ arises in this way and this gives us a one-to-one corrrespondence τ between $\mathcal{F} \times \mathcal{F}$ and $\mathcal{J}$. It follows from the definition of weight of perfect matchings and the definition of weight of permutation digraphs that

$$(\text{wt } F_1)(\text{wt } F_2) = \pm\text{wt } (D(\sigma)).$$

However, by an appropriate choice of the listing of the edges of L_1 and L_2 and an appropriate choice of the listing of the vertices in their edges, we obtain that

$$\pi(L_1)\pi(L_2)^{-1} = \sigma$$

and hence

$$\text{sign}(\pi(L_1))\text{sign}(\pi(L_2)) = \text{sign } \sigma.$$

Therefore

$$\text{wt } (F_1)\text{wt } (F_2) = \text{wt } (D(\sigma))$$

and hence $(\text{pf}(M))^2 = \det(M)$. $\square$

We now turn to Tutte's algebraic proof of the theorem characterizing graphs with a perfect matching. We first derive some consequences of Lemma 9.2.10, the identity of Jacobi, for a skew-symmetric matrix $M = [m_{ij}]$ of even order n. Let $M_{i,j}$ denote the matrix obtained from M by deleting row i and column j and let $C_{i,j} = (-1)^{i+j} \det(M_{j,i})$ be the cofactor of the element m_{ij} of $M, (i, j = 1, 2, \ldots, n)$. Each principal submatrix of M is a skew-symmetric matrix. Let $M(i_1, i_2, \ldots, i_k)$ denote the principal submatrix of M obtained by deleting rows and columns $i_1, i_2, \ldots, i_k$. If k is odd, then $\det(M(i_1, i_2, \ldots, i_k)) = 0$. If k is even, then $\det(M(i_1, i_2, \ldots, i_k)) = (\text{pf}(M(i_1, i_2, \ldots, i_k)))^2$. We note that $M_{i,i} = M(i)$. It follows from Lemma 9.2.10 and the skew-symmetry of M that for $i \neq j$,

$$\det(M) \det(M(i, j)) = \det(M(i)) \det(M(j)) - C_{i,j}C_{j,i} = (C_{i,j})^2, \quad (9.35)$$

and

$$(\det(M))^3 \det(M(i,j,k,l)) = \det \begin{bmatrix} 0 & C_{i,j} & C_{i,k} & C_{i,l} \\ -C_{i,j} & 0 & C_{j,k} & C_{j,l} \\ -C_{i,k} & -C_{j,k} & 0 & C_{k,l} \\ -C_{i,l} & -C_{j,l} & -C_{k,l} & 0 \end{bmatrix}$$

$$= (C_{i,j}C_{k,l} - C_{i,k}C_{j,l} + C_{i,l}C_{j,k})^2, \ (i,j,k,l \ \text{distinct}). \qquad (9.36)$$

Applying (9.35) and (9.36), we obtain

$$\mathrm{pf}(M)\mathrm{pf}(M(i,j,k,l)) = \pm\mathrm{pf}(M(i,j))\mathrm{pf}(M(k,l))$$
$$\pm\mathrm{pf}(M(i,k))\mathrm{pf}(M(j,l)) \pm \mathrm{pf}(M(i,l))\mathrm{pf}(M(j,k)), \qquad (9.37)$$

for all distinct i, j, k and l. The actual choice of signs in (9.37) is of no consequence in what follows.

Next we recall some notation from Chapter 2. If S is a subset of the set $V = \{1, 2, \ldots, n\}$ of vertices of the graph G of order n, then $G(V - S)$ is the induced subgraph obtained from G by removing the vertices in S and all incident edges. The set of connected components of $G(V - S)$ with an odd number of vertices is denoted by $\mathcal{C}(G; S)$ and the cardinality of $\mathcal{C}(G; S)$ is $p(G; S)$.

Theorem 9.5.3.   *The graph G has a perfect matching if and only if*

$$p(G; S) \leq |S|, \ \text{for all } S \subseteq V. \qquad (9.38)$$

Proof. The condition (9.38) is clearly a necessary condition for the existence of a perfect matching. Now assume that G does not have a perfect matching. We use Lemma 9.2.10, the identity of Jacobi, to show that there exists a set S_0 of vertices for which (9.38) does not hold. If n is odd, we may choose $S_0 = \emptyset$. We now suppose that n is even. By Theorems 9.5.1 and 9.5.2 the formal adjacency matrix $M = [m_{ij}]$ of G satisfies

$$\det(M) = \mathrm{pf}(M) = 0.$$

If $\{i, j\}$ is an edge of G, then the graph $G(V - \{i, j\})$ cannot have a perfect matching and hence $\mathrm{pf}(M(i,j)) = 0$. A vertex k of G is called a *singularity* of G provided that for all vertices $i \neq k$ the graph $G(V - \{i, k\})$ does not have a perfect matching. Suppose that there is a chain i, j, k joining distinct vertices i and j such that j is not a singularity. Then there exists a vertex l different from i, j and k such that $G(V - \{j, l\})$ has a perfect matching. Then it follows from (9.37) that

$$\mathrm{pf}(M(i,k))\mathrm{pf}(M(j,l)) = 0.$$

Since $\mathrm{pf}(M(j,l)) \neq 0$, we must have $\mathrm{pf}(M(i,k)) = 0$. Hence the graph $G(V - \{i,k\})$ does not have a perfect matching. We now conclude that if i and j are distinct vertices of G which are joined by a chain each of whose interior vertices is *not* a singularity, then the graph $G(V - \{i,j\})$ does not have a perfect matching.

Suppose that there are distinct vertices i and j such that $\{i,j\}$ is not an edge of G but $G(V - \{i,j\})$ does not have a perfect matching. Adding the edge $\{i,j\}$ to G we obtain a graph which, like G, does not have a perfect matching. It follows that there exists a graph G' of order n of which G is a (spanning) subgraph such that (i) G' does not have a perfect matching, and (ii) $G(V - \{i,j\})$ has a perfect matching if and only if $\{i,j\}$ is not an edge of $G', i \neq j$.

For each set S of vertices we have

$$p(G, S) \geq p(G', S).$$

Hence it suffices to show there exists a set S_0 of vertices for which

$$p(G', S_0) > |S|. \tag{9.39}$$

Let S_0 be the set consisting of those vertices which are singularities of G'. Then each pair of vertices, at least one of which is in S_0, forms an edge of G'. Moreover, it follows from the above arguments that the connected components of the graph $G(V - S_0)$ are complete graphs. Since G' does not have a perfect matching (9.39) holds, and the proof of the theorem is complete.   □

Lovász[1979] has observed that Theorem 9.5.1 provides a random algorithm for deciding whether a graph of even order has a perfect matching. Let the number of edges of G be m. If a randomly generated m-tuple of real numbers does not cause the pfaffian of the formal adjacency matrix of G to vanish, then with probability one G has a perfect matching. Pla[1965] (see also Little and Pla[1972]) and Gibson[1972] have used pfaffians in order to count the number of perfect matchings of a graph. Identities involving pfaffians are given in Heymans[1969] and Lieb[1968].

Exercises

1. Verify that the weight of a perfect matching depends neither on the order in which the vertices are listed in the edges nor on the order in which the edges are listed.
2. Let $\alpha = (a_1, a_2, \ldots, a_p)^T$ and $\beta = (b_1, b_2, \ldots, b_p)^T$ be two real p-vectors. Define their *wedge product* $\alpha \wedge \beta$ to be the skew-symmetric matrix of order p for which the element in position (i,j) equals $a_i b_j - a_j b_i, (1 \leq i, j \leq p)$.

Let $\alpha_1, \alpha_2, \ldots, \alpha_m, \beta_1, \beta_2, \ldots, \beta_m$ be $2m$ real $2n$-vectors. Prove that

$$\text{pf}\left(\sum_{k=1}^{m} x_k(\alpha_k \wedge \beta_k)\right)$$

is linear in each of the variables $x_1, x_2, \ldots, x_m$.

3. (Continuation of Exercise 2) Prove that there exist n pairs of vectors $\{\alpha_k, \beta_k\}$ whose union is a linearly independent set if and only if

$$\det\left(\sum_{k=1}^{m} x_k(\alpha_k \wedge \beta_k)\right)$$

is not identically 0 in the variables $x_1, x_2, \ldots, x_m$ (Lovász[1979]).

4. Prove that the square of

$$\det\begin{bmatrix} 1 & 1 & 1 & \cdots & 1 \\ x_1 & x_2 & x_3 & \cdots & x_n \\ \vdots & \vdots & \vdots & \cdots & \vdots \\ x_1^{n-1} & x_2^{n-1} & x_3^{n-1} & \cdots & x_n^{n-1} \end{bmatrix}$$

equals

$$\det\begin{bmatrix} s_0 & s_1 & s_2 & \cdots & s_{n-1} \\ s_1 & s_2 & s_3 & \cdots & s_n \\ \vdots & \vdots & \vdots & \cdots & \vdots \\ s_{n-1} & s_n & s_{n+1} & \cdots & s_{2n-2} \end{bmatrix},$$

where $s_k = x_1^k + x_2^k + \cdots + x_n^k, (k \geq 0)$.

References

A. Cayley[1854], Sur les déterminantes gauches, *Crelle's J.*, 38, pp. 93–96.

P.M. Gibson[1972], The pfaffian and 1-factors of graphs, *Trans. N.Y. Acad. Sci.*, 34, pp. 52–57.

[1972], The pfaffian and 1-factors of graphs II, *Graph Theory and Applications*, Y. Alavi, D.R. Lick and A.T. White, eds., Lec. Notes in Mathematics, No. 303, pp. 89–98, Springer-Verlag, New York.

P. Heymans[1969], Pfaffians and skew-symmetric matrices, *Proc. London Math. Soc.* (3), 19, pp. 730–768.

P.W. Kasteleyn[1961], The statistics of dimers on a lattice, *Physica*, 27, pp. 1209–1225.

[1967], Graph theory and crystal physics, *Graph Theory and Theoretical Physics* (F. Harary, ed.), Academic Press, New York, pp. 43–110.

E.H. Lieb[1968], A theorem on Pfaffians, *J. Combin. Theory*, 5, pp. 313–368.

C.H.C. Little and J.M. Pla[1972], Sur l'utilisation d'un pfaffien dans l'étude des couplages parfaits d'un graphe, *C. R. Acad. Sci. Paris*, 274, p. 447.

L. Lovász[1979], On determinants, matchings, and random algorithms, *Proceedings of Fundamentals of Computation Theory*, Akademie Verlag, Berlin, pp. 565–574.

J.M. Pla[1965], Sur l'utilisation d'un pfaffien dans l'étude des couplages parfaits d'un graphe, *C. R. Acad. Sci. Paris*, 260, pp. 2967–2970.

W.T. Tutte[1947], The factorization of linear graphs, *J. London Math. Soc.*, 22, pp. 107–111.

9.6 The Formal Laplacian Matrix

In this section we consider digraphs D of order n with vertex set

$$V = \{1, 2, \ldots, n\}$$

and we assume that D has no loops. The possible lengths of directed cycles in such a digraph D are $2, 3, \ldots, n$. A *spanning arborescence*[1] of D is a subdigraph D' of D with vertex set V with the two properties: (i) D' has no directed cycles and (ii) there exists a vertex u such that every vertex different from u has outdegree equal to one in D'. The vertex u is called the *root* of D' and we say that D' is *rooted at u*. The collection of spanning arborescences of D which are rooted at u is denoted by $\Lambda_u(D)$.

Let G be a graph with vertex set V, and let $\overleftrightarrow{G}$ denote the digraph obtained from G by replacing each edge $\{i, j\}$ with the two oppositely directed arcs (i, j) and (j, i). The adjacency matrix of the graph G equals the adjacency matrix of the digraph $\overleftrightarrow{G}$. Let u be a vertex in V. The spanning arborescences of $\overleftrightarrow{G}$ rooted at u are obtained from the spanning trees T of G by orienting the edges of T toward the vertex u. The number of spanning arborescences of $\overleftrightarrow{G}$ rooted at u is thus the number $c(G)$ of spanning trees of G (the complexity of G) and hence does not depend on the choice of vertex u. In section 2.5 we showed that the adjugate of the Laplacian matrix of G is a constant matrix with each element equal to the complexity of G.

Let

$$Z = [z_{ij}], \quad (i, j = 1, 2, \ldots, n)$$

be a matrix of order n whose elements are independent indeterminates over the field F. Let

$$A = [a_{ij}], \quad (i, j = 1, 2, \ldots, n)$$

be the adjacency matrix of the digraph D of order n. Since D has no loops, each main diagonal element of A is zero. The Hadamard product

$$M = [m_{ij}] = A * Z = [a_{ij} z_{ij}]$$

is called the *formal adjacency matrix of the digraph D*. Any generic matrix of order n with main diagonal elements equal to zero may serve as the

---

[1] In section 3.7 we defined a spanning directed tree with root r of a digraph D. In a spanning arborescence with root u the arcs are directed toward the root; in a spanning directed tree with root r the arcs are directed away from the root.

formal adjacency matrix of a digraph D of order n with no loops. The matrix

$$
L(D) = \begin{bmatrix}
\sum_{j\neq 1} m_{1j} & -m_{12} & \cdots & -m_{1n} \\
-m_{21} & \sum_{j\neq 2} m_{2j} & \cdots & -m_{2n} \\
\vdots & \vdots & \ddots & \vdots \\
-m_{n1} & -m_{n2} & \cdots & \sum_{j\neq n} m_{nj}
\end{bmatrix}
$$

is the *formal Laplacian matrix of the digraph* D. The determinant of the formal Laplacian is identically zero. The matrix obtained from $L(D)$ by setting each of the nonzero elements $m_{ij}, (i \neq j)$ equal to 1 is the *Laplacian matrix* of the digraph D. If D is the digraph $\overleftrightarrow{G}$ then the Laplacian matrix of D is the Laplacian matrix of G as defined in section 2.3.

The *matrix tree theorem* asserts that the cofactor of the uth diagonal element of the Laplacian matrix of D equals the number of spanning arborescences of D which are rooted at vertex u. This theorem was apparently first proved by Borchardt[1860] and independently by Tutte[1948]. In its more general form it gives the weight of the arborescences of D which are rooted at u. Combinatorial proofs of the theorem have been given by Orlin[1978] and Chaiken[1982]. These proofs use the inclusion-exclusion principle. We follow the combinatorial proof of Zeilberger[1985].

Let $\overleftrightarrow{K}_n$ denote the complete digraph (with no loops) of order n. The digraph D of order n is a spanning subdigraph of $\overleftrightarrow{K}_n$. Let $M = [m_{ij}]$ be the formal adjacency matrix of order n of the digraph D. The *weight of an arc* (i, j) of $\overleftrightarrow{K}_n$ is defined to be

$$
\text{wt}(i, j) = m_{ij}.
$$

Thus if (i, j) is not an arc of D then its weight is equal to zero. The *weight* $\text{wt}(H)$ *of a subdigraph* H of $\overleftrightarrow{K}_n$ is defined to be the product of the weights of its arcs. If H is not a subdigraph of D its weight is equal to zero.

Theorem 9.6.1.   *The determinant of the principal submatrix* $L(D)(u)$ *of the formal Laplacian matrix* $L(D)$ *obtained by deleting row* u *and column* u *equals*

$$
\sum_{T \in \Lambda_u(\overleftrightarrow{K}_n)} \text{wt}(T) = \sum_{T \in \Lambda_u(D)} \text{wt}(T),
$$

the sum of the weights of the spanning arborescences of D *which are rooted at* u.

Proof. Without loss of generality we assume that $u = n$. Let $\mathcal{D}$ be the set of pairs (D_1, D_2) of digraphs D_1 and D_2 such that D_1 is a (possibly empty) set of vertex-disjoint directed cycles of length at least 2 whose vertices form a subset X of $\{1, 2, \ldots, n-1\}$, and D_2 is a digraph with vertex set $V = \{1, 2, \ldots, n\}$ and with $n - 1 - |X|$ arcs, one issuing from each vertex in $\overline{X} = \{1, 2, \ldots, n-1\} - X$ (the arcs may terminate at any vertex in V). The weight of (D_1, D_2) is defined by

$$\text{wt}(D_1, D_2) = (-1)^k \text{wt}(D_1)\text{wt}(D_2),$$

where k is the number of directed cycles of D_1. It follows from formula (9.3) applied to $L(D)(n)$ that

$$\det(L(D)(n)) = \text{wt}\mathcal{D} = \sum_{(D_1, D_2) \in \mathcal{D}} \text{wt}(D_1, D_2).$$

Let $\mathcal{D}_0$ be the subset of $\mathcal{D}$ consisting of those pairs (D_1, D_2) for which at least one of D_1 and D_2 has a directed cycle. We define a mapping $\sigma : \mathcal{D}_0 \rightarrow \mathcal{D}_0$ as follows. Let (D_1, D_2) be in $\mathcal{D}_0$. Let γ be the directed cycle containing the smallest numbered vertex of all directed cycles in D_1 and D_2. If γ is a directed cycle of D_1 we remove the arcs of γ from D_1 and put them in D_2 resulting in digraphs D_1' and D_2' for which (D_1', D_2') is in $\mathcal{D}_0$. If γ is a directed cycle in D_2 then we move the arcs of γ from D_2 to D_1 and obtain a pair (D_1', D_2') in $\mathcal{D}_0$. We define $\sigma(D_1, D_2) = (D_1', D_2')$ and observe that σ is an involution which is sign-reversing on the weight. Hence

$$\sum_{(D_1, D_2) \in \mathcal{D}_0} \text{wt}(D_1, D_2) = 0$$

and therefore

$$\det(L(D)(n)) = \sum_{(D_1, D_2) \in \mathcal{D} - \mathcal{D}_0} \text{wt}(D_1, D_2).$$

But a pair (D_1, D_2) belongs to $\mathcal{D} - \mathcal{D}_0$ if and only if D_1 is an empty graph $(X = \emptyset)$ and D_2 is an arborescence rooted at vertex n. Moreover, for such (D_1, D_2) we have

$$\text{wt}(D_1, D_2) = \text{wt}(D_2),$$

and the theorem now follows. $\square$

A combinatorial interpretation of the determinant of each square sub-matrix of the formal Laplacian matrix has been given by Chen[1976] and Chaiken[1982].

Exercises

1. Prove that a digraph has a spanning arborescence if and only if it satisfies the following property: For each pair of vertices a and b there is a vertex c (possibly c equals a or b) such that there are directed walks from a to c and from b to c.

2. Let D be a digraph of order n which has no directed cycles. Prove that all spanning arborescences of D (if any) have the same root.

3. (Continuation of Exercise 2) Let D be a digraph of order n which has no directed cycles. Let u be a vertex of D such that there is a spanning arborescence of D rooted at u. Determine the sum of the weights of the spanning arborescences of D rooted at u. Show that the number of spanning arborescences of D which are rooted at u equals the product of the outdegrees of all vertices different from u.

References

C.W. Borchardt[1860], Über eine der Interpolation entsprechende Darstellung der Eliminations-Resultante, *J. Reine Angew. Math.*, 57, pp. 111–121.

S. Chaiken[1982], A combinatorial proof of the all minors matrix tree theorem, *SIAM J. Alg. Disc. Meth.*, 3, pp. 319–329.

J.B. Orlin[1978], Line-digraphs, arborescences, and theorems of Tutte and Knuth, *J. Combin. Theory, Ser. B*, 25, pp. 187–198.

H.N.V. Temperley[1981], *Graph Theory and Applications*, Ellis Horwood, Chichester.

W.T. Tutte[1948], The dissection of equilateral triangles into equilateral triangles, *Proc. Cambridge Philos. Soc.*, 44, pp. 463–482.

D. Zeilberger[1985], A combinatorial approach to matrix algebra, *Discrete Math.*, 56, pp. 61–72.

9.7   Polynomial Identities

In this section we let R denote a commutative ring with identity 1 and consider the ring $M_n(R)$ of all matrices of order n with elements in R. Let $x_1, x_2, \ldots, x_k$ be k independent *noncommuting* indeterminates over R, and let

$$f(x_1, x_2, \ldots, x_k)$$

be a polynomial in the polynomial ring $R[x_1, x_2, \ldots, x_k]$. The polynomial $f(x_1, x_2, \ldots, x_k)$ is called an *identity* of $M_n(R)$ provided that

$$f(A_1, A_2, \ldots, A_k) = O$$

for all matrices $A_1, A_2, \ldots, A_k$ in $M_n(R)$. The general theory of polynomial identities is developed in Rowen[1980]. Our goal in this section is to show how combinatorial arguments can be used to prove three important identities of $M_n(R)$. One of these is a polynomial identity known as the *standard polynomial identity* of $M_n(R)$. Another is a trace identity known as the *fundamental trace identity* of $M_n(R)$. The third is the identity known as the *Cayley–Hamilton theorem*. Although the Cayley–Hamilton theorem asserts that a certain polynomial of one variable x yields a zero matrix upon replacing x by a matrix A, the coefficients of the polynomial depend on the elements of A and thus the polynomial is not a polynomial identity as defined above.

Let $A = [a_{ij}], (i, j = 1, 2, \ldots, n)$ be a matrix in $M_n(R)$, and let x be an indeterminate over R. The *characteristic polynomial* of A is the monic polynomial in $R[x]$ of degree n given by

$$\chi_A(x) = \det(xI_n - A) = x^n + \sigma_1 x^{n-1} + \cdots + \sigma_k x^{n-k} + \cdots + \sigma_{n-1} x + \sigma_n,$$

where σ_k equals the sum of the determinants of all the principal submatrices of $-A$ of order $k, (k = 1, 2, \ldots, n)$. Rutherford[1964] first discovered a combinatorial proof of the Cayley–Hamilton theorem. This proof was rediscovered by Straubing[1983] and given an exposition by Zeilberger[1985] and Brualdi[1990].

Theorem 9.7.1. *The matrix A of order n satisfies its characteristic polynomial $\chi_A(x)$, that is,*

$$A^n + \sigma_1 A^{n-1} + \cdots + \sigma_k A^{n-k} + \cdots + \sigma_{n-1} A + \sigma_n I_n = O. \qquad (9.40)$$

Proof. We prove (9.40) by showing that each element of the matrix in the left-hand side equals 0. We assign the weight

$$\mathrm{wt}(i, j) = a_{ij}, \quad (i, j = 1, 2, \ldots, n)$$

to the arcs (i, j) of the complete digraph D_n of order n, and, as in section 9.1, we use these weights to assign weights to the directed cycles of D_n and to the permutation digraphs whose vertices form a subset of $\{1, 2, \ldots, n\}$. We also define the weight of a walk to be the product of the weights of its arcs.

It follows from (9.3) [see also (9.4)] that σ_k equals the sum of the weights of all permutation digraphs whose vertices form a subset of size k of $\{1, 2, \ldots, n\}$. The element in the (i, j) position of A^{n-k} equals the sum of the weights of all walks in D_n of length $n - k$ from vertex i to vertex j. Let Ω_{ij} be the set of all ordered pairs (γ, π) for which γ is a walk in D_n from i to j of length at most n, π is a collection of vertex disjoint directed cycles in D_n and the number of arcs of γ plus the number of arcs of π equals n. We assign a weight to each pair (γ, π) in Ω_{ij} by

$$\mathrm{wt}(\gamma, \pi) = (-1)^t \times (\text{product of all arc weights of } \gamma \text{ and of } \pi),$$

where t is the number of directed cycles in π. The element in position (i, j) of the matrix on the left-hand side of equation (9.40) equals

$$\mathrm{wt}(\Omega_{ij}) = \sum_{(\gamma, \pi) \in \Omega_{ij}} \mathrm{wt}(\gamma, \pi).$$

Hence the Cayley–Hamilton theorem asserts that

$$\mathrm{wt}(\Omega_{ij}) = 0, \quad (i, j = 1, 2, \ldots, n). \qquad (9.41)$$

Let i and j be integers with $1 \leq i, j \leq n$. We define a mapping $\tau : \Omega_{ij} \rightarrow \Omega_{ij}$ as follows. Let (γ, π) be in Ω_{ij}. Since the total number of arcs in γ and π equals the number n of vertices in D_n, either there is a vertex of γ which is also a vertex of one of the directed cycles in π or the directed walk γ contains a repeated vertex and hence "contains" a directed cycle (possibly both). We walk along γ until we first arrive at a vertex u which has been previously visited or is a vertex of a directed cycle in π. (Note that these two events cannot occur simultaneously.) In the first instance, γ contains a directed cycle γ_0 whose vertices are disjoint from the vertices of the directed cycles in π; we then "remove" γ_0 from γ and include γ_0 in π. In the second instance, π contains a directed cycle π_0 one of whose vertices is a vertex of γ; we remove π_0 from π and join it to γ. In each instance we obtain a pair (γ', π') which belongs to Ω_{ij}. We define $\tau(\gamma, \pi) = (\gamma', \pi')$. The mapping τ is sign-reversing on weight and we have $\tau(\gamma', \pi') = (\gamma, \pi)$. Hence τ is a sign-reversing involution on Ω_{ij} and hence (9.41) holds. $\square$

We next turn to the fundamental trace identity. The *standard polynomial* of degree k is the polynomial in $R[x_1, x_2, \ldots, x_k]$

$$[x_1, x_2, \ldots, x_k]_k = \sum_{\pi} (\text{sign}\pi) x_{\pi(1)} x_{\pi(2)} \cdots x_{\pi(k)}$$

where the summation extends over all permutations π of $\{1, 2, \ldots, k\}$. The standard polynomials satisfy the relations

$$[x_1, x_2, \ldots, x_k]_k = \sum_{i=1}^{k} (-1)^{i-1} x_i [x_1, \ldots, x_{i-1}, x_{i+1}, \ldots, x_k]_{k-1}, \quad (9.42)$$

and

$$[x_1, x_2, \ldots, x_k]_k = \sum_{i=1}^{k} (-1)^{k-i} [x_1, \ldots, x_{i-1}, x_{i+1}, \ldots, x_k]_{k-1} x_i, \quad (9.43)$$

where we adopt the convention that $[\emptyset]_0 = 1$. We have

$$[x_1, x_2]_2 = x_1 x_2 - x_2 x_1$$

and hence

$$\text{tr}([A_1, A_2]_2) = \text{tr}(A_1 A_2 - A_2 A_1) = 0.$$

Thus $\text{tr}[x_1, x_2]_2$ is a trace identity of $M_n(R)$ for all $n \geq 1$. More generally, it follows from the above relations that

$$\text{tr}[x_1, x_2, \ldots, x_k]_k, \quad (k \geq 2)$$

is a trace identity of $M_n(R)$ for all $n \geq 1$. These identities are sometimes called the *trivial trace identities* since they hold for all n. The following trace identity, which is not trivial in the sense just described, goes back to Frobenius[1896]. It was also discovered by Lew[1966] and Procesi[1976]. We give the combinatorial proof of Laue[1988], which is also presented in Brualdi[1990]. For $n = 2$ and $k = 3$ this identity asserts that if A_1, A_2, A_3 are matrices of order 2 with elements in R, then

$$\text{tr}(A_1 A_2 A_3) + \text{tr}(A_1 A_3 A_2) - \text{tr}(A_1)\text{tr}(A_2 A_3) - \text{tr}(A_2)\text{tr}(A_1 A_3)$$
$$-\text{tr}(A_3)\text{tr}(A_1 A_2) + \text{tr}(A_1)\text{tr}(A_2)\text{tr}(A_3) = 0.$$

Theorem 9.7.2.   *Let k and n be integers with $k > n$ and let $A_1, A_2, \ldots, A_k$ be matrices in $M_n(R)$. Then*

$$\sum_{\pi} \text{sign}(\pi) \prod_{\pi_i} \text{tr}\left(\prod_{p \in \pi_i} A_p\right) = O, \qquad (9.44)$$

where the summation extends over all permutations π of $\{1, 2, \ldots, k\}$.

Proof. First we clarify the expression on the left hand side in (9.44). The first product is over all cycles π_i of the permutation π. The second product is over all elements p of the cycle π_i taken in the cyclical order of π_i.

The ring $M_n(R)$ is actually an R-module, and the function on the left-hand side of (9.44) is an R-multilinear function of its arguments $A_1 A_2, \ldots, A_k$. Hence the validity of (9.44) for all matrices in $M_n(R)$ follows from its validity for all matrices chosen from the standard basis $\{E_{ij} : i, j = 1, 2, \ldots, n\}$ of the R-module $M_n(R)$, where E_{ij} denote the matrix in $M_n(R)$ whose only nonzero element is a 1 in position (i, j).

Let $A_t = E_{i_t j_t}, (t = 1, 2, \ldots, k)$. We define a general digraph D with vertex set $V = \{1, 2, \ldots, n\}$ in which (i, j) is an arc of multiplicity r provided (i, j) occurs r times among the elements $(i_1, j_1), (i_2, j_2), \ldots, (i_k, j_k)$. The general digraph D has n vertices and $k > n$ arcs. Let π be a permutation of $\{1, 2, \ldots, k\}$ and let π_i be a cycle of π. Then $\text{tr}(\prod_{p \in \pi_i} A_p)$ equals 1 if the arcs $(i_p, j_p), p \in \pi_i$, taken in the cyclical order of π_i, determine a directed closed trail of D and equals 0 otherwise. Hence

$$\prod_{\pi_i} \text{tr}\left(\prod_{p \in \pi_i} A_p\right)$$

equals 1 if the cycles π_i of π determine a partition θ of the arcs of D into directed closed trails, and equals 0 otherwise. (We note that each permutation π of $\{1, 2, \ldots, k\}$ determines a permutation of the arcs of D, but only some of the permutations determine a partition of the arcs into directed closed trails.) It now follows that (9.44) is equivalent to the following combinatorial statement:

(∗) If D is a general digraph of order n with $k > n$ arcs, then the number of partitions of the arcs of D into directed closed trails which correspond to even permutations of $\{1, 2, \ldots, k\}$ equals the number of partitions which correspond to odd permutations of $\{1, 2, \ldots, k\}$.

We now verify $(*)$. Since D has more arcs than vertices, there are two arcs α and β with the same initial vertex u (α and β might also have the same terminal vertex).

Let Θ be the set of partitions of the arcs of D into directed closed trails. We define a mapping $\sigma : \Theta \to \Theta$ as follows. Let θ be a partition in Θ. First suppose that α and β are in the same directed closed trail $\theta_{\alpha,\beta}$ of θ. We may assume that the arc α is written first in $\theta_{\alpha,\beta}$:

$$\alpha, \cdots, \gamma, \beta, \cdots, \delta.$$

We then replace θ in Θ with the two directed closed trails

$$\alpha, \cdots, \gamma$$

and

$$\beta, \cdots, \delta,$$

resulting in a partition θ' in Θ. Now suppose that the arcs α and β are in different directed closed trails θ_α and θ_β of θ. We may assume that α is written first in θ_α and β is written first in θ_β. We then follow θ_α with θ_β resulting in a closed directed trail $\theta_{\alpha,\beta}$. In this case we let θ' be the partition in Θ obtained by replacing θ_α and θ_β with $\theta_{\alpha,\beta}$. We define $\sigma(\theta)$ to be θ'. It follows from the construction that σ is a sign-reversing involution on Θ and $(*)$ holds.   □

Procesi[1976] proved that every trace identity is a consequence of the trace identity (9.44) and the trivial trace identities.

We now turn to the theorem of Amitsur and Levitzki[1950,1951]. This theorem asserts that the standard polynomial of degree $2n$

$$[x_1, x_2, \ldots, x_{2n}]_{2n}$$

is a polynomial identity of $M_n(R)$. This implies that matrices of order n satisfy a weakened form of the commutative law. We follow the combinatorial proof of Swan[1963,1969], an exposition of which is also given in Bollobás[1979].

Theorem 9.7.3. *For all matrices $A_1, A_2, \ldots, A_{2n}$ in $M_n(R)$, we have*

$$[A_1, A_2, \ldots, A_{2n}]_{2n} = O. \tag{9.45}$$

Proof. As in the proof of (9.44) we need only verify (9.45) for matrices $A_1, A_2, \ldots, A_{2n}$ chosen from the standard basis $\{E_{ij} : i, j = 1, 2, \ldots n\}$ of

$M_n(R)$. Let $A_t = E_{i_t j_t}, (t = 1, 2, \ldots, 2n)$. Let D be the general digraph with vertex set $V = \{1, 2, \ldots, n\}$ and the $2n$ arcs

$$\alpha_t = (i_t, j_t), \quad (t = 1, 2, \ldots, 2n).$$

Let π be a permutation of $\{1, 2, \ldots, 2n\}$. We have

$$A_{\pi(1)} A_{\pi(2)} \cdots A_{\pi(2n)} = E_{i_{\pi(1)} i_{\pi(2n)}} \qquad (9.46)$$

if

$$\alpha_{\pi(1)}, \alpha_{\pi(2)}, \ldots, \alpha_{\pi(2n)} \qquad (9.47)$$

is a directed trail in D from vertex $i_{\pi(1)}$ to vertex $j_{\pi(2n)}$. Such a trail includes each arc of D exactly once and is called an *Eulerian trail*. The product on the left-hand side of (9.46) equals a zero matrix if (9.47) is not an Eulerian trail. Each Eulerian trail in D is a permutation of the $2n$ arcs $\alpha_1, \alpha_2, \ldots, \alpha_{2n}$ of D. It is convenient to identify the Eulerian trail (9.47) with the permutation π of $\{1, 2, \ldots, 2n\}$. Thus each Eulerian trail has a *sign* which is equal to the sign of the permutation π. Let Γ_{uv} denote the set of permutations π which are Eulerian trails from vertex u to vertex v. Let

$$\epsilon(D : u, v) = \sum_{\pi \in \Gamma_{uv}} \text{sign} \, \pi.$$

It then follows that (9.45) is equivalent to the combinatorial statement:

(∗) For each pair of vertices u and v of the directed multigraph D of order n with $2n$ arcs,

$$\epsilon(D : u, v) = 0.$$

By introducing a new vertex w and the arcs (w, u) and (v, w) from w to u and v to w, respectively, we see that it suffices to prove (∗) under the additional assumption that $v = u$. In addition it suffices to verify (∗) for any one vertex u.

We verify (∗) by induction on n. If $n = 1$ then (∗) holds. Now suppose that $n > 1$. We assume that for each vertex of D the indegree equals the outdegree, for otherwise Γ_{uu} is empty and the assertion holds trivially. We distinguish three cases.

Case 1. There is a vertex $a \neq u$ with indegree and outdegree equal to 1. Let (a_1, a) be the unique arc entering a and let (a, a_2) be the unique arc issuing from a. If $a_1 = a_2$, the assertion follows by applying the induction hypothesis to the digraph obtained by removing the vertex a and the arcs (a_1, a) and (a, a_2). Now suppose that $a_1 \neq a_2$. We may assume that $u \neq a_2$. Let $\alpha_1 = (a_2, b_1), \alpha_2 = (a_2, b_2), \ldots, \alpha_t = (a_2, b_t)$ be the arcs issuing from a_2. Let D_i be the general digraph obtained from D by removing the vertex

a and the arcs $(a_1, a), (a, a_2)$ and (a_2, b_i) and adding the arc $(a_1, b_i), (i = 1, 2, \ldots, t)$. By the induction hypothesis $\epsilon(D_i : u, u) = 0$ and hence

$$\epsilon(D : u, u) = \sum_{i=1}^{t} \epsilon(D_i : u, u) = 0.$$

Case 2. There is a loop at a vertex $a \neq u$ whose indegree and outdegree equal 2. Let the arcs incident with a be $(a, a), (b, a)$ and (a, c). Let D' be the general digraph obtained from D by removing the vertex a and its incident arcs and adding the arc (b, c). Applying the induction hypothesis we obtain

$$\epsilon(D : u, u) = \epsilon(D' : u, u) = 0.$$

Since the total number of arcs of D is $2n$, if Cases 1 and 2 do not apply, then the following case applies.

Case 3. Either each vertex has indegree and outdegree equal to 2, or u has indegree and outdegree equal to 1, there is a vertex with indegree and outdegree equal to 3 and every other vertex has indegree and outdegree equal to 2. Since Case 2 does not apply, there exist vertices a and b such that $\alpha = (a, b)$ is an arc and a and b have indegree and outdegree equal to 2. Let the two arcs entering a be $\alpha_1 = (c_1, a)$ $\alpha_2 = (c_2, a)$. Let D_i be the general digraph obtained from D removing the arcs α and α_i and adding the arcs (c_i, b) and $(b, b), (i = 1, 2)$. Each Eulerian trail from u to u in D is an Eulerian trail in exactly one of D_1 and D_2. However, the general digraphs D_1 and D_2 contain Eulerian trails that do not correspond to Eulerian trails in D. Let the two arcs of D leaving b be $\beta_1 = (b, d_1)$ and $\beta_2 = (b, d_2)$. Let D'_i be the general digraph obtained from D by removing the arcs α and β_i and adding the arcs (b, b) and $(a, c_i), (i = 1, 2)$. It can be verified that the Eulerian trails from u to u in D_1 and D_2 that do not arise from Eulerian trails in D correspond exactly to the Eulerian trails in D'_1 and D'_2 from u to u. Moreover, we have

$$\epsilon(D : u, u) = \epsilon(D_1 : u, u) + \epsilon(D_2 : u, u) - \epsilon(D'_1 : u, u) - \epsilon(D'_2 : u, u).$$

The general digraphs D_1 and D_2 satisfy the requirements of Case 1 and the general digraphs D'_1 and D'_2 satisfy the requirements of Case 2. Hence $\epsilon(D : u, u) = 0$ in this case also.   □

Razmyslov[1974] has shown that the Amitsur–Levitzki theorem can be deduced from the Cayley–Hamilton theorem and indeed that every polynomial identity (in the case of characteristic zero) is a "consequence" of the Cayley–Hamilton theorem. Amitsur and Levitzki have shown that $M_n(R)$

does not satisfy the standard identity of degree $2n - 1$ and that every multilinear identity of $M_n(R)$ (again in the case of characteristic zero) of degree at most $2n$ is a multiple of the standard identity of degree $2n$.

Exercises

1. Verify the relations (9.42) and (9.43).
2. Write out (9.44) in the case that $n = 3$ and $k = 4$.
3. Let the $2n - 1$ matrices $A_1, A_2, \ldots, A_{2n-1}$ of order n be defined by: $A_1 = E_{11}, A_2 = E_{12}, A_3 = E_{22}, A_4 = E_{23}, \ldots, A_{2n-1} = E_{nn}$. Let π be a permutation of $\{1, 2, \ldots, 2n - 1\}$. Prove that $A_{\pi(1)} A_{\pi(2)} \cdots A_{\pi(2n-1)}$ equals E_{nn} if π is the identity permutation and O otherwise.
4. Prove that the standard polynomial of degree $2n - 1$ is not a polynomial identity for $M_n(R)$.
5. (Newton's formulas) Let A be a matrix of order n and let σ_k be the coefficient of x^{n-k} in the characteristic polynomial of A. Prove that

$$(-1)^k k \sigma_k = \sum_{i=1}^{k} (-1)^{i-1} \sigma_{k-i} \mathrm{tr}(A^i), \quad (1 \leq i \leq k).$$

References

S.A. Amitsur and J. Levitzki[1950], Minimal identities for algebras, *Proc. Amer. Math. Soc.*, 1, pp. 449–463.

   [1951], Remarks on minimal identities for algebras, *Proc. Amer. Math. Soc.*, 2, pp. 320–327.

B. Bollobás[1979], *Graph Theory: An Introductory Course*, Springer-Verlag, New York.

R.A. Brualdi[1990], The many facets of combinatorial matrix theory, in *Matrix Theory and Applications*, C.R. Johnson ed., Proc. Symposia Pure and Applied Math. vol. 40, Amer. Math. Soc., Providence.

F.G. Frobenius[1896], Über die Primfactoren der Gruppendeterminante, *Sitzungsber. der Königlich Preuss. Akad. Wiss. Berlin*, pp. 1343–1382.

H. Laue[1988], A graph theoretic proof of the fundamental trace identity, *Discrete Math.*, 69, pp. 197–198.

J.S. Lew[1966], The generalized Cayley–Hamilton theorem in n dimensions, *Z. Angew. Math. Phys.*, 17, pp. 650–653.

C. Procesi[1976], The invariant theory of $n \times n$ matrices, *Advances in Math.*, 19, pp. 306–381.

Yu. P. Ramyslov[1974], Trace identities of full matrix algebras over a field of characteristic zero, *Math. USSR-Isv.*, 8, pp. 727–760.

L.R. Rowen[1980], *Polynomial Identities in Ring Theory*, Academic Press, New York.

D.E. Rutherford[1964], The Cayley–Hamilton theorem for semirings, *Proc. Roy. Soc. Edinburgh Sec. A*, 66, pp. 211–215.

H. Straubing[1983], A combinatorial proof of the Cayley–Hamilton theorem, *Discrete Math.*, 43, pp. 273–279.

R.G. Swan[1963], An application of graph theory to algebra, *Proc. Amer. Math. Soc.*, 14, pp. 367–373.

[1969], Correction to "An application of graph theory to algebra," *Proc. Amer. Math. Soc.*, 21, pp. 379–380.

D. Zeilberger[1985], A combinatorial approach to matrix algebra, *Discrete Math.*, 56, pp. 61–72.

9.8   Generic Nilpotent Matrices

Let

$$M = [m_{ij}], \quad (i, j = 1, 2, \ldots, n)$$

be a generic matrix of order n whose nonzero elements are independent indeterminates over a field F, and let $D(M)$ be the digraph corresponding to M. The vertex set of $D(M)$ is $V = \{1, 2, \ldots, n\}$ and there is an arc (i, j) from vertex i to vertex j if and only if m_{ij} is an indeterminate. We assign to each arc (i, j) the *weight*

$$\mathrm{wt}(i, j) = m_{ij}.$$

The matrix M is *nilpotent* provided that there exists a positive integer k such that M^k is a zero matrix. The smallest such integer k is the *nilpotent index* of M. The matrix M is nilpotent if and only if each of its eigenvalues is equal to zero.

The digraph of the nilpotent generic matrix M does not have any directed cycles and thus $D(M)$ is an *acyclic digraph* of order n. It follows from Lemma 3.2.3 that without loss of generality we may assume that each arc of $D(M)$ is of the form (i, j) where $1 \leq i < j \leq n$. With this assumption the matrix M has 0's on and below its main diagonal and is a *strictly upper triangular matrix*. For each positive integer k the element in position (i, j) of M^k equals the sum of the weights of all directed walks from i to j of length k. Because $D(M)$ is acyclic there exists a nonnegative integer r equal to the length of its longest path (directed chain). The acyclicity of $D(M)$ implies there is no directed walk of length $r + 1$ or greater in $D(M)$ and hence M^{r+1} is a zero matrix. Because M is a generic matrix, M^r is not a zero matrix and we conclude that the nilpotent index of M is $r + 1$.

We remark that a nilpotent matrix with elements from the field F may have a digraph which is not acyclic. For example, the matrix

$$A = \begin{bmatrix} 2 & 4 \\ -1 & -2 \end{bmatrix}$$

is nilpotent of index 2 and the digraph $D(A)$ is the complete digraph of order 2.

Let A be a nilpotent matrix of order n. The Jordan canonical form

JCF(A) of A is a direct sum of (nilpotent) *Jordan blocks* of order $t \geq 1$ of the form

$$B_t = \begin{bmatrix} 0 & 1 & 0 & \cdots & 0 & 0 \\ 0 & 0 & 1 & \cdots & 0 & 0 \\ 0 & 0 & 0 & \cdots & 0 & 0 \\ \vdots & \vdots & \vdots & \ddots & \vdots & \vdots \\ 0 & 0 & 0 & \cdots & 0 & 1 \\ 0 & 0 & 0 & \cdots & 0 & 0 \end{bmatrix}.$$

The ordering of the Jordan blocks in the Jordan canonical form can be chosen arbitrarily. In order that the Jordan canonical form of A be unique, we assume that the Jordan blocks are ordered from largest to smallest. Thus there exists a partition

$$jp(A) = (n_1, n_2, \ldots, n_s)$$

of n into $s \geq 1$ parts such that

$$\mathrm{JCF}(A) = B_{n_1} \oplus B_{n_2} \oplus \cdots \oplus B_{n_s}. \tag{9.48}$$

[Recall that $(n_1, n_2, \ldots, n_s)$ is a partition of n means that $n = n_1 + n_2 + \cdots + n_s$ and $n_1 \geq n_2 \geq \cdots \geq n_s \geq 1$. In section 6.5 a partition is regarded as an infinite nonincreasing sequence of nonnegative integers with only a finite number of nonzero terms. In the partition $(n_1, n_2, \ldots, n_s)$ we have suppressed the trailing terms equal to zero.] As in section 6.5 the set of all partitions of n is denoted by P_n. We call the partition $jp(A)$ the *Jordan partition* of the nilpotent matrix A. The *length* s of the Jordan partition equals the *nullity* of A, equivalently s equals the number of linearly independent eigenvectors for the eigenvalue 0. The order n_1 of the largest Jordan block equals the nilpotent index of A. Thus the order of the largest Jordan block of the Jordan canonical form of a generic nilpotent matrix M equals $r + 1$ where r is the length of the longest path in $D(M)$. We shall show how the entire Jordan partition of M can be determined by the digraph $D(M)$.

Let D be a digraph of order n with vertex set V. Let k be a positive integer. A *k-path* of D is a set X of vertices which can be partitioned into k (possibly empty) sets $X_1, X_2, \ldots, X_k$ such that each X_i is the set of vertices of a path of D. (We allow a trivial path consisting of a single vertex.) The largest number of vertices in a k-path of D is called the *k-path number* of D and is denoted by $p_k(D)$. We define $p_0(D) = 0$ and note that there is a smallest positive integer $s \leq n$ such that

$$0 = p_0(D) < p_1(D) < \cdots < p_s(D) = \cdots = p_n(D) = \cdots = n.$$

The sequence

$$p(D) = (p_0(D), p_1(D), \ldots, p_n(D))$$

is called the *path-number sequence* of D. It is sometimes convenient to regard the path-number sequence as an infinite sequence all but a finite number of terms of which equal n. In the notation of section 6.5 the path-number sequence $p(D)$ belongs to T_n. We call the integer s the *width* of D and denote it by width(D). If $k \leq$ width(D) then a k-path can be partitioned into k but no fewer paths of D. Notice that $p_1(D)$ equals 1 plus the length of the longest path in D. Hence if M is a generic nilpotent matrix, then $p_1(D(M))$ equals the nilpotent index of M.

If D is the digraph corresponding to the Jordan canonical form matrix in (9.48), then we see that the width of D equals the number s of Jordan blocks and

$$p_k(D) = n_1 + n_2 + \cdots + n_k, \quad (k = 0, 1, \ldots, s).$$

Hence

$$n_k = p_k(D) - p_{k-1}(D), \quad (k = 1, 2, \ldots, s).$$

Gansner[1981] and Saks[1986] showed that these equations hold between the Jordan partition of a generic nilpotent matrix and the k-path numbers of its digraph. In order to derive their results we first recall some basic facts concerning the Jordan invariants of a nilpotent matrix A of order n.

Let λ be an indeterminate over the field F. The greatest common divisor of the determinants of the submatrices of order k of $\lambda I_n - A$ is called the kth *determinantal divisor* of A (strictly speaking, of $\lambda I_n - A$) and is denoted by $d_k(A : \lambda), (k = 1, 2, \ldots, n)$. We define $d_0(A : \lambda) = 1$. There exists a positive integer $t \leq n$ and integers $\ell_0, \ell_1, \ldots, \ell_n$ satisfying

$$0 = \ell_0 < \ell_1 < \cdots < \ell_t = \ell_{t+1} = \cdots = \ell_n = n$$

such that

$$d_{n-k}(A : \lambda) = \lambda^{n-\ell_k}, \quad (k = 0, 1, \ldots, n).$$

We call the sequence

$$d(A) = (\ell_0, \ell_1, \ldots, \ell_n)$$

the *divisor sequence* of A. With the convention established above $d(A)$ belongs to T_n.

The polynomials

$$d_i(A : \lambda)/d_{i-1}(A : \lambda) = \lambda^{\ell_{n-i+1}-\ell_{n-i}}, \quad (i = n-t+1, n-t+2, \ldots, n)$$

are the elementary divisors, equivalently the invariant factors, of A. The

Jordan partition of A equals the difference sequence (as defined in section 6.5) of the divisor sequence

$$\delta d(A) = (\ell_1 - \ell_0, \ell_2 - \ell_1, \ldots, \ell_t - \ell_{t-1})$$

where the difference sequence has been truncated by omitting the trailing terms equal to 0. The nullity of A is equal to t, the number of nonconstant elementary divisors of A. Thus the Jordan partition of A is determined by the divisor sequence of A. We shall show that the divisor sequence of a generic nilpotent matrix M equals the path-number sequence of its digraph, and hence that the Jordan partition is the difference sequence of the path-number sequence. First we prove three lemmas.

Lemma 9.8.1. *Let $A = [a_{ij}]$ be a matrix of order n each of whose main diagonal elements equals zero. Let k be a positive integer with $k \leq$ width$(D(M))$ and let X be a k-path of $D(A)$ of size r. Then there is a submatrix of A of order $r - k$ with term rank equal to $r - k$.*

Proof. Since $k \leq$ width$(D(M))$ the set X can be partitioned into exactly k paths γ_t joining a vertex i_t to a vertex j_t, $(t = 1, 2, \ldots, k)$. Let B be the principal submatrix of order r of A determined by the rows and columns whose indices lie in X. Let B' be the submatrix of B obtained by deleting rows $j_1, j_2, \ldots, j_k$ and columns $i_1, i_2, \ldots, i_k$. Then B' is a submatrix of order $r - k$ of A and B' has term rank equal to $r - k$. $\square$

Lemma 9.8.2. *Let T be a strictly upper triangular matrix of order n. Let α and β be subsets of $\{1, 2, \ldots, n\}$ of size r, and suppose that the submatrix $T[\alpha, \beta]$ of T determined by the rows with index in α and columns with index in β has term rank r. Then the complementary submatrix $T[\overline{\alpha}, \overline{\beta}]$ of order $n - r$ is a strictly upper triangular matrix.*

Proof. Let

$$\alpha = \{i_1, i_2, \ldots, i_r\}, \qquad \overline{\alpha} = \{k_1, k_2, \ldots, k_{n-r}\}$$

and

$$\beta = \{j_1, j_2, \ldots, j_r\}, \qquad \overline{\beta} = \{l_1, l_2, \ldots, l_{n-r}\}.$$

(The elements of each of the sets are assumed to be listed in strictly increasing order.)

If there exists an integer k with $j_k \leq i_k$, then it follows from the assumption that T is strictly upper triangular that $T[\alpha, \beta]$ has a zero submatrix of size $(r - k + 1)$ by k contradicting the assumption that $T[\alpha, \beta]$ has term rank r. Hence

$$j_k > i_k, \quad (k = 1, 2, \ldots, r).$$

This implies that

$$k_i \geq l_i, \quad (i = 1, 2, \ldots, n - r)$$

and $T[\overline{\alpha}, \overline{\beta}]$ is a strictly upper triangular matrix.   □

Lemma 9.8.3.   *Let T, α and β satisfy the assumptions of Lemma 9.8.2. Let X be a set of r nonzero elements of $T[\alpha, \beta]$ with no two on the same line. Then the arcs of the digraph $D(T)$ corresponding to the elements of X can be partitioned into $e \leq n - r$ pairwise vertex disjoint paths each of which joins a vertex in $\overline{\beta}$ to a vertex in $\overline{\alpha}$. The vertices on these paths form an $(n - r)$-path of $D(T)$ of size $e + r$.*

Proof. By Lemma 9.8.2 $T[\overline{\alpha}, \overline{\beta}]$ is a strictly upper triangular matrix. Let T' and $T[\overline{\alpha}, \overline{\beta}]'$ be the matrices obtained from T and $T[\overline{\alpha}, \overline{\beta}]$, respectively, by replacing the 0's on the main diagonal of $T[\overline{\alpha}, \overline{\beta}]$ with 1's. Let X' be the union of X and the set of diagonal positions of $T[\overline{\alpha}, \overline{\beta}]$. The elements of X' correspond to a set Γ of n arcs of the digraph $D(T')$, one entering each vertex and one issuing from each vertex. The arcs of Γ can be partitioned into directed cycles $\gamma_1, \gamma_2, \ldots, \gamma_t$. Since T is strictly upper triangular $D(T)$ has has no directed cycles, and hence the removal of the $n - r$ arcs corresponding to the main diagonal elements of $T[\overline{\alpha}, \overline{\beta}]'$ results in a collection of pairwise vertex disjoint paths of $D(T)$. The removal of each arc increases the number of paths by at most one, and the conclusions now follow.   □

We now prove the theorem of Gansner[1981] and Saks[1986].

Theorem 9.8.4.   *Let M be a generic nilpotent matrix of order n. Then the divisor sequence $d(M)$ of M equals the path-number sequence $p(D(M))$ of the digraph $D(M)$. Hence the Jordan partition of M equals the difference sequence $\delta p(D(M))$.*

Proof. Let k be an integer with $k \leq \text{width}(D(M))$. The theorem asserts that the degree $n - \ell_k$ of the determinantal divisor $d_{n-k}(M : \lambda)$ of M satisfies

$$n - \ell_k = n - p_k(D(M)). \tag{9.49}$$

It follows from Lemma 9.8.1 that there is a submatrix of M of order $p_k(D(M)) - k$ with a nonzero term in its determinant expansion and hence a submatrix C of order $n - k$ of $\lambda I_n - M$ whose determinant expansion contains a nonzero term of degree $n - p_k(D(M))$. Since M is generic, $\det(C)$ contains a nonzero term of degree $n - p_k(D(M))$. Hence

$$n - \ell_k \leq n - p_k(D(M)).$$

Now let $B = (\lambda I_n - M)[\alpha, \beta]$ be a submatrix of order $n - k$ of $\lambda I_n - M$ such that $\det(B)$ has a nonzero term of degree $n - \ell_k$. Then there exists a set γ of size $n - \ell_k$ with $\gamma \subseteq \alpha \cap \beta$ such that there is a nonzero term in the determinant expansion of the submatrix $M[\alpha - \gamma, \beta - \gamma]$ of order $\ell_k - k$ of the strictly upper triangular matrix $M[\overline{\gamma}, \overline{\gamma}]$ of order ℓ_k. We apply Lemma 9.8.3 where T is the strictly upper triangular matrix $M[\overline{\gamma}, \overline{\gamma}]$ of order ℓ_k and $r = \ell_k - k$ and conclude that there exists an e-path X of size $\ell_k - k + e$ in $D(M)$ where $e \leq k$. There are $(k - e) + (n - \ell_k)$ vertices not on this e-path. Adding any $k - e$ of them to X we obtain a k-path of size ℓ_k. Hence $p_k(D(M)) \geq \ell_k$ and thus

$$n - p_k(D(M)) \leq n - \ell_k.$$

We now conclude that $\ell_k = p_k(D(M))$ for all $k = 0, 1, \ldots, n$.   □

Corollary 9.8.5.   *The path-number sequence of an acyclic digraph is a convex sequence.*

Let A be a nilpotent matrix of order n. It is a direct consequence of the Jordan canonical form that the conjugate $(jp(A))^*$ of the Jordan partition $jp(A) = (n_1, n_2, \ldots, n_s)$ of A is the partition

$$(m_1, m_2, \ldots, m_{n_1})$$

of n in which $m_1 + m_2 + \cdots + m_k$ equals the nullity of A^k, $(k = 1, 2, \ldots, n_1)$. (Note that $A^s = O$ for $s \geq n_1$.) For a generic nilpotent matrix M of order n we have by Theorem 9.8.4 that the difference sequence of the conjugate of the path-number sequence equals the sequence of nullities of powers of M.

For a matrix which is not a generic nilpotent matrix the path-number sequence need not equal the divisor sequence. This is because of the possibility of cancellation of terms in taking determinants. The general situation is discussed in Brualdi[1987] and also in Brualdi[1985] where the above ideas are used to derive the Jordan partition of the tensor product of two matrices in terms of the Jordan partition of the individual matrices.

We conclude this section by showing how the existence of the Jordan canonical form can be derived using some of the ideas of the theory of digraphs. The idea for this derivation goes back to Turnbull and Aitken[1932] and the derivation was given an exposition in Brualdi[1987].

Let A be a complex matrix of order n. The Jordan canonical form of A is a matrix $\mathrm{JCF}(A)$ similar to A which is a direct sum of Jordan blocks

$$aI_t + B_t = \begin{bmatrix} a & 1 & 0 & \cdots & 0 & 0 \\ 0 & a & 1 & \cdots & 0 & 0 \\ 0 & 0 & a & \cdots & 0 & 0 \\ \vdots & \vdots & \vdots & \ddots & \vdots & \vdots \\ 0 & 0 & 0 & \cdots & a & 1 \\ 0 & 0 & 0 & \cdots & 0 & a \end{bmatrix}$$

each having a unique eigenvalue a. The existence of the Jordan canonical form can be easily reduced to consideration of nilpotent matrices in the following way.

By Jacobi's theorem there is an upper triangular matrix T which is similar to A. The eigenvalues of A are the n diagonal elements of T and T can be chosen so that equal eigenvalues of A occur consecutively on the main diagonal of T. Suppose that A has two unequal eigenvalues. Then we may write

$$T = \begin{bmatrix} T_1 & X \\ O & T_2 \end{bmatrix}$$

where T_1 is an upper triangular matrix each of whose elements on the main diagonal equals a constant a and T_2 is an upper triangular matrix with no element on its main diagonal equal to a. The matrix T is similar to the matrix obtained from T by replacing X by a zero matrix. This can be seen as follows. Adding a multiple h of column i of T to column j and then the multiple $-h$ of row j to row i replaces T with a matrix similar to T. By means of *elementary similarities* of this type we may make each element of X equal to 0. We do this row by row, starting with the last row of X, and treating each element in the current row beginning with its first. We may now treat T_2 as we did T and continue like this until we arrive at a matrix T' which is similar to T (and hence to A) where T' is a direct sum of upper triangular matrices T_i' each with a constant main diagonal. The reduction of A to Jordan canonical form will be complete once the matrices T_i' have been reduced to Jordan canonical form. Thus we need only consider upper triangular matrices with a constant main diagonal. By subtracting a multiple of an identity matrix we may assume that the main diagonal elements equal 0. Therefore we have left to show that a nilpotent upper triangular matrix N of order n has a Jordan canonical form. We use induction on n.

If $n = 1$ then N is in Jordan canonical form. If $n = 2$, then

$$N = \begin{bmatrix} 0 & b \\ 0 & 0 \end{bmatrix}$$

and N can be brought to one of the two Jordan canonical forms

$$\begin{bmatrix} 0 & 0 \\ 0 & 0 \end{bmatrix} (b = 0), \qquad \begin{bmatrix} 0 & 1 \\ 0 & 0 \end{bmatrix} (b \neq 0)$$

by a similarity transformation. Now suppose $n > 2$. We apply the induction hypothesis to the leading principal submatrix of order $n - 1$ of N, and conclude that N is similar to a matrix

$$N' = \begin{bmatrix} & & & * \\ & N_1 & & \vdots \\ & & & * \\ 0 & \cdots & 0 & 0 \end{bmatrix}, \tag{9.50}$$

where N_1 is a matrix of order $n - 1$ in Jordan canonical form. Each row of the matrix (9.50) contains at most two nonzero elements.

We now show that by similarity transformations we can eliminate all nonzero elements in the last column of (9.50) which are in a row containing a 1 of N_1. Suppose that there is a nonzero element h in row i of the last column, and that row i of N_1 contains a 1. Since N_1 is in Jordan canonical form this 1 is in column $i + 1$. The elementary similarity transformation which adds $-h$ times column $i + 1$ to column n and h times row n to row $i + 1$ replaces h by 0 and does not alter any other element of the matrix (9.50). Hence we may apply a finite number of similarity transformations of this kind and obtain a matrix N' of the form (9.50) where N_1 is a matrix in Jordan canonical form and each row of N' contains at most one nonzero element.

If the last column of N' contains only 0's, then N' is in Jordan canonical form. Suppose now that there is at least one nonzero element in column n of N'. We show that by similarity transformations we can eliminate all but one nonzero element in column n without changing any other element of N'. We consider the digraph $D(N')$ with vertex set $\{1, 2, \ldots, n\}$. This digraph consists of $t \geq 0$ vertex disjoint paths $\gamma_1, \ldots, \gamma_t$, and $s \geq 1$ paths $\pi_1, \ldots, \pi_s$ which are vertex disjoint except for the fact that they all terminate at vertex n. Moreover the set of t paths and the set of s paths have no vertex in common. If $t > 0$, then the inductive assumption applies to the principal submatrix of N' obtained by deleting the rows and columns corresponding to the vertices of γ_1. We now assume that $t = 0$ so that $D(N')$ consists of paths $\pi_1, \pi_2, \ldots, \pi_s$ of lengths $\ell_1, \ell_2, \ldots, \ell_s$, respectively, all terminating at vertex n. We now show that by a sequence of similarity transformations we can eliminate all the nonzero elements in column n of N' except for one corresponding to an arc of a longest path, again without changing any other element of N'. Let h be a nonzero element in the last column n of N corresponding to the last arc of a longest path π_k. We may simultaneously permute rows and columns if necessary and assume that π_k corresponds to the last block in the Jordan canonical form N_1. The similarity transformation which multiplies the last row by h and the last column by h^{-1} allows us to assume that $h = 1$. Let

$$\pi_k : i_0 \to \cdots \to i_{\ell_k - 1} \to i_{\ell_k} = n.$$

Let

$$\pi_r : j_0 \to \cdots \to j_{\ell_r-1} \to j_{\ell_r} = n$$

be any other path with $r \neq k$. Let p be the nonzero element in position (j_{ℓ_r-1}, n) of N'. We successively perform the following elementary similarity transformations:

(i) Add $-p$ times row i_{ℓ_k-1} to row i_{ℓ_r-1} and add p times column i_{ℓ_r-1} to column i_{ℓ_k-1}.

(ii) Add $-p$ times row i_{ℓ_k-2} to row j_{ℓ_r-2} and add p times column j_{ℓ_r-2} to column i_{ℓ_k-2}.

$\vdots$

(ℓ_r) Add $-p$ times row $i_{\ell_k-\ell_r}$ to row j_0 and add p times column j_0 to column $i_{\ell_k-\ell_r}$.

The result of this sequence of similarity transformations is to replace p with 0 and leave all other elements of N' unchanged. We may repeat until all the nonzero elements in column n of N', other than the nonzero element corresponding to the arc of the longest path π_k, have been replaced with 0. The digraph of the resulting matrix N'' consists of the path π_k and the paths π_r' obtained by deleting the last arc of $\pi_r, (r \neq k)$. It follows that the matrix N'' is similar to our given matrix A and is in Jordan canonical form with Jordan blocks of sizes $\ell_1, \ldots, \ell_{k-1}, \ell_k + 1, \ell_{k+1}, \ldots, \ell_s$.

In closing we remark that the above argument only establishes the existence (and not the uniqueness) of the Jordan canonical form.

Exercises

1. Construct an acyclic digraph D which does not have a 2-path γ of maximal size $p_2(D)$ such that γ can be partitioned into two 1-paths one of which has size $p_1(D)$. Thus a 2-path (and more generally a k-path) cannot always be obtained by successively choosing 1-paths of maximal size (Stanley[1980]).

2. Show by example that Theorem 9.8.4 does not hold in general for nilpotent matrices.

3. Let T be a tree of order n and let T^* be the digraph obtained by assigning a direction to each edge of T. Let A be the (nilpotent) adjacency matrix of T^*. Prove that the Jordan partition of A equals the difference sequence $\delta p(T^*)$ (Brualdi[1987]).

4. Determine the Jordan canonical form of the matrix

$$\begin{bmatrix} 0 & 1 & 0 & 0 & 0 & 0 & 0 & 0 & 0 & 1 \\ 0 & 0 & 1 & 0 & 0 & 0 & 0 & 0 & 0 & 1 \\ 0 & 0 & 0 & 0 & 0 & 0 & 0 & 0 & 0 & 1 \\ 0 & 0 & 0 & 0 & 1 & 0 & 0 & 0 & 0 & 1 \\ 0 & 0 & 0 & 0 & 0 & 0 & 0 & 0 & 0 & 1 \\ 0 & 0 & 0 & 0 & 0 & 0 & 1 & 0 & 0 & 1 \\ 0 & 0 & 0 & 0 & 0 & 0 & 0 & 1 & 0 & 1 \\ 0 & 0 & 0 & 0 & 0 & 0 & 0 & 0 & 1 & 1 \\ 0 & 0 & 0 & 0 & 0 & 0 & 0 & 0 & 0 & 1 \\ 0 & 0 & 0 & 0 & 0 & 0 & 0 & 0 & 0 & 0 \end{bmatrix}.$$

References

R.A. Brualdi[1985], Combinatorial verification of the elementary divisors of tensor products, *Linear Alg. Applics.*, 71, pp. 31–47.

  [1987], The Jordan canonical form: an old proof, *Amer. Math. Monthly*, 94, pp. 257–267.

  [1987], Combinatorially determined elementary divisors, *Congressus Numerantium*, 58, pp. 193–216.

R.A. Brualdi and K.L. Chavey[1991], Elementary divisors and ranked posets with application to matrix compounds, *Linear Multilin. Alg.*, to be published.

E.R. Gansner[1981], Acyclic digraphs, Young tableaux and nilpotent matrices, *SIAM J. Alg. Disc. Meth.*, 2, pp. 429–440.

M.E. Saks[1986], Some sequences associated with combinatorial structures, *Discrete Math.*, 59, pp. 135–166.

H. Schneider[1986], The influence of the marked reduced graph of a nonnegative matrix on the Jordan form and on related properties: A survey, *Linear Alg. Applics.*, 84, pp. 161–189.

R.P. Stanley[1980], Weyl groups, the hard Lefschetz theorem, and the Sperner property, *SIAM J. Alg. Disc. Methods*, 1, pp. 168–184.

MASTER REFERENCE LIST

C.M. Ablow and J.L. Brenner[1963], Roots and canonical forms for circulant matrices, *Trans. Amer. Math. Soc.*, 107, pp. 360–376.

R. Ahlswede and D.E. Daykin, The number of values of combinatorial functions, *Bull. London Math. Soc.*, 11, pp. 49–51.

A.V. Aho, J.E. Hopcroft and J.D. Ullman[1975], *The Design and Analysis of Computer Algorithms*, Addison-Wesley, Reading, Mass.

N. Alon, R.A. Brualdi and B.L. Shader[1991], Multicolored trees in bipartite decompositions of graphs, *J. Combin. Theory, Ser. B*, to be published.

S.A. Amitsur and J. Levitzki[1950], Minimal identities for algebras, *Proc. Amer. Math. Soc.*, 1, pp. 449–463.

[1951], Remarks on minimal identities for algebras, *Proc. Amer. Math. Soc.*, 2, pp. 320–327.

L.D. Andersen and A.J.W. Hilton[1983], Thanks Evans!, *Proc. London Math. Soc.* (3), 47, pp. 507–522.

I. Anderson[1971], Perfect matchings of graphs, *J. Combin. Theory*, 10, pp. 183–186.

R.P. Anstee[1982], Properties of a class of (0,1)-matrices covering a given matrix, *Canad. J. Math.*, 34, pp. 438–453.

M. Aschbacher[1971], The non-existence of rank three permutation groups of degree 3250 and subdegree 57, *J. Algebra*, 19, pp. 538–540.

M.J. Atallah[1982], Finding the cycle index of an irreducible, nonnegative matrix, *SIAM J. Computing*, 11, pp. 567–570.

K.B. Athreya, C.B. Pranesachar and N.M. Singhi[1980], On the number of latin rectangles and chromatic polynomial of $L(K_{r,s})$, *Europ. J. Combinatorics*, 1, pp. 9–17.

K. Balasubramanian[1990], On transversals in latin squares, *Linear Alg. Applics.*, 131, pp. 125–129.

S.E. Bammel and J. Rothstein[1975], The number of 9×9 latin squares, *Discrete Math.*, 11, pp. 93–95.

E. Bannai and T. Ito[1973], On finite Moore graphs, *J. Fac. Sci. Univ. Tokyo*, 20, pp. 191–208.

L. Bassett, J. Maybee and J. Quirk[1983], Qualitative economics and the scope of the correspondence principle, *Econometrica*, 26, pp. 544–563.

V.J. Baston[1982], A Marica-Schönheim theorem for an infinite sequence of finite sets, *Math. Z.*, 179, pp. 531–533.

C. Berge[1958], Sur le couplage maximum d'un graphe, *C.R. Acad. Sciences* (Paris), 247, pp. 258–259.

[1976], *Graphs and Hypergraphs*, North Holland, Amsterdam.

S. Beslin and S. Ligh[1989], Greatest common divisor matrices, *Linear Alg. Applics.*, 118, pp. 69–76.

T. Beth[1983], Eine Bermerkung zur Abschätzung der Anzahl orthogonaler lateinischer Quadrate mittels Siebverfahren, *Abh. Math. Sem. Hamburg*, 53, pp. 284–288.

N. Biggs[1974], *Algebraic Graph Theory*, Cambridge Tracts in Mathematics No. 67, Cambridge University Press, Cambridge.

G. Birkhoff[1946], Tres observaciones sobre el algebra lineal, *Univ. Nac. Tucumán Rev. Ser. A*, pp. 147–151.

B. Bollobás[1979], *Graph Theory*, Springer-Verlag, New York.

V.I. Bol'shakov[1986], Upper values of a permanent in Λ_n^k, *Combinatorial analysis, No. 7 (Russian)*, Moskov. Gos. Univ., Moscow, pp. 92–118 and 164–165.

[1986], The spectrum of the permanent on Λ_n^k, *Proceedings of the All-Union seminar on discrete mathematics and its applications, Moscow (1984), (Russian)*, Moskov. Gos. Univ. Mekh.-Mat. Fak., Moscow, pp. 65–73.

J.A. Bondy and U.S.R. Murty[1976], *Graph Theory with Applications*, North-Holland, New York.

C.W. Borchardt[1860], Über eine der Interpolation entsprechende Darstellung der Eliminations-Resultante, *J. Reine Angew. Math.*, 57, pp. 111–121.

R.C. Bose[1938], On the application of the properties of Galois fields to the problem of construction of hyper-Graeco-Latin squares, *Sankhyā*, 3, pp. 323–338.

R.C. Bose and S.S. Shrikhande[1959], On the falsity of Euler's conjecture about the non-existence of two orthogonal Latin squares of order $4t + 2$, *Proc. Nat. Acad. Sci. U.S.A.*, 45, pp. 734–737.

[1960], On the composite of balanced incomplete block designs, *Canad. J. Math.*, 12, pp. 177–188.

[1960], On the construction of sets of mutually orthogonal Latin squares and the falsity of a conjecture of Euler, *Trans. Amer. Math. Soc.*, 95, pp. 191–209.

R.C. Bose, S.S. Shrikhande and E.T. Parker[1960], Further results on the construction of mutually orthogonal Latin squares and the falsity of Euler's conjecture, *Canad. J. Math.*, 12, pp. 189–203.

A. Brauer[1947], Limits for the characteristic roots of a matrix II, *Duke Math. J.*, 14, pp. 21–26.

A. Brauer and I.C. Gentry[1968], On the characteristic roots of tournament matrices, *Bull. Amer. Math. Soc.*, 74, pp. 1133–1135.

R.K. Brayton, D. Coppersmith and A.J. Hoffman[1974], Self-orthogonal latin squares of all orders $n \neq 2, 3, 6$, *Bull. Amer. Math. Soc.*, 80, pp. 116–118.

L.M. Brégman[1973], Certain properties of nonnegative matrices and their permanents, *Dokl. Akad. Nauk SSSR*, 211, pp. 27–30. (*Soviet Math. Dokl.*, 14, pp. 945–949.)

W.G. Bridges[1971], The polynomial of a non-regular digraph, *Pacific J. Math.*, 38, pp. 325–341.

[1972], The regularity of x^3-graphs, *J. Combin. Theory, Ser. B*, 12, pp. 174–176.

W.G. Bridges and R.A. Mena[1981], x^k-digraphs, *J. Combin. Theory, Ser. B*, 30, pp. 136–143.

A. Brouwer, A. Cohen and A. Neumaier[1989], *Distance Regular Graphs*, Springer-Verlag, Berlin.

A.E. Brouwer, A.J. de Vries and R.M.A. Wieringa[1978], A lower bound for the length of partial transversals in a latin square, *Nieuw Archief voor Wiskunde* (3), XXVI, pp. 330–332.

R.A. Brualdi[1966], Term rank of the direct product of matrices, *Canad. J. Math.*, 18, pp. 126–138.

[1966], Permanent of the direct product of matrices, *Pacific J. Math.*, 16, pp. 471–482.

[1971], Matchings in arbitrary graphs, *Proc. Cambridge Phil. Soc.*, 69, pp. 401–407.

[1976], Combinatorial properties of symmetric non-negative matrices, *Teorie Combinatoire*, Toma II, Accademia Nazionale dei Lincei, Roma, pp. 99–120.

[1977], *Introductory Combinatorics*, Elsevier Science Publishers, New York.

[1979], Matrices permutation equivalent to irreducible matrices and applications, *Linear Multilin. Alg.*, 7, pp. 1–12.

[1979], The diagonal hypergraph of a matrix (bipartite graph), *Discrete Math.*, 27, pp. 127–147.

[1980], On the diagonal hypergraph of a matrix, *Annals of Discrete Math.*, 8, pp. 261–264.

[1982], Matrices, eigenvalues, and directed graphs, *Linear Multilin. Alg.*, 11, pp. 143–165.

[1985], Combinatorial verification of the elementary divisors of tensor products, *Linear Alg. Applics.*, 71, pp. 31–47.

[1987], The Jordan canonical form: an old proof, *Amer. Math. Monthly*, 94, pp. 257–267.

[1987], Combinatorially determined elementary divisors, *Congressus Numerantium*, 58, pp. 193–216.

[1988], Counting permutations with restricted positions: permanents of (0,1)-matrices. A tale in four parts. In The 1987 Utah State University Department of Mathematics Conference Report by L. Beasley and E.E. Underwood, *Linear Alg. Applics.*, 104, pp. 173–183.

[1988], In Memoriam Herbert J. Ryser 1923-1985, *J. Combin. Theory, Ser. A*, 47, pp. 1–5.

[1990], The many facets of combinatorial matrix theory, in *Matrix Theory and Applications*, C.R. Johnson ed., Proc. Symposia Pure and Applied Math. vol. 40, Amer. Math. Soc., Providence.

[1992], The symbiotic relationship of combinatorics and matrix theory, *Linear Alg. Applics.*, to be published.

R.A. Brualdi and K.L. Chavey[1991], Elementary divisors and ranked posets with application to matrix compounds, *Linear Multilin. Alg.*, to be published.

R.A. Brualdi and J. Csima[1986], Extending subpermutation matrices in regular classes of matrices, *Discrete Math.*, 62, pp. 99–101.

[1991], Butterfly embedding proof of a theorem of König, *Amer. Math. Monthly*, to be published.

R.A. Brualdi and P.M. Gibson[1977], Convex polyhedra of doubly stochastic matrices I. Applications of the permanent function, *J. Combin. Theory, Ser. A*, 22, pp. 194–230.

R.A. Brualdi, J.L. Goldwasser and T.S. Michael[1988], Maximum permanents of matrices of zeros and ones, *J. Combin. Theory, Ser. A*, 47, pp. 207–245.

R.A. Brualdi, F. Harary and Z. Miller[1980], Bigraphs versus digraphs via matrices, *J. Graph Theory*, 4, pp. 51–73.

R.A. Brualdi and M.B. Hedrick[1979], A unified treatment of nearly reducible and nearly decomposable matrices, *Linear Alg. Applics.*, 24, pp. 51–73.

R.A. Brualdi and M. Lewin[1982], On powers of nonnegative matrices, *Linear Alg. Applics.*, 43, pp. 87–97.

R.A. Brualdi and B. Liu[1991], Fully indecomposable exponents of primitive matrices, *Proc. Amer. Math. Soc.*, to be published.

  [1990], Hall exponents of Boolean matrices, *Czech. Math. J.*, 40 (115), pp. 663–674.

  [1990], Generalized exponents of primitive directed graphs, *J. Graph Theory*, 14, pp. 483–499.

R.A. Brualdi and T.S. Michael[1989], The class of 2-multigraphs with a prescribed degree sequence, *Linear Multilin. Alg.*, 24, pp. 81–102.

R.A. Brualdi and M. Newman[1965], Some theorems on the permanent, *J. Res. National Bur. Stands.*, 69B, pp. 159–163.

R.A. Brualdi, S.V. Parter and H. Schneider[1966], The diagonal equivalence of a nonnegative matrix to a stochastic matrix, *J. Math. Anal. Applics.*, 16, pp. 31–50.

R.A. Brualdi and J.A. Ross[1980], On the exponent of a primitive, nearly reducible matrix, *Math. Oper. Res.*, 5, pp. 229–241.

  [1981], Matrices with isomorphic diagonal hypergraphs, *Discrete Math.*, 33, pp. 123–138.

R.A. Brualdi and B.L. Shader[1990], Matrix factorizations of determinants and permanents, *J. Combin. Theory, Ser. A*, 54, pp. 13–134.

  [1991], On converting the permanent into the determinant and sign-nonsingular matrices, *Applied Geometry and Discrete Mathematics* (P. Gritzmann and B. Sturmfels, eds.), to be published.

R.H. Bruck[1951], Finite nets. I. Numerical invariants, *Canad. J. Math.*, 3, pp. 94–107.

  [1963], Finite nets. II. Uniqueness and imbedding, *Pacific J. Math.*, 13, pp. 421–457.

T. Brylawski[1975], Modular constructions for combinatorial geometries, *Trans. Amer. Math. Soc.*, 203, pp. 1–44.

D. de Caen and D. Gregory[1987], On the decomposition of a directed graph into complete bipartite subgraphs, *Ars Combinatoria*, 23B, pp. 139–146.

D. de Caen and D.G. Hoffman[1989], Impossibility of decomposing the complete graph on n points into $n-1$ isomorphic complete bipartite graphs, *SIAM J. Disc. Math*, 2, pp. 48–50.

P.J. Cameron[1978], Strongly regular graphs, *Selected Topics in Graph Theory* (L.W. Beineke and R.J. Wilson, eds.), Academic Press, New York, pp. 337–360.

P.J. Cameron, J.-M. Goethals and J.J. Seidel[1978], Strongly regular graphs having strongly regular subconstituents, *J. Algebra*, 55, pp. 257–280.

P.J. Cameron, J.M. Goethals, J.J. Seidel and E.E. Shult[1976], Line graphs, root systems, and elliptic geometry, *J. Algebra*, 43, pp. 305–327.

P.J. Cameron and J.H. van Lint[1975], *Graph Theory, Coding Theory and Block Designs*, London Math. Soc. Lecture Note Series No.19, Cambridge University Press, Cambridge.

P. Camion[1965], Characterization of totally unimodular matrices, *Proc. Amer. Math. Soc.*, 16, pp. 1068–1073.

P. Cartier and D. Foata[1969], *Problèmes Combinatoires de Commutation et Réarrangement*, Lec. Notes in Math. No. 85, Springer, Berlin.

A. Cayley[1854], Sur les déterminantes gauches, *Crelle's J.*, 38, pp. 93–96.

[1889], A theorem on trees, *Quarterly J. Math.*, 23, pp. 376–378.

S. Chaiken[1982], A combinatorial proof of the all minors matrix tree theorem, *SIAM J. Alg. Disc. Meth.*, 3, pp. 319–329.

C.Y. Chao[1977], On a conjecture of the semigroup of fully indecomposable relations, *Czech. Math. J.*, 27, pp. 591–597.

C.Y. Chao and T. Wang[1987], On the matrix equation $A^2 = J$, *J. Math. Res. and Exposition*, 2, pp. 207–215.

C.Y. Chao and M.C. Zhang[1983], On the semigroup of fully indecomposable relations, *Czech. Math. J.*, 33, pp. 314–319.

G. Chaty and M. Chein[1976], A note on top down and bottom up analysis of strongly connected digraphs, *Discrete Math.*, 16, pp. 309–311.

W.C. Chu[1985], Some algebraic identities concerning determinants and permanents, *Linear Alg. Applics.*, 116, pp. 35–40.

F.R.K. Chung, P. Diaconis, R.L. Graham and C.L. Mallows[1981], On the permanents of complements of the direct sum of identity matrices, *Advances in Applied Math.*, 2, pp. 121–137.

V. Chungphaisan[1974], Conditions for sequences to be r-graphic, *Discrete Math.*, 7, pp. 31–39.

C.L. Coates[1959], Flow graph solutions of linear algebraic equations, *IEEE Trans. Circuit Theory*, CT-6, pp. 170–187.

C.C. Colburn[1984], The complexity of completing partial latin squares, *Discrete Applied Math.*, 8, pp. 25–30.

S.A. Cook[1971], The complexity of theorem proving procedures, *Proc. 3rd ACM Symp. on Theory of Computing*, pp. 151–158.

A.B. Cruse[1974], On embedding incomplete symmetric latin squares, *J. Combin. Theory, Ser. A*, 16, pp. 18–22.

J. Csima[1979], On the plane term rank of three dimensional matrices, *Discrete Math.*, 28, pp. 147–152.

D.M. Cvetković[1975], The determinant concept defined by means of graph theory, *Mat. Vesnik*, 12, pp. 333–336.

D.M. Cvetković, M. Doob and H. Sachs[1982], *Spectra of Graphs – Theory and Application*, 2nd ed., Deutscher Verlag der Wissenschaften, Berlin, Academic Press, New York.

D. Cvetković, M. Doob, I. Gutman and A. Torĝasev[1988], *Recent Results in the Theory of Graph Spectra*, Annals of Discrete Mathematics No. 36, Elsevier Science Publishers, New York.

R.H. Damerell[1973], On Moore graphs, *Proc. Cambridge Phil. Soc.*, 74, pp. 227–236.

[1983], On Smetaniuk's construction for latin squares and the Andersen-Hilton theorem, *Proc. London Math. Soc.* (3), 47, pp. 523–526.

P.J. Davis[1979], *Circulant Matrices*, John Wiley & Sons, New York.

D.E. Daykin and R. Häggvist[1981], Problem No. 6347, *Amer. Math. Monthly*, 88, p. 446.

D.E. Daykin and L. Lovász[1976], The number of values of a Boolean function, *J. London Math. Soc.*, 12 (2), pp. 225–230.

E.V. Denardo[1977], Periods of connected networks, *Math. Oper. Res.*, 2, pp. 20–24.

J. Dénes and A.D. Keedwell[1974], *Latin Squares and Their Applications*, Academic Press, New York.

J. Donald, J. Elwin, R. Hager and P. Salamon[1984], A graph-theoretic upper bound on the permanent of a nonnegative matrix, *Linear Alg. Applics.*, 61, pp. 187–198.

M. Doob[1984], Applications of graph theory in linear algebra, *Math. Mag.*, 57, pp. 67–76.

A.L. Dulmage and N.S. Mendelsohn[1958], Coverings of bipartite graphs, *Canad. J. Math.*, 10, pp. 517–534.

[1958], Some generalizations of the problem of distinct representatives, *Canad. J. Math.*, 10, pp. 230–241.

[1959], A structure theory of bipartite graphs of finite exterior dimension, *Trans. Roy. Soc. Canad.*, Third Ser. Sec. III 53, pp. 1–3.

[1962], The exponent of a primitive matrix, *Canad. Math. Bull.*, 5, pp. 642–656.

[1963] The characteristic equation of an imprimitive matrix, *SIAM J. Appl. Math.*, 11, pp. 1034–1045.

[1964], Gaps in the exponent set of primitive matrices, *Illinois J. Math.*, 8, pp. 642–656.

[1964], The exponents of incidence matrices, *Duke Math. J.*, 31, pp. 575–584.

[1967], Graphs and matrices, *Graph Theory and Theoretical Physics* (F. Harary, ed.), Academic Press, New York, pp. 167–227.

J. Edmonds[1967], Systems of distinct representatives and linear algebra, *J. Res. Nat. Bur. Standards*, 71B, pp. 241–245.

G.P. Egoryčev[1981], A solution of van der Waerden's permanent problem, *Dokl. Akad. Nauk SSSR*, 258, pp. 1041–1044. (*Soviet Math. Dokl.*, 23, pp. 619–622.)

P. Erdös and T. Gallai[1960], Graphs with prescribed degrees of vertices, *Mat. Lapok*, 11, pp. 264–274 (in Hungarian).

P. Erdös, D.R. Hickerson, D.A. Norton and S.K. Stein[1988], Has every latin square of order n a partial transversal of size $n - 1$?, *Amer. Math. Monthly*, 95, pp. 428–430.

P. Erdös, A. Rényi and V.T. Sós[1966], On a problem of graph theory, *Studia Sci. Math. Hungar.*, 1, pp. 215–235.

R. Euler, R.E. Burkhard and R. Grommes[1986], On latin squares and the facial structure of related polytopes, *Discrete Math.*, 62, pp. 155–181.

T. Evans[1960], Embedding incomplete latin squares, *Amer. Math. Monthly*, 67, pp. 958–961.

S. Even[1979], *Graph Algorithms*, Computer Science Press, Potomac, Maryland.

C.J. Everett and P.R. Stein[1973], The asymptotic number of (0,1)-matrices with zero permanent, *Discrete Math.*, 6, pp. 29–34.

D.I. Falikman[1981], A proof of van der Waerden's conjecture on the permanent of a doubly stochastic matrix, *Mat. Zametki*, 29, pp. 931–938. (*Math. Notes*, 29, pp. 475–479.)

M. Fiedler[1973], Algebraic connectivity of graphs, *Czech. Math. J.*, 23, pp. 298–305.

[1975], A property of eigenvectors of nonnegative symmetric matrices and its application to graph theory, *Czech. Math. J.*, 25, pp. 619–633.

[1990], Absolute connectivity of trees, *Linear Multilin. Alg.*, 26, pp. 86–106.

S. Fiorini and R.J. Wilson[1977], *Edge-colorings of graphs*, Pitman, London.

J. Folkman and D.R. Fulkerson[1969], Edge colorings in bipartite graphs, *Combinatorial Mathematics and Their Applications* (R.C. Bose and T. Dowling, eds.), University of North Carolina Press, Chapel Hill, pp. 561–577.

L.R. Ford, Jr. and D.R. Fulkerson[1962], *Flows in Networks*, Princeton University Press, Princeton.

T.H. Foregger[1975], An upper bound for the permanent of a fully indecomposable matrix, *Proc. Amer. Math. Soc.*, 49, pp. 319–324.

G. Frobenius[1896], Über die Primfactoren der Gruppendeterminante, *Sitzungsber. der Königlich Preuss. Akad. Wiss. Berlin*, pp. 1343–1382.

[1879], Theorie der linearen Formen mit ganzen Coefficienten, *J. für reine und angew. Math.*, 86, pp. 146–208.

[1917], Über zerlegbare Determinanten, *Sitzungsber. Preuss. Akad. Wiss. Berlin*, pp. 274–277.

[1912], Über Matrizen aus nicht negativen Elementen, *Sitzungsber. Preuss. Akad. Wiss. Berlin*, pp. 476–457.

D.R. Fulkerson[1960], Zero-one matrices with zero trace, *Pacific J. Math.*, 10, pp. 831–836.

[1964], The maximum number of disjoint permutations contained in a matrix of zeros and ones, *Canad. J. Math.*, 10, pp. 729–735.

[1970], Blocking Polyhedra, *Graph Theory and Its Applications* (B. Harris, ed.), Academic Press, New York, pp. 93–111.

D.R. Fulkerson, A.J. Hoffman and M.H. McAndrew[1965], Some properties of graphs with multiple edges, *Canad. J. Math.*, 17, pp. 166–177.

D. Gale[1957], A theorem on flows in networks, *Pacific J. Math.*, 7, pp. 1073–1082.

T. Gallai[1963], Neuer Beweis eines Tutte'schen Satzes, *Magyar Tud. Akad. Közl.*, 8, pp. 135–139.

E.R. Gansner[1981], Acyclic digraphs, Young tableaux and nilpotent matrices, *SIAM J. Alg. Disc. Meth.*, 2, pp. 429–440.

F.R. Gantmacher[1959], *The Theory of Matrices*, vol. 2, Chelsea, New York.

A.M.H. Gerards[1989], A short proof of Tutte's characterization of totally unimodular matroids, *Linear Alg. Applics.*, 114/115, pp. 207–212.

I. Gessel[1985], Counting three-line latin rectangles, *Proc. Colloque de Combinatoire Énumérative*, UQAM, pp. 106–111.

[1987], Counting latin rectangles, *Bull. Amer. Math. Soc. (New Series)*, 16, pp. 79–82.

P.M. Gibson[1966], A short proof of an inequality for the permanent function, *Proc. Amer. Math. Soc.*, 17, pp. 535–536.

[1971], Conversion of the permanent into the determinant, *Proc. Amer. Math. Soc.*, 27, pp. 471–476.

[1972], A lower bound for the permanent of a (0,1)-matrix, *Proc. Amer. Math. Soc.*, 33, pp. 245–246.

[1972], The pfaffian and 1-factors of graphs, *Trans. N.Y. Acad. Sci.*, 34, pp. 52–57.

[1972], The pfaffian and 1-factors of graphs II, *Graph Theory and Applications* (Y. Alavi, D.R. Lick and A.T. White, eds.), Lec. Notes in Mathematics, No. 303, pp. 89–98, Springer-Verlag, New York.

J.-M. Goethals and J.J. Seidel[1967], Orthogonal matrices with zero diagonal, *Canad. J. Math.*, 19, pp. 1001–1010.

[1970], Strongly regular graphs derived from combinatorial designs, *Canad. J. Math.*, 22, pp. 597–614.

R.L. Graham and H.O. Pollak[1971], On the addressing problem for loop switching, *Bell System Tech. J.*, 50, pp. 2495–2519.

[1972], On embedding graphs in squashed cubes, *Lecture Notes in Math*, vol. 303, Springer-Verlag, New York, pp. 99–110.

R. Grone[1991], On the geometry and Laplacian of a graph, *Linear Alg. Applics.*, to be published.

B. Grone and R. Merris[1987], Algebraic connectivity of trees, *Czech. Math. J.*, 37, pp. 660–670.

R. Guérin[1968], Existence et propriétés des carrés latins orthogonaux II, *Publ. Inst. Statist. Univ. Paris*, 15, pp. 215–293.

R.P. Gupta[1967], On basis diagraphs, *J. Comb. Theory*, 3, pp. 16–24.

[1967], A decomposition theorem for bipartite graphs, *Théorie des Graphes Rome I.C.C.* (P. Rosenstiehl, ed.), Dunod, Paris, pp. 135–138.

[1974], On decompositions of a multigraph with spanning subgraphs, *Bull. Amer. Math. Soc.*, 80, pp. 500–502.

[1978], An edge-coloration theorem for bipartite graphs with applications, *Discrete Math.*, 23, pp. 229–233.

H. Hadwiger, H. Debrunner and V. Klee[1964], *Combinatorial Geometry in the Plane*, Holt, Rinehart and Winston, New York.

W. Haemers[1979], *Eigenvalue Techniques in Design and Graph Theory*, Mathematisch Centrum, Amsterdam.

R. Häggvist[1978], A solution to the Evans conjecture for latin squares of large size, *Combinatorics*, Proc. Conf. on Combinatorics, Kesthely (Hungary) 1976, János Bolyai Math. Soc. and North Holland, pp. 495–513.

S.L. Hakimi[1962], On realizability of a set of integers as degrees of the vertices of a linear graph I, *J. Soc. Indust. Appl. Math.*, 10, pp. 496–506.

M. Hall, Jr.[1945], An existence theorem for latin squares, *Bull. Amer. Math. Soc.* 51, pp. 387–388.

[1948], Distinct representatives of subsets, *Bull. Amer. Math. Soc.*, 54, pp. 922–926.

[1952], A combinatorial problem on abelian groups, *Proc. Amer. Math. Soc.*, 3, pp. 584–587.

[1986], *Combinatorial Theory*, 2nd edition, Wiley, New York,

P. Hall[1935], On representatives of subsets, *J. London Math. Soc.*, 10, pp. 26–30.

H. Hanani[1970], On the number of orthogonal latin squares, *J. Combin. Theory*, 8, pp. 247–271.

F. Harary[1959], A graph theoretic method for the complete reduction of a matrix with a view toward finding its eigenvalues, *J. Math. and Physics*, 38, pp. 104–111.

[1962], The determinant of the adjacency matrix of a graph, *SIAM Review*, 4, pp. 202–210.

[1967], Graphs and matrices, *SIAM Review*, 9, pp. 83–90.

[1969], *Graph Theory*, Addison-Wesley, Reading, Mass.

F. Harary, R. Norman and D. Cartwright[1965], *Structural Models*, Wiley, New York.

F. Harary and E.M. Palmer[1973], *Graphical Enumeration*, Academic Press, New York.

D.J. Hartfiel[1970], A simplified form for nearly reducible and nearly decomposable matrices, *Proc. Amer. Math. Soc.*, 24, pp. 388–393.

[1971], On constructing nearly decomposable matrices, *Proc. Amer. Math. Soc.*, 27, pp. 222–228.

[1975], A canonical form for fully indecomposable (0,1)-matrices, *Canad. Math. Bull*, 18, pp. 223–227.

D.J. Hartfiel and R. Loewy[1984], A determinantal version of the Frobenius-König theorem, *Linear Multilin. Alg.*, 16, pp. 155–165.

V. Havel[1955], A remark on the existence of finite graphs (in Hungarian), *Časopis Pěst. Mat.*, 80, pp. 477–480.

B.R. Heap and M.S. Lynn[1964], The index of primitivity of a non-negative matrix, *Numer. Math.*, 6, pp. 120–141.

M. Hedrick and R. Sinkhorn[1970], A special class of irreducible matrices—the nearly reducible matrices, *J. Algebra*, 16, pp. 143–150.

J.R. Henderson[1975], Permanents of (0,1)-matrices having at most two 0's per line, *Canad. Math. Bull.*, 18, pp. 353–358.

P. Heymans[1969], Pfaffians and skew-symmetric matrices, *Proc. London Math. Soc.* (3), 19, pp. 730–768.

D.G. Higman[1971], Partial geometries, generalized quadrangles and strongly regular graphs, *Atti di Conv. Geometria Combinatoria e sue Applicazione* (A. Barlotti, ed.), Perugia, pp. 263–293.

A.J. Hoffman[1960], On the uniqueness of the triangular association scheme, *Ann. Math. Statist.*, 31, pp. 492–497.

[1960], Some recent applications of the theory of linear inequalities to extremal combinatorial analysis, *Proc. Symp. in Applied Mathematics*, vol. 10, Amer. Math. Soc., pp. 113–127.

[1963], On the polynomial of a graph, *Amer. Math. Monthly*, 70, pp. 30–36.

[1967], Research problem 2-11, *J. Combin. Theory*, 2, p. 393.

[1970], $-1 - \sqrt{2}$? *Combinatorial Structures and Their Applications*, Gordon and Breach, New York, pp. 173–176.

[1977], On graphs whose least eigenvalue exceeds $-1-\sqrt{2}$, *Linear Alg. Applics.*, 16, pp. 153–165.

A.J. Hoffman and J.B. Kruskal[1956], Integral boundary points of convex polyhedra, *Annals of Math. Studies*, No. 38, Princeton University Press, Princeton, pp. 223–246.

A.J. Hoffman and M.H. McAndrew[1965], The polynomial of a directed graph, *Proc. Amer. Math. Soc.*, 16, pp. 303–309.

A.J. Hoffman and D.K. Ray-Chaudhuri, On a spectral characterization of regular line graphs, unpublished manuscript.

J.C. Holladay and R.S. Varga[1958], On powers of non-negative matrices, *Proc. Amer. Math. Soc.*, 9, pp. 631–634.

R. Horn and C.R. Johnson[1985], *Matrix Analysis*, Cambridge University Press, Cambridge.

X.L. Hubaut[1975], Strongly regular graphs, *Discrete Math.*, 13, pp. 357–381.

C.G. Jacobi[1841], De determinantibus functionalibus, *J. Reine Angew. Math.*, 22, pp. 319–359.

K. Jones, J.R. Lundgren, N.J. Pullman and R. Rees[1988], A note on the covering numbers of $K_n - K_m$ and complete t-partite graphs, *Congressus Num.*, 66, pp. 181–184.

D.D. Joshi[1987], *Linear Estimation and Design of Experiments*, Wiley, New York.

A.-A.A. Jucys[1976], The number of distinct latin squares as a group-theoretical constant, *J. Combin. Theory, Ser. A*, 20, pp. 265–272.

W.B. Jurkat and H.J. Ryser[1966], Matrix factorizations of determinants and permanents, *J. Algebra*, 3, pp. 1–27.

   [1967], Term ranks and permanents of nonnegative matrices, *J. Algebra*, 5, pp. 342–357.

   [1968], Extremal configurations and decomposition theorems, *J. Algebra*, 8, pp. 194–222.

R. Kallman[1982], A method for finding permanents of 0,1 matrices, *Maths. of Computation*, 38, pp. 167–170.

I. Kaplansky[1939], A generalization of the 'Problème des recontres,' *Amer. Math. Monthly*, 46, pp. 159–161.

   [1943], Solution of the "Problème des ménages," *Bull. Amer. Math. Soc.*, 49, pp. 784–785.

   [1944], Symbolic solution of certain problems in permutations, *Bull. Amer. Math. Soc.*, 50, pp. 906–914.

R.M. Karp[1972], Reducibility among combinatorial problems, In *Complexity of Computer Calculations* (R.E. Miller and J.W. Thatcher, eds.), Plenum, New York, pp. 85–104.

P.W. Kasteleyn[1961], The statistics of dimers on a lattice, *Physica*, 27, pp. 1209–1225.

   [1967], Graph theory and crystal physics, *Graph Theory and Theoretical Physics* (F. Harary, ed.), Academic Press, New York, pp. 43–110.

J.G. Kemeny and J.L. Snell[1960], *Finite Markov Chains*, Van Nostrand, Princeton.

D.G. Kendall[1969], Incidence matrices, interval graphs and seriation in archaeology, *Pacific J. Math.*, 28, pp. 565–570.

F. King and K. Wang[1985], On the g-circulant solutions to the matrix equation $A^m = \lambda J$, II, *J. Combin. Theory, Ser. A*, 38, pp. 182–186.

V. Klee, R. Ladner and R. Manber[1983], Signsolvability revisited, *Linear Alg. Applics.*, 59, pp. 131–157.

D.E. Knuth[1970], Notes on central groupoids, *J. Combin. Theory*, 8, pp. 376–390.

   [1974], Wheels within wheels, *J. Combin. Theory, Ser. B*, 16, pp. 42–46.

K.K. Koksma[1969], A lower bound for the order of a partial transversal, *J. Combin. Theory*, 7, pp. 94–95.

J. Komlós[1967], On the determinant of (0,1)-matrices, *Studia Sci. Math. Hung.*, 2, pp. 7–21.

D. König[1916], Über Graphen and ihre Anwendung auf Determinantentheorie und Mengenlehre, *Math. Ann.*, 77, pp. 453–465.

[1936], *Theorie der endlichen und unendlichen Graphen*, Leipzig, reprinted by Chelsea[1960], New York.

J.P.S. Kung[1984], Jacobi's identity and the König-Egerváry theorem, *Discrete Math.*, 49, pp. 75–77.

C.W.H. Lam[1975], A generalization of cyclic difference sets I, *J. Combin. Theory, Ser. A*, 19, pp. 51–65.

[1975], A generalization of cyclic difference sets II, *J. Combin. Theory, Ser. A*, 19, pp. 177–191.

[1977], On some solutions of $A^k = dI + \lambda J$, *J. Combin. Theory, Ser. A*, 23, pp. 140–147.

C.W.H. Lam and J.H. van Lint[1978], Directed graphs with unique paths of fixed length, *J. Combin. Theory, Ser. B*, 24, pp. 331–337.

C.W.H. Lam, L.H. Thiel and S. Swierzc[1989], The nonexistence of finite projective planes of order 10, *Canad. J. Math.*, XLI, pp. 1117–1123.

H. Laue[1988], A graph theoretic proof of the fundamental trace identity, *Discrete Math.*, 69, pp. 197–198.

J.S. Lew[1966], The generalized Cayley-Hamilton theorem in n dimensions, *Z. Angew. Math. Phys.*, 17, pp. 650–653.

M. Lewin[1971], On exponents of primitive matrices, *Numer. Math.*, 18, pp. 154–161.

[1971], On nonnegative matrices, *Pacific. J. Math.*, 36, pp. 753–759.

[1974], Bounds for the exponents of doubly stochastic matrices, *Math. Zeit.*, 137, pp. 21–30.

M. Lewin and Y. Vitek[1981], A system of gaps in the exponent set of primitive matrices, *Illinois J. Math.*, 25, pp. 87–98.

Chang Li-Chien[1959], The uniqueness and non-uniqueness of the triangular association scheme, *Sci. Record. Peking Math.* (New Ser.), 3, pp. 604–613.

[1960], Associations of partially balanced designs with parameters $v = 28$, $n_1 = 12$, $n_2 = 15$, and $p_{11}^2 = 4$, *Sci. Record. Peking Math.*, (New Ser.), 4. pp. 12–18.

E.H. Lieb[1968], A theorem on Pfaffians, *J. Combin. Theory*, 5, pp. 313–368.

C.C. Lindner[1970], On completing latin rectangles, *Canad. Math. Bull.*, 13, pp. 65–68.

J.H. van Lint[1985], $(0, 1, *)$ distance problems in Combinatorics, *Surveys in Combinatorics* (I. Anderson, ed.), London Math. Soc. Lecture Notes 103, Cambridge University Press, pp. 113–135.

J.H. van Lint and J.J. Seidel[1966], Equilateral point sets in elliptic geometry, *Nederl. Akad. Wetensch. Proc*, Ser. A, 69 (=*Indag. Math.*, 28), pp. 335–348.

C.H.C. Little[1975], A characterization of convertible (0,1)-matrices, *J. Combin. Theory, Ser. B*, 18, pp. 187–208.

C.H.C. Little and J.M. Pla[1972], Sur l'utilisation d'un pfaffien dans l'étude des couplages parfaits d'un graphe, *C. R. Acad. Scie. Paris*, 274, p. 447.

B. Liu[1990], A note on the exponents of primitive (0,1)-matrices, *Linear Alg. Applics.*, 140, pp. 45–51.

[1991], New results on the exponent set of primitive, nearly reducible matrices, *Linear Alg. Applics*, to be published.

B. Liu, B. McKay, N. Wormwald and K.M. Zhang[1990], The exponent set of symmetric primitive (0,1)-matrices with zero trace, *Linear Alg. Applics*, 133, pp. 121–131.

L. Lovász[1970], Subgraphs with prescribed valencies, *J. Combin. Theory*, 8, pp. 391–416.

[1979], On determinants, matchings, and random algorithms, *Proceedings of Fundamentals of Computation Theory*, Akademie Verlag, Berlin, pp. 565–574.

L. Lovász and M.D. Plummer[1977], On minimal elementary bipartite graphs, *J. Combin. Theory, Ser. A*, 23, pp. 127–138.

R.D. Luce[1952], Two decomposition theorems for a class of finite oriented graphs, *Amer. J. Math.*, 74, pp. 701–722.

S.L. Ma and W.C. Waterhouse[1987], The g-circulant solutions of $A^m = \lambda J$, *Linear Alg. Applics.*, 85, pp. 211–220.

P.A. MacMahon[1915], *Combinatory Analysis*, vol. 1, Cambridge University Press, Cambridge.

H.F. MacNeish[1922], Euler squares, *Ann. Math.*, 23, pp. 221–227.

W. Mader[1973], Grad und lokaler Zusammenhang in endlichen Graphen, *Math. Ann.*, 205, pp. 9–11.

H.B. Mann[1942], The construction of orthogonal latin squares, *Ann. Math. Statist.*, 13, pp. 418–423.

M. Marcus and H. Minc[1961], On the relation bewteen the determinant and the permanent, *Illinois J. Math.*, 5, pp. 376–381.

[1963], Disjoint pairs of sets and incidence matrices, *Illinois J. Math*, 7, pp. 137–147.

[1964], *A Survey of Matrix Theorey and Matrix Inequalities*, Allyn and Bacon, Boston.

J. Marica and J. Schönheim[1969], Differences of sets and a problem of Graham, *Canad. Math. Bull.*, 12, pp. 635–637.

J.S. Maybee[1981], Sign solvability, in *Computer assisted analysis and model simplification* (H. Greenberg and J. Maybee, eds.), Academic Press, New York.

P.K. Menon[1961], Method of constructing two mutually orthogonal latin squares of order $3n + 1$, *Sankhyā A*, 23, pp. 281–282.

D. Merriell[1980], The maximum permanent in Λ_n^k, *Linear Multilin. Alg.*, 9, pp. 81–91.

H. Minc[1963], Upper bound for permanents of (0,1)-matrices, *Bull. Amer. Math. Soc.*, 69, pp. 789–791.

[1967], A lower bound for permanents of (0,1)-matrices, *Proc. Amer. Math. Soc.*, 18, pp. 1128–1132.

[1969], On lower bounds for permanents of (0,1)-matrices, *Proc. Amer. Math. Soc.*, 22, pp. 117–123.

[1969], Nearly decomposable matrices, *Linear Alg. Applics.*, 5, pp. 181–187.

[1973], (0,1)-matrices with minimal permanents, *Israel J. Math.*, 77, pp. 27–30.

[1974], An unresolved conjecture on permanents of (0,1)-matrices, *Linear Multilin. Alg.*, 2, pp. 57–64.

[1974], The structure of irreducible matrices, *Linear Multilin. Algebra*, 2, pp. 85–90.

[1978], *Permanents*, Addison-Wesley, Reading, Mass.

L. Mirsky[1968], Combinatorial theorems and integral matrices, *J. Combin. Theory*, 5, pp. 30–44.

[1971], *Transversal Theory*, Academic Press, New York.

B. Mohar[1988], The Laplacian spectrum of graphs, *Preprint Series Dept. Math. University E.K. Ljubljana*, 26, pp. 353–382.

J.W. Moon and L. Moser[1966], Almost all (0,1)-matrices are primitive, *Studia Scient. Math. Hung.*, 1, pp. 153–156.

J.W. Moon and N.J. Pullman[1967], On the powers of tournament matrices, *J. Combin. Theory*, 3, pp. 1–9.

T. Muir[1882], On a class of permanent symmetric functions, *Proc. Roy. Soc. Edinburgh*, 11, pp. 409–418.

[1897], A relation between permanents and determinants, *Proc. Roy. Soc. Edinburgh*, 22, pp. 134–136.

K. Murota[1989], On the irreducibility of layered mixed matrices, *Linear Multilin. Alg.*, 24, pp. 273–288.

J.R. Nechvatal[1981], Asymptotic enumeration of generalized latin rectangles, *Utilitas Math.*, 20, pp. 273–292.

A. Nijenhuis[1976], On permanents and the zeros of rook polynomials, *J. Combin. Theory, Ser. A*, 21, pp. 240–244.

A. Nijenhuis and H. Wilf[1970], On a conjecture of Ryser and Minc, *Nederl. Akad. Wetensch. Proc., Ser. A*, 73 (=*Indag. Math.*, 32), pp. 151–158.

[1978], *Combinatorial Algorithms*, second ed., Academic Press, New York.

G.N. de Oliveira[1974], Note on the characteristic roots of tournament matrices, *Linear Alg. Applics.*, 8, pp. 271–272.

W.E. Opencomb[1984], On the intricacy of combinatorial problems, *Discrete Math.*, 50, pp. 71–97.

J. Orlin[1977], Contentment in graph theory: covering graphs with cliques, *Indag. Math.*, 39, pp. 406–424.

[1978], Line-digraphs, Arborescences, and theorems of Tutte and Knuth, *J. Combin. Theory, Ser. B*, 25, pp. 187–198.

P.A. Ostrand[1970], Systems of distinct representatives II, *J. Math. Anal. Applics.*, 32, pp. 1–4.

A. Ostrowski[1937], Über die Determinanten mit überwiegender Hauptdiagonale, *Comm. Math. Helv.*, 10, pp. 69–96.

[1951], Über das Nichtverschwinden einer Klassen von Determinanten und die Lokalisierung der charakteristischen Wurzeln von Matrizen, *Compositio Math.*, 9, pp. 209–226.

M.W. Padberg[1976], A note on the total unimodularity of matrices, *Discrete Math.*, 14, pp. 273–278.

L.J. Paige[1947], A note on finite abelian groups, *Bull. Amer. Math. Soc.*, 53, pp. 590–593.

E.T. Parker[1959], Orthogonal Latin squares, *Proc. Nat. Acad. Sci.*, 45, pp. 859–862.

[1959], Construction of some sets of mutually orthogonal Latin squares, *Proc. Amer. Math. Soc.*, 10, pp. 946–949.

[1990], private communication.

G.W. Peck[1984], A new proof of a theorem of Graham and Pollak, *Discrete Math.*, 49, pp. 327–328.

H. Perfect1966], Symmetrized form of P. Hall's theorem on distinct representatives, *Quart. J. Math. Oxford* (2), 17, pp. 303–306.

J.M. Pla[1965], Sur l'utilisation d'un pfaffien dans l'étude des couplages parfaits d'un graphe, *C. R. Acad. Sci. Paris*, 260, pp. 2967–2970.

H. Poincaré[1901], Second complément à l'analysis situs, *Proc. London Math. Soc.*, 32, pp. 277–308.

G. Pólya[1913], Aufgabe 424, *Arch. Math. Phys.* (3), 20, p. 271.

C.R. Pranesachar[1981], Enumeration of latin rectangles via SDR's, *Combinatorics and Graph Theory* (S.B. Rao, ed.), Lecture Notes in Math., 885, Springer-Verlag, Berlin and New York, pp. 380–390.

C. Procesi[1976], The invariant theory of $n \times n$ matrices, *Advances in Math.*, 19, pp. 306–381.

V. Pták[1958], On a combinatorial theorem and its applications to nonnegative matrices, *Czech. Math. J.*, 8, pp. 487–495.

V. Pták and J. Sedláček[1958], On the index of imprimitivity of non-negative matrices, *Czech. Math. J.*, 8, pp. 496–501.

R. Rado[1967], On the number of systems of distinct representatives, *J. London Math. Soc.*, 42, pp. 107–109.

D. Raghavarao[1971], *Construction and combinatorial problems in design of experiments*, Wiley, New York (reprinted[1988] by Dover, Mineola, NY).

Yu. P. Ramyslov[1974], Trace identities of full matrix algebras over a field of characteristic zero, *Math. USSR-Isv.*, 8, pp. 727–760.

J. Riordan[1946], Three-line latin rectangles-II, *Amer. Math. Monthly*, 53, pp. 18–20.

V. Romanovsky[1936], Recherches sur les Chains de Markoff, *Acta. Math.*, 66, pp. 147–251.

D. Rosenblatt[1957], On the graphs and asymptotic forms of finite boolean relation matrices and stochastic matrices, *Naval Research Logistics Quarterly*, 4, pp. 151–167.

J.A. Ross[1980], *Some problems in combinatorial matrix theory*, Ph.D. thesis, University of Wisconsin, Madison.

[1982], On the exponent of a primitive, nearly reducible matrix II, *SIAM J. Alg. Disc. Meth.*, 3, pp. 385–410.

L.R. Rowen[1980], *Polynomial Identities in Ring Theory*, Academic Press, New York.

D.E. Rutherford[1964], The Cayley-Hamilton theorem for semirings, *Proc. Roy. Soc. Edinburgh Sec. A*, 66, pp. 211–215.

H.J. Ryser[1951], A combinatorial theorem with an application to latin rectangles, *Proc. Amer. Math. Soc.*, 2, pp. 550–552.

[1957], Combinatorial properties of matrices of 0's and 1's, *Canad. J. Math.*, 9, pp. 371–377.

[1960], Matrices of zeros and ones, *Bull. Amer. Mat. Soc.*, 66, pp. 442–464.

[1963], *Combinatorial Mathematics*, Carus Mathematical Monograph No. 14, Math. Assoc. of Amer., Washington, D.C.

[1967], Neuere Probleme in der Kombinatorik (prepared by D.W. Miller), *Vorträge über Kombinatorik*, Oberwolfach, pp. 69–91.

[1969], Combinatorial configurations, *SIAM J. Appl. Math.*, 17, pp. 593–602.

[1970], A generalization of the matrix equation $A^2 = J$, *Linear Alg. Applics.*, 3, pp. 451–460.

[1972], A fundamental matrix equation for finite sets, *Proc. Amer. Math. Soc.*, 34, pp. 332–336.

[1973], Analogs of a theorem of Schur on matrix transformations, *J. Algebra*, 25, pp. 176–184.

[1973], Indeterminates and incidence matrices, *Linear Multilin. Alg.*, 1, pp. 149–157.

[1974], Indeterminates and incidence matrices, in *Combinatorics*, Math. Centre Tracts Vol. 55, Mathematisch Centrum, Amsterdam, pp. 3–17.

[1975], The formal incidence matrix, *Linear Multilin. Alg.*, 3, pp. 99–104.

[1982], Set intersection matrices, *J. Combin. Theory, Ser. A*, 32, pp. 162–177.

[1984], Matrices and set differences, *Discrete Math.*, 49, pp. 169–173.

H. Sachs[1967], Über Teiler, Faktoren und charakteristische Polynome von Graphen II, *Wiss. Z. Techn. Hochsch. Ilmenau*, 13, pp. 405–412.

M. Saks[1986], Some sequences associated with combinatorial structures, *Discrete Math.*, 59, pp. 135–166.

H. Schneider[1977], The concepts of irreducibility and full indecomposability of a matrix in the works of Frobenius, König and Markov, *Linear Alg. Applics.*, 18, pp. 139–162.

[1986], The influence of the marked reduced graph of a nonnegative matrix on the Jordan form and on related properties: A survey, *Linear Alg. Applics.*, 84, pp. 161–189.

A. Schrijver[1978], A short proof of Minc's conjecture, *J. Combin. Theory, Ser. A*, 25, pp. 80–83.

[1983], Bounds on permanents, and the number of 1-factors and 1-factorizations of bipartite graphs, *Surveys in Combinatorics*, Cambridge University Press, Cambridge, pp. 107–134.

[1986], *Theory of Linear and Integer Programming*, Wiley, New York.

A. Schrijver and W.G. Valiant[1980], On lower bounds for permanents, *Indag. Math.*, 42, pp. 425–427.

A. Schuchat[1984], Matrix and network models in archaeology, *Math. Magazine*, 57, pp. 3–14.

S. Schwarz[1973], The semigroup of fully indecomposable relations and Hall relations, *Czech. Math. J.*, 23, pp. 151–163.

A.J. Schwenk and R.J. Wilson[1978], On the eigenvalues of a graph, *Selected Topics in Graph Theory* (L.W. Beineke and R.J. Wilson, eds.), Academic Press, New York, pp. 307–336.

J. Sedláček[1959], O incidenčnich matiach orientirovaných grafò, *Časop. Pěst. Mat.*, 84, pp. 303–316.

J.J. Seidel[1968], Strongly regular graphs with (-1,1,0) adjacency matrix having eigenvalue 3, *Linear Alg. Applics.*, 1, pp. 281–298.

[1969], Strongly regular graphs, *Recent Progress in Combinatorics* (W.T. Tutte, ed.), Academic Press, New York, pp. 185–198.

[1974], Graphs and two-graphs, Proceedings of the Fifth Southeastern Conference on Combinatorics, Graph Theory and Computing, Congressus Numerantium X, *Utilitas Math.*, Winnipeg, pp. 125–143.

[1976], A survey of two-graphs, *Teorie Combinatorie*, Tomo I, (B. Segre, ed.), Accademia Nazionale dei Lincei, Rome, pp. 481–511.

[1979], Strongly regular graphs, *Surveys in Combinatorics, Proc. 7th British*

Combinatorial Conference, London Math. Soc. Lecture Note Ser. 38 (B. Bollobás, ed.), Cambridge University Press, Cambridge.

P.D. Seymour[1980], Decomposition of regular matroids, *J. Combin. Theory, Ser. B*, 28, pp. 305–359.

[1982], Applications of the regular matroid decomposition, *Colloquia Math. Soc. János Bolyai*, No. 40 Matroid Theory, Szeged (Hungary), pp. 345–357.

P. Seymour and C. Thomassen[1987], Characterization of even directed graphs, *J. Combin. Theory, Ser. B*, 42, pp. 36–45.

C.E. Shannon[1949], A theorem on coloring the lines of a network, *J. Math. Phys.*, 28, pp. 148–151.

J.Y. Shao[1985], On a conjecture about the exponent set of primitive matrices, *Linear Alg. Applics.*, 65, pp. 91–123.

[1985], On the exponent of a primitive digraph, *Linear Alg. Applics.*, 64, pp. 21–31.

[1985], Matrices permutation equivalent to primitive matrices, *Linear Alg. Applics.*, 65, pp. 225–247.

[1987], The exponent set of symmetric primitive matrices, *Scientia Sinica Ser. A*, vol. XXX, pp. 348–358.

P.W. Shor[1982], A lower bound for the length of a partial transversal in a latin square, *J. Combin Theory, Ser. A*, 33, pp. 1–8.

S.S. Shrikhande[1959], The uniqueness of the L_2 association scheme, *Ann. Math. Statist.*, 30, pp. 781–798.

[1961], A note on mutually orthogonal latin squares, *Sankhyā* A, 23, pp. 115–116.

R. Sinkhorn and P. Knopp[1969], Problems involving diagonal products in nonnegative matrices, *Trans. Amer. Math. Soc.*, 136, pp. 67–75.

B. Smetaniuk[1981], A new construction for latin squares I. Proof of the Evans conjecture, *Ars Combinatoria*, 11, pp. 155–172.

[1982], A new construction on latin squares-II: The number of latin squares is strictly increasing, *Ars Combinatoria*, 14, pp. 131–145.

H.J.S. Smith[1876], On the value of a certain arithmetical determinant, *Proc. London Math. Soc.*, 7, pp. 208–212.

J.H. Smith[1970], Some properties of the spectrum of a graph, *Combinatorial Structures and Their Applications* (R. Guy, N. Sauer and J. Schönheim, eds.), Gordon and Breach, New York, pp. 403–406.

R.P. Stanley[1980], Weyl groups, the hard Lefschetz theorem, and the Sperner property, *SIAM J. Alg. Disc. Methods*, 1, pp. 168–184.

[1986], *Enumerative Combinatorics*, vol. I, Wadsworth & Brooks/Cole, Monterrey.

H. Straubing[1983], A combinatorial proof of the Cayley-Hamilton theorem, *Discrete Math.*, 43, pp. 273–279.

R.G. Swan[1963], An application of graph theory to algebra, *Proc. Amer. Math. Soc.*, 14, pp. 367–373.

[1969], Correction to "An application of graph theory to algebra," *Proc. Amer. Math. Soc.*, 21, pp. 379–380.

G. Szegö[1913], Zu Aufgabe 424, *Arch. Math. Phys.* (3), 21, pp. 291–292.

R.E. Tarjan[1972], Depth first search and linear graph algorithms, *SIAM J. Computing*, 1, pp. 146–160.

G. Tarry[1900,1901], Le problème de 36 officeurs, *Compte Rendu de l'Association Française pour l'Avancement de Science Naturel*, 1, pp. 122–123, and 2, pp. 170–203.

O. Taussky[1949], A recurring theorem on determinants, *Amer. Math. Monthly*, 10, pp. 672–676.

H.N.V. Temperley[1964], On the mutual cancellation of cluster integrals in Mayer's fugacity series, *Proc. Phys. Soc.*, 83, pp. 3–16.

[1981], *Graph Theory and Applications*, Ellis Horwood, Chichester.

C. Thomassen[1981], A remark on the factor theorems of Lovász and Tutte, *J. Graph Theory*, 5, pp. 441–442.

[1986], Sign-nonsingular matrices and even cycles in directed graphs, *Linear Alg. Applics.*, 75, pp. 27–41.

J. Touchard[1934], Sur un problème de permutations, *C. R. Acad. Sci. Paris*, 198, pp. 631–633.

W.T. Tutte[1947], The factorization of linear graphs, *J. London Math. Soc.*, 22, pp. 107–111.

[1948], The dissection of equilateral triangles into equilateral triangles, *Proc. Cambridge Philos. Soc.*, 44, pp. 463–482.

[1952], The factors of graphs, *Canadian J. Math.*, 4, pp. 314–328.

[1958], A homotopy theorem for matroids, I and II, *Trans. Amer. Math. Soc.*, 88, pp. 144–174.

[1978], The subgraph problem, *Discrete Math.*, 3, pp. 289–295.

[1981], Graph factors, *Combinatorica*, 1, pp. 79–97.

[1984], Graph Theory, *Encyclopedia of Mathematics and Its Applications*, vol. 21, Addison-Wesley, Reading, Mass.

H. Tverberg[1982], On the decomposition of K_n into complete bipartite graphs, *J. Graph Theory*, 6, pp. 493–494.

L.G. Valiant[1979], Completeness classes in algebra, *Proc. 11th Ann. ACM Symp. Theory of Computing*, pp. 249–261.

[1979], The complexity of computing the permanent, *Theoret. Comput. Sci.*, 8, pp. 189–201.

R.S. Varga[1962], *Matrix Iterative Analysis*, Prentice-Hall, Englewood Cliffs, N.J.

D. Vere-Jones[1984], An identity involving permanents, *Linear Alg. Applics.*, 63, pp. 267–270.

[1988], A generalization of permanents and determinants, *Linear Alg. Applics.*, 111, pp. 119–124.

V.G. Vizing[1964], On an estimate of the chromatic class of a p-graph (in Russian), *Diskret. Analiz.*, 3, pp. 25–30.

[1965], The chromatic class of a multigraph (in Russian), *Cybernetics*, 3, pp. 32–41.

W. Vogel[1963], Bermerkungen zur Theorie der Matrizen aus Nullen und Einsen, *Archiv der Math.*, 14, pp. 139–144.

J. von zur Gathen[1987], Permanent and determinant, *Linear Alg. Applics.*, 96, pp. 87–100.

[1987], Feasible arithmetic computations: Valiant's hypothesis, *J. Symbolic Comp.*, 4, pp. 137–172.

M. Voorhoeve[1979], A lower bound for the permanents of certain (0,1)-matrices, *Indag. Math.*, 41, pp. 83–86.

A.J. de Vries and R.M.A. Wieringa[1978], Een ondergrens voor de lengte van een partiele transversaal in een Latijns vierkant, preprint.

K. Wang[1980], On the matrix equation $A^m = \lambda J$, *J. Combin. Theory, Ser. A*, 29, pp. 134–141.

[1981], A generalization of group difference sets and the matrix equation $A^m = dI + \lambda J$, *Aequationes Math.*, 23, pp. 212–222.

[1982], On the g-circulant solutions to the matrix equation $A^m = \lambda J$, *J. Combin. Theory, Ser. A*, 33, pp. 287–296.

N. White (ed.)[1986], Theory of Matroids, *Encyclopedia of Maths. and Its Applics.*, Cambridge University Press, Cambridge.

[1987], Unimodular matroids, Combinatorial Geometries (N. White, ed.), *Encyclopedia of Maths. and Its Applics.*, Cambridge University Press, Cambridge.

H. Wielandt[1958], Unzerlegbare, nicht negative Matrizen, *Math. Zeit.*, 52, pp. 642–645.

H.S. Wilf[1968], A mechanical counting method and combinatorial applications, *J. Combin. Theory*, 4, pp. 246–258.

[1986], *Algorithms and Complexity*, Prentice-Hall, Englewood Cliffs, N.J.

R.J. Wilson[1985], *Introduction to Graph Theory*, 3rd edition, Longman, Harlow, Essex.

R.M. Wilson[1974], Concerning the number of mutually orthogonal latin squares, *Discrete Math.*, 9, pp. 181–198.

[1974], A few more squares, *Congressus Numerantium*, No. X, pp. 675–680.

D.E. Woolbright[1978], An $n \times n$ latin square has a transversal with at least $n - \sqrt{n}$ distinct symbols, *J. Combin. Theory, Ser. A*, 24, pp. 235–237.

D.E. Woolbright and H.-L. Fu[1987], The rainbow theorem of 1-factorization, preprint.

S. Yang and G.P. Barker[1988], On the exponent of a primitive, minimally strong digraph, *Linear Alg. Applics.*, 99, pp. 177–198.

D. Zeilberger[1985], A combinatorial approach to matrix algebra, *Discrete Math.*, 56, pp. 61–72.

K.M. Zhang[1987], On Lewin and Vitek's conjecture about the exponent set of primitive matrices, *Linear Alg. Applics.*, 96, pp. 101–108.

L. Zhu[1982], A short disproof of Euler's conjecture concerning orthogonal latin squares, *Ars Combinatoria*, 14, pp. 47–55.

INDEX

Printed in the United States
By Bookmasters